The Care of Life

The Care of Life

Transdisciplinary Perspectives in Bioethics and Biopolitics

Edited by Miguel de Beistegui, Giuseppe Bianco and Marjorie Gracieuse

ROWMAN &
LITTLEFIELD
INTERNATIONAL
London • New York

Published by Rowman & Littlefield International, Ltd.
Unit A, Whitacre Mews, 26-34 Stannary Street, London SE11 4AB
www.rowmaninternational.com

Rowman & Littlefield International, Ltd. is an affiliate of Rowman & Littlefield
4501 Forbes Boulevard, Suite 200, Lanham, Maryland 20706, USA
With additional offices in Boulder, New York, Toronto (Canada), and Plymouth (UK)
www.rowman.com

British Library Cataloguing in Publication Information Available
A catalogue record for this book is available from the British Library

ISBN: HB 978-1-78348-037-1
ISBN: PB 978-1-78348-038-8

Library of Congress Cataloging-in-Publication Data

The care of life : transdisciplinary perspectives in bioethics and biopolitics / [edited by] Miguel de Beistegui, Giuseppe Bianco and Marjorie Gracieuse.
pages cm
Includes index.
ISBN 978-1-78348-037-1 (cloth : alk. paper) — ISBN 978-1-78348-038-8 (pbk. : alk. paper) — ISBN 978-1-78348-039-5 (electronic)
1. Bioethics—Philosophy. 2. Biopolitics—Philosophy. 3. Interdisciplinary approach to knowledge. I. Beistegui, Miguel de, 1966– editor. II. Bianco, Giuseppe, editor. III. Gracieuse, Marjorie, editor.
QH332.C368 2015
174.2—dc23
2014028213

Printed in the United States of America

Contents

Acknowledgements vii

Introduction: Life: *Who* Cares? 1
Giuseppe Bianco, with Miguel de Beistegui

Part I: "Bioethics" and "Biopolitics" 23

1 A Brief History of Bioethics 25
Guillaume Le Blanc

2 Between (Bio)-Politics and (Bio)-Ethics: What Life? 33
Luca Paltrinieri

3 What Is Vital? 47
Frédéric Worms

Part II: Norms and Normality 61

4 The Return of Vitalism: Canguilhem, Bergson and the Project of a Biophilosophy 63
Charles T. Wolfe and Andy Wong

5 Life and Objective Norms: Canguilhem in the Context of Contemporary Meta-ethics 79
Sander Werkhoven

6 Critical and Political Stakes of a Philosophy of Norms: Part I: Towards a Critical Philosophy of Norms 95
Julien Pieron

7 Critical and Political Stakes of a Philosophy of Norms: Part II: Theory of Norms and Social Criticism 109
Florence Caeymaex

8 The Racial Politics of Life Itself: Goldstein, Uexküll, Canguilhem and Fanon 121
Robert Bernasconi

9 History and the Politics of "Life" 135
Claudia Stein and Roger Cooter

Part III: Pathology and Ageing **149**

10 Life of Pain: Remarks about Negativity and Effort in Georges Canguilhem 151
Giuseppe Bianco

11 Human Life and Subjectivity: Learning from Foucault 165
Piergiorgio Donatelli

12 Ethics of Care and Face Transplants: After Levinas, Deleuze and Guattari 181
Marjorie Gracieuse

13 Ageing and Longevity 199
Paul-Antoine Miquel

14 Generational Change as a Vehicle for Radical Conceptual Change: The Case for Periodic Rejuvenation 211
Steve Fuller

Part IV: Desire and Pleasure **225**

15 Sexuality and Liberalism 227
Patrick Singy

16 Desire Within and Beyond Biopolitics 241
Miguel de Beistegui

17 The Pragmatics of Desire and Pleasure: Rethinking Somatic Powers with Foucault, Deleuze and Guattari 261
Marjorie Gracieuse and Nicolae Morar

18 Revenge of the Tender Pervert: On a Troubling Concept That Refuses to Go Away 279
Hector Kollias

19 *Jouissance* in Lacan's Seminar XX: Progelomena to a New Reading of Strange Enjoyment and Being an Angel 295
Lorenzo Chiesa

Index 309

About the Contributors 319

Acknowledgements

We would like to thank the Leverhulme Trust for its generous support. Most of the chapters gathered in this volume were originally presented at one of the three workshops organized by the Research Project in Bioethics and Biopolitics at the University of Warwick between 2011 and 2013, which was funded by the Leverhulme Trust. We're also very grateful to our editor, Sarah Campbell, for her trust, patience and good humour in the last few months.

Introduction

Life: Who *Cares?*

Giuseppe Bianco, with Miguel de Beistegui

How can we account for the fact that, in the last twenty years at least, "life" has become a privileged category with which the humanities in general, and philosophy in particular, have approached virtually all epistemological, political, ethical and even ontological problems? And what kind of history does the concept of "life" have, both inside and outside the humanities? Our goal, in this brief introduction, is to mention some of the reasons why, according to us, we seem to have reached a moment in history where every distinction and opposition is made no longer in relation to life, but within it, and where life is a problem inextricably theoretical and practical, ontological and political. These reasons are, no doubt, contingent, and our aim is not to reconstruct a grand narrative regarding the meaning and direction of that phenomenon. Rather, we wish to identify some of the signs that allow us to recognize it as a key development, and open the critical space within which it is discussed in the various chapters of this volume.

BIRTH AND DEATH OF A CONCEPT

In *The Logic of Life* (1970), the French biologist and historian of the life sciences François Jacob declared that life was no longer interrogated in the laboratories.[1] At roughly the same time, and more boldly still, his colleague Jacques Monod claimed that, while the secret of life had once seemed inaccessible, it was, by then, mostly solved.[2] With such claims, the two scientists who, together with André Lwoff, were awarded the 1965 Nobel Prize for their work on molecular biology, wanted to stress the trend towards the reduction of biological phenomena to the laws governing the inanimate world and implicit in molecular biology. They interpreted life as a "code" or a "message" inscribed in every living being and reproduced through the self-copying of the DNA fibre. The "discovery" of the genome by James Watson, Maurice Wilkins and Francis Crick, who were awarded the Nobel Prize in 1962, seemed like the definitive way to deprive life of all the qualities that were separating it from inanimate matter. The old question "What is life?" now sounded obsolete: molecular biology was finally able to provide the definitive proofs to the answer that, in 1944,[3] Erwin Schrödinger had already given, signalling in the process what Michel Morange once called the "Twilight of life".[4]

Monod and Jacob further claim that, as this question progressively faded away, it was no longer possible to consider the various versions of "vitalism" as viable orientations in biology:[5] contrary to what the (now discredited) vitalism of the eighteenth century claimed,[6] life could no longer be seen as a specific "type" of matter. In addition, and contrary to what philosophers such as Nietzsche and Bergson, or biologists such as Hans Driesch, once claimed, life could no longer be seen as a transformative power opposed or at least irreducible to the laws governing inanimate matter. In the end, it was the very concept of life itself that turned out to be useless, or at least the trace of a false problem. Paraphrasing Lévi-Strauss, we could say that, in the same way that anthropology had "dissolved" the notion of "man" with the help of structural linguistics, molecular biology was dissolving the common notion of "life" with the help of advances made in information theory and chemistry.[7] It is not by chance that Lévi-Strauss mentioned the discoveries of genetics in *The Savage Mind*:[8] according to him, they represented proof of the effectiveness of the sciences of information applied to both human and biological phenomena. Throughout the 1960s and, especially, 1970s, the paradigm of language—evident in the use and abuse of notions such as "message", "code", "system" and "structure"—dominated the field of the humanities. As a counterpart of this enthusiasm, little use was made of categories such as agency, individuality, subjectivity and purposiveness.

Roughly at the same time—in 1966, to be precise—Michel Foucault published a book that sketched an "archeology of the human sciences".[9] Inspired by the historical epistemology of Gaston Bachelard and secretly animated by Martin Heidegger, the French philosopher asserted that the notion of life was, like that of man, a recent conceptual creation, which dated from the beginning of the nineteenth century. Until then, he wrote, "life did not exist. Only living beings existed".[10] What Foucault wanted to stress was that, far from being what Kripke calls a "rigid designator", the notion of life indicated different classes of phenomena at different moments in history.[11] Georges Canguilhem, a philosopher and historian of the life sciences who supervised Foucault's PhD on the history of madness, later confirmed his student's view and argued that "the birth . . . of the concept of life"[12] dated from the beginning of the nineteenth century. In Aristotle, "life" is to be found everywhere and, as a result, doesn't have a clear specificity. The situation is quite different in Descartes: given the fact that his mechanism reduces everything to extension and matter, there is no room whatever for "life". The concept of life as we understand it today emerged through a series of discourses, experiments and institutions, which we need to say a few words about.

If, after the discovery of DNA, and in the eyes of positivist biologists such as Jacob, Monod and Lwoff, the concept of life was no longer necessary, Foucault's claim is that it did not yet exist in the eighteenth century. The life of life itself unfolds between that *no longer*, which marks the "death" of a concept,[13] and that *not yet*, which signals its birth. Where Jacob and Monod claimed that life was a useless concept in biology, Foucault was more nuanced and suggested that the historical transcendental, or the *episteme*, which dominated Western culture for 150 years—and in which a specific conception of life, labour and language played a major role—was probably destined to change. Like "man", "life" was destined to disappear and be "erased, like a face drawn in sand at the edge of the sea".[14] In Foucault's visionary evocation of a future disappearance of "man" at the very end of *The Order of Things* one can clearly hear the echo of Nietzsche's prophecy about its overcoming. But, just as in Nietzsche, the prophecy concerning the overcoming of man is essentially ambiguous: the notion of life was not about to disappear entirely, but was destined to go through mutations, in the biological and medical fields as well as in the humanities.

THE FORGETTING OF LIFE

The slow "disappearance" of the notion of life from the laboratories was mirrored within the humanities. Taken together, the emergence of the life sciences in the nineteenth century, the growing importance of the theory of evolution, which inflicted a second "narcissistic wound" to the human psyche, and the great influence of the metaphor of the organism,[15] triggered a twofold effect in much of philosophy and the social sciences. On the one hand, the results of biology made possible a philosophy of nature for which life was nothing but a higher degree of organization of inanimate matter, and one that, as such, could be understood by the classical schemas of mechanics. On the other hand, finality, normativity and the will were located in a single and *exceptional* living being—namely, the human.[16]

To this epistemological climate, we should add the historical and political context of the 1930s and 1940s, which was largely dominated by the ideological use fascism made of the romantic philosophy that conceives of life as an obscure force[17] that transcends reason and the values of democracy and humanism. As a result, life became a highly problematic and suspicious notion, the use of which required the utmost caution. Most philosophers, psychologists, anthropologists and sociologists, as well as a number of biologists, resisted the temptation to approach life as a phenomenon endowed with specific features and that is required to be treated differently from inanimate matter.[18] It was as if, to protect the values of humanism, scientific progress and democracy, one needed to deprive life of any potentiality, agency or originality, and ensure that the human intellect, which was given an exceptional place within nature, be able to control and dominate life itself.

The situation in France was, in that respect, symptomatic. As part of the process of autonomization of the discipline, French sociology opposed social Darwinism and organicism, as well as the legacy of Spencer's and Comte's social positivism; psychology emphasized the moment of humanization in the course of evolution and the peculiarity of human cognition; beginning in the 1930s, psychoanalysis and theorists such as Politzer and Lacan developed a critique of notions such as "psychical energies" and, more generally, Freud's metapsychology, with the aim of condemning their realism, vitalism and irrationalism. And in French philosophy, the critique of positivism was accompanied by a lack of interest in the factual development of biology, as shown in Léon Brunschvicg's and Emile Meyerson's epistemologies.

After the "Lysenko affair"[19]—a Soviet self-proclaimed agronomist and biologist who, in the name of "proletarian science", pretended to criticize the results of "bourgeois" biology—philosophers and scientist influenced by Marxism avoided speculations concerning the supposed dialectical character of life and embraced the rationalist tradition of Cartesian mechanism. If, during the 1940s, Marxist biologists such as John Burdon Sanderson Haldane and Marcel Prenant showed a certain enthusiasm for the project of a "proletarian biology", which followed the principles of dialectical materialism, the 1960s witnessed the publication of a book by another Marxist biologist, Ernest Kahane, the title of which, *La vie n'existe pas!*,[20] did not leave any room for a biology other than mechanistic and reductionist.

The situation in Germany was not altogether different. While the philosophical anthropology of philosophers like Max Scheler, Arnold Gehlen and Helmuth Plessner revealed a strong interest in biology, the situation changed in the immediate aftermath of the Second World War. To be sure, Ernst Cassirer fell under the influence of Jakob von Uexküll's notion of *Bauplan* and emphasized the specificity of organisms and their evolution in the posthumously published fourth volume of his *Philosophy of Symbolic Forms* and later in the books *An Essay on Man* and *The Problem of Knowledge*.[21] In doing so, he distinguished himself from Hermann Cohen and Paul Natorp, his teachers at Marburg, who were indifferent to the specificity in question. As for Heidegger, we could say that, despite his early (and rather Bergsonian) definitions of phenomenology as "the absolute *sympathy with life* that is identical with life-experience (*Erlebnis*)"[22] and of who we are as "factical life" (*faktische Leben*), his neat separation between the human *Dasein*, who is "rich in world", and animals, who are "poor in world", is the sign of a systematic effort to distinguish human existence from biological life, and to subordinate any discussion of life to the *metaphysical* question regarding the singular status of Dasein.[23] Similarly, Husserl's notion of the *Lebenswelt* has nothing to do with biology, and one should agree with Jean-Marie Schaeffer's claim that the insistence on the importance of corporeity in phenomenology gives an epistemic privilege to the self-inspection of consciousness.[24] It's only with Merleau-Ponty's later thought, and his 1956–1958 lecture courses on *Nature* in particular,[25] as well as Hans Jonas's *The Phenomenon of Life* and *The Imperative of Responsibility*, that phenomenology became interested in thinking the singularity of life.[26] More recently, this tradition has been developed by Renaud Barbaras.[27]

The situation in the Anglophone world, and in the UK in particular, was somewhat different, but yielded a similar suspicion regarding the category of life. Unlike his colleague Alfred North Whitehead who, by extending the category of organism, introduced purposiveness in every aspect of reality, Bertrand Russell, who had already denounced Bergsonism as an anti-intellectualism and an "imaginative and poetic view of the world",[28] wrote in 1925 that "vitalism as a philosophy, and evolutionism, show . . . a lack of sense of proportion and logical relevance".[29] At the same time, the logical positivism of the Vienna Circle, which was mainly made up of physicists, mathematicians and philosophers interested in those scientific disciplines, showed the same lack of interest in the life sciences. In the 1920s, Moritz Schlick, whose antipathy for Bergson was not a secret,[30] also rejected *en bloc* the vitalism implicit in Hans Driech's theory of *entelechy*.[31] According to Schlick, the causal closure of physical space-time implies that there cannot be causes of organic processes which are not themselves spatial. As a result, the laws of biology can, in principle, be reduced to those of physics. Ten years later, Philipp Frank, another member of the Circle, followed the same path.[32] Even if this refutation of "ontological" vitalism did not necessarily imply a reductionist conception of biological phenomena, analytic philosophers such as Ernst Nagel not only based their research on an attempt to give deflationary accounts of teleological explanations in biology[33] but also showed no interest in biology's most recent investigations. The case of Karl Popper, who, as late as 1974, claimed that Darwinism was not a scientific enterprise,[34] is also paradigmatic. In 1969, the zoologist and philosopher Ernst Mayr complained that books with "philosophy of science" in their titles were misleading and should be retitled "philosophy of physics".[35] It was only in the 1970s and 1980s that analytic philosophy of science became more interested in biology, initially focusing on the theory of evolution and then, in the 1990s, integrating the most recent results of biology.[36] Exemplary, in that respect, is David Hull's 1974 *Philosophy of Biological Science*: while displaying an obvious interest in biology from a philosophical point of view, it also claims (symptomatically, in our view) that "both scientists and philosophers take ontological reduction for granted" and that "organisms are nothing but atoms, and that is that".[37]

BIOPHILOSOPHY

Two years after the publication of Monod's, Jacob's and Lwoff's[38] books, Georges Canguilhem, whose philosophy of norms is an inspiration for many of the chapters gathered in this volume, examined the different approaches to life proposed by the three biologists.[39] He showed his appreciation for the work of Jacob, which he compared to that of Michel Foucault.[40] Both the biologist and the philosopher, in his view, criticize linear accounts of time and try to reconstruct the history of the life sciences by drawing on anonymous conceptual correspondences. But Canguilhem also points to a blind spot in Jacob's work: while trying to give an answer to the old question about the meaning of life, he also ignores the *subject* who is posing the question. According to the scientists' interpretation, DNA is to be understood as a code,[41] or a message, but one that is impersonal, has neither sender nor addressee, and is without will or scope. "Code" or "message" are simple metaphors,[42] the *raison d'être* of which has to be understood in relation to the human subject who seeks to find a meaning and who is unsatisfied with the meaning that he finds.[43] By underlying the change of language and the use of metaphors in biology, Canguilhem wanted to stress the centrality of the living being who is the subject of the knowledge of life, and of its valorization. The works of Jacob, Monod and Lwoff are not, in fact, deprived of any normative aspect. On the contrary, they contain elements that are incompatible with the rigorous scientism they advocate: Lwoff stresses the originality and the specificity of "the biological order"; Monod introduces the category of teleonomy and wants to deduce an ethics from biology; and Jacob emphasizes the irreducibility of the living being to matter, and the need to understand it through a series of existential categories such as openness, freedom, integration, death and desire. According to Canguilhem, the meaning of those categories and structures, which science objectifies, requires another discipline—namely, philosophy. The French philosopher makes the case for another critical discipline, one that would be able to articulate the necessary conflict between the objectifying act that is proper to knowledge and the auto-normativity of the living being in search for meaning: "outside of the laboratory . . . he concluded—love, birth and death continue to propose to living beings, sons of order and chaos, the immemorial figures of those questions that the science of life no longer poses to life".[44]

The importance of the living being as the absolute centre of knowledge and action, of science and ethics, was present in Canguilhem's thought from the start—that is, from the time of his PhD in medicine, published in 1943 as *The Normal and the Pathological.*[45] This book constitutes an exception in the panorama of twentieth-century Western philosophy, and especially in French thought, in that it strongly criticizes the then established ideas of the organism as a simple machine, of diseases as something quantifiable, and of the parallel between organisms and societies.[46] As Sander Werkhoven and Julien Pieron, each in his own way, explain in their chapters, Canguilhem proposes an idea of the organism as a being endowed with a certain normative power. Therefore, the idea of normality is based on a statistic: it means the highest frequency of cases of a certain type.[47]

In 1947,[48] in his review of Raymond Ruyer's *Éléments de psychobiologie*, Canguilhem lamented the absence of a "biological philosophy" or a "biophilosophy" in France: he opposed this philosophy, which puts life at the centre of our experience, to a "philosophy of biology" in the sense of a disengaged reflection on the methods of the science of life.[49] As Giuseppe Bianco shows in his chapter, the philosopher was following in the footsteps of Kurt Goldstein,[50] and of a certain "romantic" and holistic tradition proper to German *Lebensphilosophie*:[51] it is, he claims, of the utmost importance to think of life not as an "empire within an empire", or as a particular kind of matter subtracted to the laws of physics, but as the absolute centre of our experience of the world, and this means of our ethics, politics and science. According to Canguilhem, the tendency to treat the living as a simple result of a series of physical interactions was the result of a choice of values that privileges conservation, regularity and prevision over the "spontaneous production" which, according to him, is proper to life itself. He located the paradigm of this deliberate will to ignore the specificity of life in Cartesian mechanism, and emphasized the importance of scientific vitalism in the movement towards the development and independence of biology in the nineteenth century. Canguilhem praised Bergson for being the only French philosopher who located life at the centre of his interests and tried to inscribe knowledge and technology in the movement of life itself. As Charles T. Wolfe and Andy Wong show in their chapter, Canguilhem viewed his "vitalism" not as a metaphysics—although he was not entirely immune to that temptation, evident in his insistence on "Life" as a kind of transhistorical notion—or as an admission of the limits of scientific reason, or even as an invitation to rely on an extra-scientific and intuitive access to Life, but rather as the acknowledgement of the originality of the living and of its role in the construction of concepts used to gain access to reality. This thread basically consisted in combining historical and scientific information about the life sciences with critical, normative and even speculative considerations. As Wolfe and Wong emphasize, it found a limited success in France: only four or five thinkers, who are currently being rediscovered, tried to deal, from a "vitalist" point of view, with the conceptual foundations of the life sciences.

BIOETHICS AND BIOPOLITICS

Since the 1960s,[52] a number of technological breakthroughs have taken place in the field of biology and clinical medicine: new devices offer the possibility of keeping alive patients who otherwise would have died; artificial fertilization and new reproduction techniques reconfigure the relationships between parents and children; contraceptives, prenatal testing, safe abortions and developments in genetics give increased choices about the number and kinds of children we have; transplants and the possibility of sex change provoke discussions about identity. These breakthroughs, and the issues they raise, go hand in hand with a growing concern about the power exercised by doctors and medical institutions, and give birth to issues about patients' rights and the rights of the community as a whole to be involved in decisions that concern them. In 1970, the American chemist Van Rensselaer Potter[53] proposed the term "bioethics" to designate a "science of survival" in the ecological sense, an interdisciplinary study aimed at ensuring the preservation of the biosphere. This field of research never became widely established, but the term bioethics was used to refer to the growing interest in the ethical issues arising from the progresses in biology and medicine. This discipline was constituted as an extension of a reflection that began with medical and nursing ethics: Joseph Fletcher's *Morals and Medicine*, published in 1954, is considered to be the first work in that field.

As Guillaume Le Blanc shows in his chapter, and by anchoring his analysis in Foucault's *Birth of the Clinic*, this double origin of bioethics is the sign of a contradiction that has affected the "care" of life since the beginning of the century, but especially after World War II, when the Nuremberg Code established the necessity of human consent in medicine. The first usage of the term shows that one of the ancestors of bioethics is ecology. The term emerged in the nineteenth century, but was instituted as a discipline in the twentieth century. It needs to be understood in the light of the transformation of the attitude towards life introduced by Darwin and Malthus, and according to which human beings share something with the rest of the animal kingdom, and belong to an ecosystem or to "nature", which they need to take care of. In that respect, ecology postulates the existence of an all-comprehensive system called "life". On the other hand, the birth of bioethics also constituted a reaction to the transformation of medicine. To the extent that it considers pathology as a lower degree of normality, medicine is no longer characterized by the value of health, but by that of normality, defined as the maximal statistic frequency of a certain configuration of human life. Medicine is then extended to all aspects of human life and its main objective is to restore man's so-called natural norms. Bioethics—as medical ethics—is then a reaction against this extension, which forgets the importance of human subjectivity. Therefore, as Le Blanc points out, the contradiction is located between the two origins of bioethics and its two contradictory objectives: the first, derived from ecology, seeks to defend the entirety of "life", whereas the second focuses on the defence of the human from the abuses of the medical apparatus.[54]

This contradiction, however, cannot be understood without introducing a further concept, that of biopolitics.[55] The emergence of modern medicine and ecology during the nineteenth century, and that of bioethics in the aftermath of the Second World War, has to be read against the backdrop of the transformation in the nature of power which, according to Foucault, began to take place towards the beginning of the nineteenth century.[56] Until then, he claims, power was essentially defined as *sovereign* power, or the power of the sword. It consisted in the sovereign's power or right to kill—that is, to "take life or let live". It's sovereign power thus defined that began to be complemented, rather than simply replaced, by a different, and in fact opposite, kind of right—namely, the right to live. The new kind of power, which still governs us today, consists in "making live and letting die". It acts no longer on the body, which sovereign power can (still) kill, discipline, punish, but on "man-as-living-being", or "man-as-species". As an object of power and a problem of government, human beings now constitute a mass (a population) affected and in fact *characterized* by natural phenomena such as birth, death, reproduction, illness and so on. As a result of this historical development, the nature of politics itself has changed, and life itself, in the words of Paltrinieri, has become "an unalterable element at the base of political action". Politics has become *biopolitics*: it is concerned with birth and mortality rates, fertility, longevity, morbidity and hygiene. Those new phenomena and concerns resulted in the development of medicine, the main function of which was public hygiene, as well as of institutions aimed at "coordinating medical care, centralising power, and normalising knowledge".[57]

This means that the object or target of politics has, to a large extent, become the well-being of the population as a whole, understood precisely as a *living* entity. The well-being in question includes the right to a decent life and good health, and requires the creation of the system of health care, social services and social security we have become accustomed to.[58] But it also, and by the same token, introduces new modalities of power, new institutions and new discourses—new ways, that is, of taking hold of human life, of shaping and organizing it, transforming it, extending and amplifying it, or, on the contrary, limiting it, containing it, and even exterminating it.[59] On the one hand, state power is now concerned with the health, safety and security of its population. At times, it acts on life directly, and even against the will of those most immediately concerned, as in the case of the artificially prolonged life in hospital, which can rob the individual of the very decency of life the system also seeks to establish. At other times, though, and simultaneously, its action is more indirect, and reaches the most intimate spheres of life. Through various incentives, the diffusion of information and advice, and methods of self-regulation, it seeks to influence the way in which we live, understand life itself, and relate to ourselves. In other words, it generates norms of behaviour and operates like a normalizing (and not just normative) force, as Pieron and Caeymaex show in their chapters, and Kollias explores in relation to sexuality. It requires and produces at the same time a new type of subject. It develops new technologies of government through freedom, responsibility, interests and desires, as Beistegui and Singy show in their respective chapters. Thus, the life that's at stake within biopower is not exclusively biological, but is also an object of economic governmentality. On the other hand, we shouldn't forget that biopower also paves the way for an understanding of populations in terms of race and, through the biologization of politics, for state racism, eugenics and genocide (a topic that Bernasconi addresses in his chapter). The political battle, then, is one for the very definition of life itself, as Gracieuse and Morar argue, and may involve forms of resistance to, but also reconfigurations of, biopolitics itself.

This struggle involves also issues of sexuality and gender and the redefinition of what is the relation between, on the one hand, human reproduction and human sexual desire (and pleasure) and, on the other hand, desire and the production and consumption of goods in the broader sense. As such, the notions and, we would claim, experience of desire and pleasure, which the fourth and final part of this volume investigates, can be situated at the junction of two discourses and practices, with their respective histories: that of sexuality, psychiatry and psychoanalysis, and that of political economy. Those two types of discourses and the experiences they formalize have largely defined who we are today, and they continue to shape our subjectivity. During the 1930s, Wilhelm Reich, an Austrian psychoanalyst, was the first to try to merge the theories of Freud with the analysis brought by Marxism in publications such as *The Mass Psychology of Fascism* (1933) and *The Sexual Revolution* (1936). Around the same time (in 1936, to be precise), the members of the Institut für Sozialforschung, founded in the University of Frankfurt in 1923, published a volume, *Studien über Autorität und Familie* (*Studies on Authority and Family*), in which, like Reich, they explored the role of the family in the service of producing an authoritarian structure. Especially noteworthy is Erich Fromm's essay, which consists of a summary of the basic concepts of psychoanalysis and suggestions on how they can be applied to a broad understanding of social dynamics. This thread, commonly called "Freudo-Marxism", disappeared in the 1940s, but reemerged in the figure of another member of the Frankfurt School and former student of Heidegger, Herbert Marcuse, who analysed the role of sexuality in capitalist societies in works such as *Eros and Civilization* (1955).[60] Marcuse criticized the hypothesis, formulated by Freud in *Civilization and Its Discontents*, that the suppression of instincts and of the energy produced by sexuality is necessary to the progress of society. He displaced the conflict between Eros and work (or the reality principle) to the one between Eros and alienated labour.

This tradition was taken up in France only at the end of the 1960s, especially in the aftermath of the events of May 1968, and following the demands for the liberation of sexuality. Until that time, psychoanalysis and social critique had been kept separate, mostly as a result of the critique of the communist party's condemnation of the former as part of "bourgeois ideology". Towards the beginning of the 1970s, Jacques Lacan, who was then in contact with Louis Althusser, began using Marxian concepts in his own seminar. To his quadripartite theory of discourse (the discourses of the hysteric, the master, the analyst, and the university) he thus added a fifth discourse, the "discourse of the capitalist". The problematization of desire—at the junction between social critique and psychoanalysis—by Foucault, Lyotard and Deleuze and Guattari arose in that context. For different reasons, Foucault,[61] Lyotard,[62] and Deleuze and Guattari[63] all reject the Freudo-Marxist ideas of the repression of desire as responsible for the ills of contemporary society, the "liberation" of desire as the solution to those ills and the celebration of pleasure as the natural outcome of that liberation. In the first volume of his *History of Sexuality*, Foucault insists on the fact that, since the middle of the nineteenth century, desire, far from being suppressed and condemned to silence, has been a constant source of attention, description and subjected to highly sophisticated processes of confession and veridiction, many of which were inherited from Christian practices of "alethurgy" and "pastoral power". As for Deleuze and Guattari, they claim that, far from repressing desire, capitalism, like fascism, is a product of it. In fact, they insist, desire is itself essentially productive, not lacking in anything (*pace* Lacan), and not oriented towards pleasure as towards its ultimate goal. Capitalism is the dominant *economic* organization of desire, in the same way that fascism was its more strictly *political* organization. Desire, they claim, is the manifestation of life itself, the way in which life, which they conceive as inorganic, insofar as it generates and exceeds the organism, is able to produce individualized subjects and objects.

As a result of being inscribed in a socioeconomic history, read through the framework of a broader social critique and, in the case of Deleuze and Guattari, inserted in a broader ontology, the basic concepts of psychoanalysis dealing with human desire and sexuality took a new dimension. At roughly the same time, psychoanalysis and, more generally, the sciences of the psyche were also going through other, maybe even more radical, transformations.

THE LIFE OF THE BRAIN

In *The Physiology of Truth* (2002),[64] the neurophysiologist Jean-Pierre Changeux declared that everything that once belonged to the field of the spiritual, the transcendent and the immaterial was in the process of being materialized, naturalized. Twenty years earlier, in his controversial *The Neuronal Man: The Biology of Mind,*[65] he had already presented the view, which he shared with other scientists, according to which the brain needed to be understood from the point of view of evolution, artificial intelligence and molecular biology. There is, he claims in that book, no such thing as the "mind", and nothing that can count as specifically psychic; there are only neurons, synapses, electricity and chemistry. The human brain, or, as he calls it, "the cerebral machine", is nothing but "an assembly of neurons".[66]

In the twenty years that separated the publication of those two books, the field of neurophysiology saw an explosion of studies seeking to explain the reflexive, decisional, emotional and, sometimes, even social functions of the human mind by appealing to the morphology of the brain. But for the radical and exceptional case of John Eccles and a few other pioneers, the majority of those scientists, particularly Daniel Dennett and Paul and Patricia Churchland, rejected the category of the "mental" in favour of a "materialist eliminative" approach, which reduces mental states to a certain state of the brain. This approach, which aims to naturalize the human as a whole, had a deep influence on the humanities and gave birth to what Charles T. Wolfe recently called a "brain theory".[67] Where the philosophy of neuroscience limits itself to an epistemological reflection on the discipline, and "neurophilosophers"[68] use the results of the neurosciences to solve a number of classical philosophical questions through a reductive method, other researchers stress the compatibility between the philosophical framework emerging from the results of neuroscience with the phenomenological tradition that focuses on action and embodiment. As Claudia Stein and Roger Cooter point out in their chapter, this influence of cognitive neuroscience went as far as to affect the discourse of some historians, who affirm the possibility of deciphering a "deep history" contained in our brains. This project, like many appropriations of the discourses of neuroscience by the humanities, takes for granted the results of neuroscience and the "spontaneous philosophy of scientists", as Althusser would have called it. The result is that, paradoxically, most of those projects lack a historical and critical dimension.

By contrast, and as early as 1981, Georges Canguilhem criticized cognitive neuroscience, which he saw as a new phrenology.[69] According to the philosopher, not only does it conflate a reality (the brain) with a model (the computer), but it also ignores the set of values and the social implications implicit in the choice of this very model. Initially, scientists ignored this type of epistemological reflection, based on the same critical approach that characterized Canguilhem's critique of the mechanical models adopted in biology. At the same time, so-called continental philosophy remained entrenched in a refusal to take into account the progresses of science.

In the last few years, however, the project of the neurosciences has drawn fundamental and critical interrogations. The non-eliminative theories of enactivism, autopoiesis, plasticity and emergentism generated questions about the status and specificity of the brain and of the models, metaphors and concepts used to understand it. Those questions echo other interrogations concerning the kind of *life* that allows for the knowledge of life by the living human being, the models that research institutions are using and the possibilities of a critique of those models. In so-called continental philosophy, Paul Ricœur attempted a conversation with Changeux as early as 1998, while Deleuze and Guattari's last book, *What Is Philosophy?*, ends with a paradigmatic chapter entitled "From Chaos to the Brain".[70] Finally, Catherine Malabou's *What Should We Do with Our Brain?*[71] can be considered as one of the latest attempts to discuss the results of the naturalizing project of neuroscience, to which she adds an historical and critical dimension.

THE REBIRTH OF LIFE?

The rise of bioethics as an independent discipline, the concept of biopolitics as a key notion to understand the intertwining between knowledge and power since the nineteenth century and the rebirth of a biophilosophy all seem to indicate that, in the humanities, the concept of Life represents today, to borrow Sartre's famous expression, "the unsurpassable horizon of our time".[72] On the other hand, and as Michel Morange shows in a recent book,[73] the blurring of the informational interpretation of genetics, the critique of the notion of program and the new interdisciplinary advances in biology of the last thirty or forty years have provoked biologists themselves to raise anew the question of the origin and features of life. Since the 1990s, aspects of biology normally considered separated are converging towards a "synthetic" perspective involving the three aspects of evolution, development and cognition.

Contrary to Jacob's declaration, with which we began, contemporary discourses of life seem to proliferate in various contexts, inside, but especially outside the laboratory. By defining and assessing critically the specific meaning of life at stake in bioethics and biopolitics, and by focusing on the debates around norms and values, as well as specific aspects of life today, from ageing, suffering and dying to desire, pleasure and sexuality, this volume seeks to give a sense of why and how life has become an all-encompassing problem—why all questions, especially ethical and political ones, have become vital questions.

Most, but not all, of the contributions contained in this volume take their point of departure in a certain "continental tradition" that we could characterize as a critical philosophy of norms, or a critical biophilosophy, which draws on the works of philosophers such as Henri Bergson, Geroges Canguilhem, Michel Foucault,[74] Jacques Lacan, Georges Bataille, Bruno Latour, Etienne Balibar, Gilles Deleuze and Félix Guattari. Earlier and often shorter versions of most of the chapters gathered in this volume were presented in the three workshops, held at the University of Warwick, around the theme of bioethics and biopolitics, between 2011 and 2014. We are extremely grateful to the Leverhulme Trust, who funded the workshops and the research, presented in this volume, carried out by G. Bianco and M. Gracieuse during those years.

The volume is divided in four parts. Part I explores the origins, foundations and presuppositions of the concepts of bioethics and biopolitics through a series of methodologically specific investigations and critical engagements. Part II focuses on the concepts of norms, normativity and normality, from which the discourses of biology and medicine, and the various institutions on which they rely, are indissociable. Part III shows how this normative and normalizing environment affects the way in which phenomena that all human beings are confronted with—namely, disease, identity (in the case of Gracieuse), pain (in the case of Bianco) and ageing (in the case of Miquel and Fuller)—are thought about scientifically, but also dealt with in a practical, medical environment. Finally, part IV explores the ways in which, through the discourses of psychiatry and psychoanalysis, as well as the law and political economy, the areas of desire and pleasure are themselves subjected to normative and normalizing processes, which provide a window into the state of the contemporary subject. All four parts raise the question not only of the good or most useful concepts to adopt in the attempt to think life, or the "Age of Life", but also of the good or decent life. And it's within the context of this double approach—theoretical on the one hand, ethical and political on the other—that the idea of *care* (in all its ambiguous and delicate dimensions) is given centre stage.

BIBLIOGRAPHY

Agamben, Giorgio. *Homo Sacer: Sovereign Power and Bare Life*. Translated by Daniel Heller-Roazen. Stanford, CA: Stanford University Press, 1998.

Barbaras, Renaud. *Introduction à une phénoménologie de la vie*. Paris: Vrin, 2008.

———. *La vie lacunaire*. Paris: Vrin, 2011.

Biopolitics: An Advanced Introduction. Translated by Erik Frederick Trump. New York: New York University Press, 2011.

Brandon, Robert N. *Concepts and Methods in Evolutionary Biology*. Cambridge: Cambridge University Press, 1996.

Canguilhem, Georges. *Knowledge of Life*. Translated by Stefanos Geroulanos and Daniela Ginsburg. New York: Fordham University Press, 2008.

———. "The Brain and Thought". *Radical Philosophy* 148, no. 7 (2008).
———. "Le concept et la vie". In *Études d'histoire et de philosophie des sciences.* Paris: Vrin, 1968, 335–64.
———. "Logique du vivant et histoire de la biologie". *Sciences* 71 (1971): 20–25.
———. "Note sur la situation faite en France à la philosophie biologique". *Revue de métaphysique et de morale* 52 (1947): 322–32.
———. "The Death of Man, or Exhaustion of the Cogito?" In *The Cambridge Companion to Foucault*, Translated by Catherine Porter, edited by Gary Gutting. New York: Cambridge University Press, 1995.
———. "Vie". In *Encyclopædia Universalis, Encyclopedia Universalis*, t. 23 (1989): 530–46.
———. *The Normal and the Pathological.* Translated by Carolyn R. Fawcett and Robert S. Cohen. New York: Zone Books, 1998.
———. *Writings on Medicine.* Translated by Ste Geroulanos and Todd Meyers. New York: Fordham University Press, 2012.
Cassirer, Ernst. *An Essay on Man: An Introduction to a Philosophy of Human Culture.* New Haven, CT: Yale University Press, 1944.
———. *The Philosophy of Symbolic Forms: The Metaphysics of Symbolic Form.* Vol. 4. Translated and edited by John Michael Krois and Donald P. Verene. New Haven, CT: Yale University Press, 1997.
———. *The Problem of Knowledge. Philosophy, Science, and History since Hegel.* Translated by William H. Woglom and Charles W. Hendel. New Haven, CT: Yale University Press, 1950.
Changeux, Jean-Pierre. *Neuronal Man: The Biology of Mind.* Translated by Laurence Garey. Princeton, NJ: Princeton University Press, 1997.
———. *The Physiology of Truth. Neuroscience and Human Knowledge.* Translated by Malcolm DeBevoise. Cambridge, MA: Harvard University Press, 2009.
Charles T. Wolfe (edited by). *Brain Theory: Essays in Critical Neurophilosophy.* London: Palgrave Macmillan, 2014.
Churchland, Patricia Smith. *Brain-Wise: Studies in Neurophilosophy.* Cambridge, MA: MIT Press, 2002.
———. *Neurophilosophy: Toward a Unified Science of the Mind-Brain.* Cambridge, MA: MIT Press, 1986.
Crick, Francis. *Of Molecules and Men.* Seattle: University of Washington Press, 1968.
Deleuze, Gilles, and Felix Guattari. *Anti-Oedipus.* Translated by Robert Hurley, Mark Seem and Helen R. Lane. London: Athlone Press, 1984.
———. *What Is Philosophy?* Translated by Janis Tomlinson and Graham Burchell. New York: Columbia University Press, 1996.
Derrida, Jacques. *The Animal That Therefore I Am (More to Follow).* Translated by David Willis and edited by Marie-Louis Mallet. New York: Fordham University Press, 2008.
Derrida, Jacques. *The Beast and the Sovereign.* Translated by Geoffrey Bennington. Chicago: University of Chicago Press, 2009.
Esposito, Roberto. "Totalitarianism or Biopolitics? Concerning a Philosophical Interpretation of the Twentieth Century". *Critical Inquiry* 34(4): 633–44.
———. *Bios: Biopolitics and Philosophy.* Translated by Timothy Campbell. Minneapolis: University of Minnesota Press, 2008.
Foucault, Michel. *Society Must Be Defended: Lectures at the Collège de France, 1975–1975.* Translated by David Macey. London: Penguin, 2003.
———. *The History of Sexuality, Volume 1: An Introduction.* Translated by Robert Hurley. London: Allen Lane, 1979.
———. *The Order of Things.* Translated by Adam Scheridan. London: Tavistock, 1970.
Frank, Philipp. *The Law of Causality and Its Limits.* Edited and translated by Marie Neurath and Robert S. Cohen. Dordrecht: Kluwer, 1998.
Geroulanos, Stefanos, and Todd Meyers. *Experimente im Individuum. Kurt Goldstein und die Frage des Organismus.* Berlin: August Verlag, 2014.

Giroux, Élodie. *Après Canguilhem: définir la santé et la maladie*. Paris: Presses Universitaires de France, 2010.

Goldstein, Kurt. *The Organism: A Holistic Approach to Biology Derived from Pathological Data in Man*. New York: Zone Books, 1995.

Guillo, Dominique. *Les figures de l'organisation. Sciences de la vie et sciences sociales au xixe siècle*. Paris: Presses Universitaires de France, 2003.

Hans-Jörg Rheinberger. *An epistemology of the concrete: Twentieth-century histories of life*. Durham, NC: Duke University Press, 2010.

Heidegger, Martin. *The Fundamental Concepts of Metaphysics: World, Finitude, Solitude*. Translated by William McNeill and Nicholas Walker. Bloomington: Indiana University Press, 1995.

Henry, Michel. *Philosophy and Phenomenology of the Body*. Translated by Girard J. Etzkorn. New York: Springer, 1975.

Hull, David. "Philosophy and Biology". In *Contemporary Philosophy: A New Survey*, vol. 2, edited by Guttorm Fløistad. The Hague: Martinus Nijhoff, 1982.

Hull, David. *Philosophy of Biological Science*. Englewood Cliffs, NJ: Prentice Hall, 1974.

Jacob, François. *The Logic of Life: A History of Heredity*. Translated by Betty E. Spillman. Princeton, NJ: Princeton University Press, 1993.

Jonas, Hans. *The Imperative of Responsibility: In Search of Ethics for the Technological Age*. Translated by Hans Jonas and David Herr. Chicago: University of Chicago Press, 1984.

Jonas, Hans. *The Phenomenon of Life: Toward a Philosophical Biology*. New York: Harper & Row, 1966.

Kahane, Ernest. *La vie n'existe pas!* Paris: Éditions rationalistes, 1962.

Kuhse, Helga, and Peter Singer, eds. *A Companion to Bioethics*. London: Blackwell, 2003.

Lecourt, Dominique. *Proletarian Science? The Case of Lysenko*. Translated by Ben Brewster. London: New Left Books, 1977.

Lévi-Strauss, Claude. *The Savage Mind*. Translated by George Wiedenfeld and Nicolson Ltd. Oxford: Oxford University Press, 1966.

Lwoff, André *The Biological Order*. Cambridge, MA: MIT Press, 1965.

Lyotard, Jean-François. *Libidinal Economy*. Translated by Ian Hamilton Grant. Bloomington: Indiana University Press, 1993.

Malabou, Catherine. *What Should We Do with Our Brain?* Translated by Sebastian Rand. New York: Fordham University Press, 2008.

Marcuse, Herbert. *Eros and Civilization*, London: Routledge, 1987.

Mayr, Ernst. "Footnotes on the Philosophy of Biology". *Philosophy of Science* 36, no. 2 (1969): 197–202.

Merleau-Ponty, Maurice. *Nature: Course Notes from the Collège de France*. Translated by Robert Vallier. Evanston, IL: Northwestern University Press, 2003.

———. *Phenomenology of Perception*. Translated by Colin Smith and Kegan Paul. London: Routledge, 1962.

Monod, Jacques. *Chance and Necessity: An Essay on the Natural Philosophy of Modern Biology*. Translated by Austryn Wainhouse. New York: Alfred A. Knopf, 1971.

Morange, Michel. *La vie expliquée? 50 ans après la double hélice*. Paris: Odile Jacob, 2007.

Muhle, Maria. *Eine Genealogie der Biopolitik. Der Lebensbegriff bei Foucault und Canguilhem*. München: Fink, 2013.

Nagel, Ernst. "Teleology Revisited". *Journal of Philosophy* 74 (1977): 261–301.

———. *The Structure of Science*. New York: Harcourt, Brace and World, 1961.

Pichot, André. *Histoire de la notion de vie*. Paris: Gallimard, 1993.

Plutynski, Anya, and Sahotra Sarkar. "Introduction" to *A Companion to the Philosophy of Biology*. London: Blackwell, 2008.

Popper, Karl. "Scientific Reduction and the Essential Incompleteness of All Science". In *Studies in the Philosophy of Biology: Reduction and Related problems*, edited by Francisco Jose Ayala and Theodosius Grigorievich Dobzhansky. Berkeley: University of California Press, 1974.

Potter, Van Rensselaer. "Bioethics, Science of Survival". *Biology and Medicine* 14 (1970): 127–53.

Ramberg, Peter J. "The Death of Vitalism and the Birth of Organic Chemistry: Wöhler's Urea Synthesis and the Disciplinary Identity of Organic Chemistry". *Ambix* 47, no. 3 (2000).

Rose, Nikolas. *Governing the Present: Administering Economic, Social and Personal Life*. Cambridge: Polity Press, 2008.

———. *The Politics of Life Itself: Biomedicine, Power and Subjectivity in the Twenty First Century*. Princeton, NJ: Princeton University Press, 2007.

Russell, Bertrand. "The Philosophy of Henri Bergson". *The Monist* 22 (1912): 321–47.

———. *What I Believe*. In *The Basic Writings of Bertrand Russell*. London: Routledge, 2004.

Sartre, Jean-Paul. *Questions de Méthode*. Paris: Gallimard, 1960.

Schaeffer, Jean-Marie. *La fin de l'exception humaine*. Paris: Gallimard, 1997.

Schlanger, Judith. *Les métaphores de l'organisme*. Paris: Harmattan, 1995.

Schlick, Moritz. "Philosophy of Organic Life". In *Readings in the Philosophy of Science*, edited by Herberg Feigl and May Brodbeck. New York: Appleton-Century Crofts, 1953.

Schlick, Moritz. *Allgemeine Erkenntnislehre*. Edited by Hans J. Wendel and Friedrich O. Engler. Vienna: Springer, 2007.

Schrödinger, Erwin. *What Is Life?* Cambridge: Cambridge University Press, 1944.

Shostak, Stanley. *Death of Life: Legacy of Molecular Biology*. London: Palgrave Macmillan, 1998.

Sternhell, Zeev. *Les anti-Lumières: Une tradition du xviii*[e] *siècle à la Guerre froide*. Paris: Fayard, 2006.

Vollmer, Gerhard. *Biophilosophie*. Reclam: Stuttgart 1995.

NOTES

1. François Jacob, *The Logic of Life: A History of Heredity*, trans. Betty E. Spillman (Princeton, NJ: Princeton University Press, 1993), 299.

2. Jacques Monod, *Chance and Necessity: An Essay on the Natural Philosophy of Modern Biology*, trans. Austryn Wainhouse (New York: Alfred A. Knopf, 1971), 211.

3. Erwin Schrödinger, *What Is Life?* (Cambridge: Cambridge University Press, 1944).

4. Cf. Michel Morange, *La vie expliquée? 50 ans après la double hélice* (Paris: Odile Jacob, 2007), chap. "Le crépuscule de la vie".

5. Crick clearly stated his intolerance to vitalism in *Of Molecules and Men* (Seattle: University of Washington Press, 1968): "To those of you who may be vitalists, I would make this prophecy: what everyone believed yesterday and you believe today, only cranks will believe tomorrow" (99).

6. Cf. Peter J. Ramberg, "The death of vitalism and the birth of organic chemistry: Wöhler's urea synthesis and the disciplinary identity of organic chemistry" (*Ambix* 47, no. 3 [2000]).

7. "So I accept the characterization of aesthete in so far as I believe the ultimate goal of the human sciences to be not to constitute, but to dissolve man": Claude Lévi-Strauss, *The Savage Mind*, trans. George Wiedenfeld and Nicholson (Oxford: Oxford University Press, 1966), 247.

8. Lévi-Strauss, *The Savage Mind*, 176.

9. Michel Foucault, *The Order of Things*, trans. Adam Scheridan (London: Tavistock, 1970).

10. Foucault, *The Order of Things*, 139.

11. For a history of the concept, see André Pichot, *Histoire de la notion de vie* (Paris: Gallimard, 1993).

12. Georges Canguilhem, "Vie", in *Encyclopædia Universalis, Encyclopedia Universalis*, t. 23 (1989): 546.
13. See Stanley Shostak, *Death of Life: Legacy of Molecular Biology* (London: Palgrave Macmillan, 1998).
14. Foucault, *The Order of Things*, 387.
15. See, for instance, Judith Schlanger, *Les métaphores de l'organisme* (Paris: Harmattan, 1995), and Dominique Guillo, *Les figures de l'organisation. Sciences de la vie et sciences sociales au xix e siècle* (Paris: Presses universitaires de France, 2003).
16. Cf. Jean-Marie Schaeffer, *La fin de l'exception humaine* (Paris: Gallimard, 1997).
17. See, for instance, Zeev Sternhell, *Les anti-Lumières: Une tradition du xviii e siècle à la Guerre froide* (Paris: Fayard, 2006).
18. Canguilhem was conscious of those risks. Cf. Georges Canguilhem, "Aspects of vitalism", in *Knowledge of Life*, trans. Stefanos Geroulanos and Daniela Ginsburg (New York: Fordham University Press, 2008).
19. See, for example, Dominique Lecourt, *Proletarian Science? The Case of Lysenko*, trans. Ben Brewster (London: New Left Books, 1977).
20. Ernest Kahane, *La vie n'existe pas!* (Paris: Éditions rationalistes, 1962).
21. Ernst Cassirer, *The Philosophy of Symbolic Forms*,Vol. 4, *The Metaphysics of Symbolic Form*, trans. and ed. John Michael Krois and Donald P. Verene (New Haven, CT: Yale University Press, 1997), 1–109; *An Essay on Man: An Introduction to a Philosophy of Human Culture* (New Haven, CT: Yale University Press, 1944); *The Problem of Knowledge. Philosophy, Science, and History since Hegel*, trans. William H. Woglom and Charles W. Hendel (New Haven, CT: Yale University Press, 1950).
22. Martin Heidegger, *Towards the Definition of Philosophy*, trans. Ted Sadler (London: Athlone, 2000), 92.
23. See Martin Heidegger, *The Fundamental Concepts of Metaphysics: World, Finitude, Solitude*, trans. William McNeill and Nicholas Walker (Bloomington: Indiana University Press, 1995), part 2. Heidegger's profound suspicion regarding the very concept of life also accounts for his condemnation of the Nazis' use of the concepts of "blood" and their "race-ideology", which he interprets as effects of the metaphysics of subjectivity. See M. Heidegger, *Gesamtausgabe Band 69: Die Geschichte des Seyns* (Frankfurt am Main: Vittorio Klostermann, 1998), 70.
24. Maurice Merleau-Ponty, *Phenomenology of Perception*, trans. Colin Smith and Kegan Paul (London: Routledge, 1962); Michel Henry, *Philosophy and Phenomenology of the Body*, trans. Girard J. Etzkorn (New York: Springer, 1975).
25. Maurice Merleau-Ponty, *Nature: Course Notes from the Collège de France*, trans. Robert Vallier (Evanston, IL: Northwestern University Press, 2003).
26. Hans Jonas, *The Imperative of Responsibility: In Search of Ethics for the Technological Age*, trans. Hans Jonas and David Herr (Chicago: University of Chicago Press, 1984); *The Phenomenon of Life: Toward a Philosophical Biology* (New York: Harper & Row, 1966).
27. Renaud Barbaras, *Introduction à une phénoménologie de la vie* (Paris: Vrin, 2008); *La vie lacunaire* (Paris: Vrin, 2011).
28. Bertrand Russell, "The Philosophy of Henri Bergson", *The Monist* 22 (1912): 321–47.
29. Bertrand Russell, *What I Believe*, in *The Basic Writings of Bertrand Russell* (1925, London: Routledge, 2004), 348.
30. Cf. Moritz Schlick, *Allgemeine Erkenntnislehre*, ed. Hans J. Wendel and Friedrich O. Engler (Vienna: Springer, 2007), sect. 17.
31. Moritz Schlick, "Philosophy of Organic Life", in *Readings in the Philosophy of Science*, ed. Herberg Feigl and May Brodbeck (New York: Appleton-Century Crofts, 1953), 523–36.
32. Philipp Frank, *The Law of Causality and Its Limits* [1932], ed. and trans. Marie Neurath and Robert S. Cohen (Dordrecht: Kluwer, 1998).
33. Ernst Nagel, *The Structure of Science* (New York: Harcourt, Brace and World, 1961), and "Teleology Revisited", *Journal of Philosophy* 74 (1977): 261–301.

34. Karl Popper, "Scientific reduction and the essential incompleteness of all science", in *Studies in the Philosophy of Biology: Reduction and Related Problems*, ed. Francisco Jose Ayala and Theodosius Grigorievich Dobzhansky (Berkeley: University of California Press, 1974), 259–84.

35. Ernst Mayr, "Footnotes on the Philosophy of Biology", *Philosophy of Science* 36, no. 2 (1969): 197–202.

36. See Sahotra Sarkar and Anya Plutynski, "Introduction" to *A Companion to the Philosophy of Biology* (London: Blackwell, 2008).

37. David Hull, "Philosophy and Biology", Guttorm Fløistad (ed.), *Contemporary Philosophy: A New Survey*, vol. 2 (The Hague: Martinus Nijhoff), 282, quoted by Wolfe and Wong. After Hull's *Philosophy of Biological Science* (Englewood Cliffs, NJ: Prentice Hall, 1974) the field developed quite rapidly. Robert N. Brandon could say that when he was a graduate during the late 1970s, "I knew five philosophers of biology: Marjorie Grene, David Hull, Michael Ruse, Mary Williams and William Wimsatt" (Robert N. Brandon, *Concepts and Methods in Evolutionary Biology* [Cambridge, Cambridge University Press, 1996], xii–xiii).

38. André Lwoff, *The Biological Order* (Cambridge, MA: MIT Press, 1965).

39. Georges Canguilhem, "Logique du vivant et histoire de la biologie" *Sciences* 71 (1971): 20–25.

40. In 1966 Canguilhem wrote two articles: the first one was a review of Michel Foucault's *Order of Things* ("The death of man, or exhaustion of the cogito?", in *The Cambridge Companion to Foucault*, trans. Catherine Porter, ed. Gary Gutting [New York: Cambridge University Press, 1995, 71–91]); the second was an article assessing the new discoveries in genetics ("Le concept et la vie", in *Études d'histoire et de philosophie des sciences* [Paris: Vrin, 1968], 335–64).

41. For this, see "Le concept et la vie", in which Canguilhem speaks about DNA as a code or a *logos*, a "material *a priori*" inscribed in life.

42. See also Michel Morange, *La vie expiqueée?*, 33.

43. In "Le concept et la vie", Canguilhem underlined to what extent but just changed *language*: if during several centuries the language was the one of mechanics, during the 1960s the dominating language was the one of linguistics and cybernetics.

44. Canguilhem, "Logique du vivant et histoire de la biologie", 25.

45. Georges Canguilhem, *The Normal and the Pathological*, trans. Carolyn R. Fawcett and Robert S. Cohen (New York: Zone Books, 1998), 91.

46. See especially Georges Canguilhem, "The problem of regulations in the organism and in society" [1956], in *Writings on Medicine*, trans. Stefanos Geroulanos and Todd Meyers (New York: Fordham University Press, 2012).

47. For a reading comparing the position of Canguilhem with the recent positions emerging from the debate between "naturalists" and "normativists", see Élodie Giroux, *Après Canguilhem : définir la santé et la maladie* (Paris: Presses Universitaires de France, 2010).

48. Georges Canguilhem, "Note sur la situation faite en France à la philosophie biologique", *Revue de métaphysique et de morale* 52 (1947): 324.

49. See Charles T. Wolf and Andy Wong's chapter for the difference. Gerhard Vollmer used the term though with a slightly different meaning (Gerhard Vollmer, *Biophilosophie* [Reclam: Stuttgart, 1995]).

50. Kurt Goldstein, *The Organism: A Holistic Approach to Biology Derived from Pathological Data in Man* (New York: Zone Books, 1995).

51. Cf. Stefanos Geroulanos, Todd Meyers, *Experimente im Individuum. Kurt Goldstein und die Frage des Organismus* (Berlin: August Verlag, 2014).

52. See Helga Kuhse and Peter Singer, eds., *A Companion to Bioethics* (London: Blackwell, 2003).

53. Van Rensselaer Potter, "Bioethics, Science of Survival", *Biology and Medicine* 14 (1970): 127–53.

54. One can formulate the hypothesis that the rise of "animal rights" and anti-"specism" movements belongs in the problematic that gave rise to ecology and bioeth-

ics. The question of the status of animality, connected to a critique of anthropocentrism, has been at the centre of the reflection of several authors, including Derrida. See, for instance, Jacques Derrida, *The Beast and the Sovereign* (2008), trans. Geoffrey Bennington (Chicago: University of Chicago Press, 2009), and *The Animal That Therefore I Am (More to Follow)*, trans. David Willis and ed. Marie-Louis Mallet (New York: Fordham University Press, 2008).

55. For an overview, see, for example, *Biopolitics: An Advanced Introduction*, trans. Erik Frederick Trump (New York: New York University Press, 2011).

56. See Michel Foucault, *Society Must Be Defended: Lectures at the Collège de France, 1975–1975*, trans. David Macey (London: Penguin, 2003), 17 March 1976.

57. Foucault, *Society Must Be Defended*, 244.

58. See Nikolas Rose, *The Politics of Life Itself: Biomedicine, Power and Subjectivity in the Twenty First Century* (Princeton, NJ: Princeton University Press, 2007), and *Governing the Present: Administering Economic, Social and Personal Life* (Cambridge: Polity Press, 2008).

59. Here, we can refer only obliquely to the ongoing debates and disagreements, within biopolitics itself, regarding the place and meaning of state and institutional racism, and of genocide in particular. Should we, as Foucault does, see those phenomena as a direct consequence of "life" having become the object of power? Or should we, as Agamben claims, understand modern biopolitics on the basis of the Nazi concentration camp? See Giorgio Agamben, *Homo Sacer: Sovereign Power and Bare Life*, trans. Daniel Heller-Roazen (Stanford, CA: Stanford University Press, 1998). A third—and somewhat intermediary—position is that of Roberto Esposito, as presented through the paradigm of "immunization" in "Totalitarianism or Biopolitics? Concerning a Philosophical Interpretation of the Twentieth Century", *Critical Inquiry* 34(4): 633–44, as well as in *Bios: Biopolitics and Philosophy*, trans. Timothy Campbell (Minneapolis: University of Minnesota Press, 2008), chapter 4, "Thanatopolitics". Nikolas Rose's position is different, in that, in the sixth chapter of *The Politics of Life Itself*, he claims that biopolitics can free itself from racism. Those questions are at the heart of Bernasconi's chapter.

60. Herbert Marcuse, *Eros and Civilization* (London: Routledge, 1987).

61. Michel Foucault, *The History of Sexuality: An Introduction* (1976), vol. 1, trans. Robert Hurley (London: Allen Lane, 1979).

62. Jean-François Lyotard, *Libidinal Economy*, trans. Ian Hamilton Grant (1972; Bloomington: Indiana University Press, 1993).

63. Gilles Deleuze and Felix Guattari, *Anti-Oedipus*, trans. Robert Hurley, Mark Seem and Helen R. Lane (London: Athlone Press, 1984).

64. Jean-Pierre Changeux, *The Physiology of Truth : Neuroscience and Human Knowledge*, trans. Malcolm DeBevoise (Cambridge, MA: Harvard University Press, 2009).

65. Jean-Pierre Changeux, *Neuronal Man: The Biology of Mind*, trans. Laurence Garey (Princeton, NJ: Princeton University Press, 1997).

66. Changeux, *Neuronal Man: The Biology of Mind*, 135.

67. Charles T. Wolfe, *Brain Theory: Essays in Critical Neurophilosophy* (London: Palgrave Macmillan, 2014).

68. See, for instance, Patricia Smith Churchland's *Neurophilosophy: Toward a Unified Science of the Mind-Brain* (Cambridge, Massachusetts, 1986), or *Brain-Wise: Studies in Neurophilosophy* (Cambridge, MA: MIT Press, 2002).

69. Georges Canguilhem, "The Brain and Thought" (1981), *Radical Philosophy* 148, no. 7 (2008).

70. Gilles Deleuze and Felix Guattari, *What Is Philosophy?*, trans. Janis Tomlinson and Graham Burchell (New York: Columbia University Press, 1996).

71. Catherine Malabou, *What Should We Do with Our Brain?*, trans. Sebastian Rand (New York: Fordham University Press, 2008).

72. Jean-Paul Sartre, *Questions de Méthode* (Paris: Gallimard, 1960), 44.

73. M. Morange, *La vie expliquée?*. See also Hans-Jörg Rheinberger's *An Epistemology of the Concrete: Twentieth-Century Histories of Life* (Durham, NC: Duke University Press, 2010).

74. For a possible genealogy of biopolitics, going from Bergson to Canguilhem and beyond, see Maria Muhle, *Eine Genealogie der Biopolitik. Der Lebensbegriff bei Foucault und Canguilhem* (München: Fink, 2013).

Part I

"Bioethics" and "Biopolitics"

ONE

A Brief History of Bioethics

Guillaume Le Blanc

WHY BIOETHICS HAS NOT ALWAYS EXISTED

I want to begin by stressing that we did not always think the questions of disease and health in terms of bioethics. There were days, not so far from us, when human life was not reflected in the vocabulary of bioethics. It seems that we already know everything about bioethics. Where does it come from? Why has it appeared as a peculiar way of questioning the power of medicine? Bioethics is everywhere and seems to be a new way of governing human life within medical power and biology. A very strong explanation consists in the claim that bioethics is an effect of the transformation of life produced by medicine. In particular, because of the disappearance of frontiers between nature and culture, and so far as life can be modified by science, the medical is disoriented and, it would seem, in need of a new moral discourse, on which new points of reference could be grounded. Also, bioethics is viewed as a new moral consciousness which, like every event, has historical roots. In every book of bioethics, we find this first chapter, which speaks of glorious fathers and mothers. For example, *Florence Nightingale Pledge*, written by the English nurse Lystra Gretter in 1893; the professional ethics of the American Medical Association, published in 1847; and even Thomas Percival's *Medical Ethics*, published in England in 1793, are most often viewed as the first utterances of a bioethics. I believe this view of history to be misled. In fact, we are dealing with a myth according to which bioethics has always existed. In my view, we need to approach the question of bioethics from another angle, and understand that a genuine history of bioethics means that there is a birth certificate of bioethics.

Bioethics does not belong to an earlier history because it is a new and radical event in the way of treating life. Foucault has helped us problematize a history based on abrupt changes. I would like to refer to a specific page of *Birth of the Clinic,* in which Foucault identifies a break between the old medicine of the eighteenth century and the new medicine of the nineteenth century.[1] Medicine is no longer characterized by the value of health, but by that of normality. If there is such a break between two kinds of medicine, how could we claim that they belong to a same space of knowledge? How could they belong to a same form of power? There are, in fact, important transformations in the spaces of both knowledge and power. The birth of the clinic is a new event in the order of knowledge and power. It signifies a new experience of power based on a new knowledge, biology. The bodies are auscultated, individualized within a new framework, that of the hospital. To be sure, the aim, in treating patients, remains the recovery of health. But the *value* of health now unfolds in the medicine of normality. And as a result of the emergence of normality, a new meaning of health appears. According to Georges Canguilhem's seminal *The Normal and the Pathological,*[2] Broussais, followed by Auguste Comte, elaborated a new way of conceiving of diseases, according to which there is no *qualitative* difference between the normal and the pathological. The pathological is a simple variation of the normal, an excessive reaction or, on the contrary, a failure to react. This means that health can no longer be annulled by a merely external entity. Health is henceforth threatened by a dysfunction of the normal itself, and the pathology is located at the heart of the normal. It is no longer an external threat; it becomes an internal possibility. In this context, the reference to health ceases to have the same meaning. It ceases to be a common value that lies outside the field of medicine, and becomes a medical construction underpinned by the identity of the normal and the pathological.

Thanks to the reversibility of the normal and the pathological, the field of medicine can henceforth conquer an appropriate space. Medicine is all the more successful that it is fixed within the limits of nature: normality, which, evidently, signals an abstraction, nonetheless indicates a natural frequency. The medicine of normality is then one in which the identity of the normal and the pathological testifies to the unity of life in its different manifestations, and fixes the medical intervention within the limits of a restoration of the natural norms, for the body as well as the mind. We need to pay specific attention to the fact that the birth of the clinic reveals a strange paradox: where it is carried out at the bedside of the patient, it remains that the medical glance is interested in an individualized body only to the extent that it can conquer the pathology that lives inside it. In addition to the clinical relation between the doctor and the patient, there is in fact another, more hidden medical relation—namely, that between the doctor and the pathology itself. This, in turn, means that the existence of the patient is bracketed. As René Leriche puts it in *Philos-*

ophie de la chirurgie,[3] the patient as such is someone who can only give false clues about his or her body and thus compromise the purity of the diagnosis. We should keep in mind that, under such conditions, bioethics can't even exist, since it's supposed to be connected with the consciousness of the sick person, and particularly with the consciousness of the dangers inherent to medicine.

TOWARDS AN ARCHAEOLOGY OF BIOETHICS

If we want to witness the birth of bioethics, we need to pay close attention to the Nuremberg Code, which was written in 1947 with the explicit purpose of avoiding the inhumanity of the experiments led by the Nazis during the Second World War. It introduces the category of consent, which became extremely influential. Its first article stipulates that "The voluntary consent of the human subject is absolutely essential". Furthermore, it must be informed. Before agreeing to the experimentation, the person must know the nature of the experiment, its length and its purpose. The experiment should benefit the person, but also society as whole (Article 2). Bioethics seems to originate from this new consciousness that someone might have about the dangers of a given experiment, and that a population might also have about the benefits of the experiment in question. The reference to consent creates a bridge between the interests of the patient and the interests of medical research, and involves a new logic in which the dangers of medicine itself are recognized. In this new history, marked by the reference to consent, a new character comes to light: the sick person herself. This promotion of the sick person is an event, when compared with the earlier history, described, for example, by Foucault in *Birth of the Clinic*.[4] But we can give two interpretations to such a promotion. On the one hand, we can claim that each sick person is an individuality, and one who, as such, medical power must respect, and in relation to whom it needs to disappear *qua* power, and exist only in order to restore the health of a given life. On the other hand, we should also realize that we can be threatened by the medical art itself: medical power can kill or damage bodies.

There are, in fact, two different episodes. In the first one, medicine is viewed as a clinic—that is, as a kind of relation between the doctor and the patient, in which the most important thing is to preserve the singularity and the dignity of the patient. In the second one, medicine is viewed as a strong power that has the ability to turn against the patient himself. In fact, in the second type of episode, it seems that the medical practice not only fights against the dangers of illness but also generates specific dangers that are due to the extension of the medical world. Each episode has, so to speak, its author. Canguilhem's reflections belong to the first one and the critic of the medical power proposed by Foucault to the second. I consider that an archaeology of bioethics refers to the second episode, and not the first. Let me explain.

What are the features of the first sequence? Contrary to Broussais and Comte's assertions, the normal and the pathological are not phenomena of the same order and are placed in two different experiences of life. Furthermore, the sick person needs to be attended to before, and ahead of, the illness. As Canguilhem asserts, "It is always the relation to the individual patient through the intermediary of clinical practice, which justifies the qualification of pathological".[5] It's always the relation to the sick person which, by means of the clinic, involves the diagnosis of a pathology. We feel sick first, before we are diagnosed as really sick, and this feeling comes from the loss, whether partial or total, of our normativity. Thus, being ill consists in the loss of normativity, and absolutely not in the loss of normality. In addition, that which, according to Canguilhem, limits medical power is, in a way, nothing other than a philosophy of life, according to which by life as such is described in terms of its own creativity.

By contrast, in the second sequence, the use of medicine is no longer thought to be entirely beneficial for the patient. It becomes clear that medicine is not only a clinic but also a power. And such a power can be used against the patient himself. From another point of view, the relation between the normal and the pathological plays a minor role when compared to the extension of the pathological field. Such an extension can no longer be viewed as an experience of life led by the patient, but needs to be interpreted as a generalization of the pathological risks. In 1974, Foucault gave three talks entitled "Crisis of Medicine or Crisis of Anti-Medicine?"[6] Initially, they were meant as a review of Illich's book, *Medical Nemesis*.[7] In those lectures, Foucault points out that our societies are regulated by the generalization of pathology, so that medicine is now everywhere and has penetrated every sphere of life. This generalization can be attributed to the end of the clinical model and to the construction of a new category, that of "the medicalizable". As a result of this medical expansion, Foucault claims, the traditional boundaries defined by the sick person and his pathologies tend to disappear. What we have instead is an economic field of health, in which medicine is required to produce health. The status of life has changed in this second scenario. Life is not only the life of someone but also, and above all else, a reality in itself. We therefore expect a certain valorization of life and we know that Foucault gives it a name: biopolitics. In this new context, life has become an economic challenge. Medicine is required to maintain life in all its forms. As a result, new kinds of risks arise, which have to do with the medicalization of life as a whole and the multiplication of therapeutic interventions. Every form of life, every organ becomes a target for medicine. Today, it's impossible to imagine a single aspect of life that's not represented by a medical domain. By definition, medicine has no field outside itself.

Let me now say a few words about the archaeology of bioethics. First, bioethics is a kind of reaction against this extension of medicine. Second, if every single field of life falls under the control of medicine, and if a certain therapeutic relentlessness belongs to modern medicine, it becomes necessary to erect limits in order to deny medicine an absolute power over life: we need bioethics to formulate moral arguments against all kinds of medical interventions on life. Third, it is particularly tempting to understand bioethics as an extension of biopolitics. But I think this would be a mistake, since bioethics would then be nothing other than a reaction against biopolitics. But the question we're confronted with is the following: how can bioethics operate as a critique of biopolitics if it does not speak under the first scenario—that is to say, under the figure of the respect of the patient's dignity and integrity? Fourth, something more important than the dignity of the patient is at stake in bioethics, and that is the promotion of life as a *value*: the defence of the dignity of the patient tends to be the strongest argument in favour of the value of life itself. From this point of view, and contrary to what I seemed to be saying at the beginning of this chapter, the bioethical gesture does not simply involve a strong break between natural life and artificial life. Rather, it contributes to the valorization of some forms of life as natural, and to the rejection of some other forms of life as artificial.

THE TWO BIRTHS OF BIOETHICS

We have to pay attention to both conditions, the absolute value of life and the absolute value of the patient, if we want to understand the two births of bioethics, which in fact refer to two different origins. To begin with, the word "bioethics" has an ecological meaning. It was used by Van Rensselaer Potter in a famous article from 1970, "Bioethics, the Science of Survival", and developed in a book the year after: *Bioethics: A Bridge to the Future.*[8] But it is also, and at the same time, used in a medical context, and with the creation of the "Joseph and Rose Kennedy Institute for the Study of Human Reproduction and Bioethics" in 1971. One of its three sections received the name of "Center for Bioethics". Let me note, in passing, that this medical turn was already prepared by Paul Ramsay in 1970 when he published his *The Patient as Person.*[9] From the start, the word "bioethics" was the site of a conflict between an ecological meaning and a medical one. How can we interpret this conflict within bioethics? Van Rensselaer Potter himself acknowledges it an article from 1987, called "Aldo Leopold's Land Ethic Revisited: Two Kinds of Bioethics".[10] In that article, he deplores the reduction of the original meaning of bioethics and its strict application to the field of medicine. By contrast, he pays attention to the two different meanings. Although medical bioethics is addressed to human persons in order to protect their existence against medical power, bioethics in the second sense concerns every living person, and even the ecosystem as a whole. Furthermore, ecological bioethics involves a strong discussion about the interdependency of all living beings and is concerned with protecting the future. By contrast, medical bioethics argues in favour of the independence of every person and takes place in the present.

It can seem surprising that, at the beginning of its history, bioethics was an ecological concept. As such, it amounted to a new government of nature, and the regulation of medicine was just a part of a general regulation of life on earth. In fact, Van Rensselaer Potter was truly inspired by Aldo Leopold and his idea of a "land ethic". In the third part of *A Sand County Almanac,*[11] first published after his death in 1949, Aldo Leopold wrote a speech for the defence of an ethics, the main function of which was to define the correct relation that we can develop with other living beings, and even the earth itself. In reality, this ethics should enable us to gain a new perspective on life, and on the earth. The earth does not belong to human beings. On the contrary, we should realize that human beings belong to the earth. The love of this belonging is the main idea behind Leopold's "land ethic". On one level, we are faced with the earth as a community: that's the main idea of ecology. On another level, we are faced with the idea that we should love the earth and treat her with respect, and that's the main idea of an ethics.

We understand now why there is a conflict between different forms of bioethics. According to Potter, if bioethics is an extension of ecological ethics, how is bioethics, understood in that sense, compatible with *medical* bioethics, understood as the government of human life, which tends to forget the reference to nonhuman nature?

BIBLIOGRAPHY

Canguilhem, Georges. *The Normal and the Pathological.* Translated by Carolyn R. Fawcett and Robert S. Cohen. New York, Zone Books, 1998.

Foucault, Michel. "The Crisis of Medicine or the Crisis of Antimedicine?". Translated by Edgar C. Knowlton et al., *Foucault Studies* 1 (2004): 5–19.

———. *The Birth of the Clinic: An Archaeology of Medical Perception.* Translated by Sheridan Smith. London: Routledge, 2003.

Illitch, Ivan. *Medical Nemesis: The Expropriation of Health.* London: Random House. 1976.

Leopold, Aldo. *A Sand County Almanac.* New York: Oxford University Press, 2001.

Leriche, René. *La philosophie de la chirurgie.* Paris: Flammarion, 1951.

Potter, Van Rensselaer. "Aldo Leopold's Land Ethics Revisited: Two Kinds of Bioethics". *Perspectives in Biology and Medicine* 30 (1987): 157–69.

———. "Bioethics, the Science of Survival". *Perspectives in Biology and Medicine* 14 (1970): 127–53.

———. *Bioethics: A Bridge to the Future.* Englewood Cliffs, NJ: Prentice-Hall, 1971.

Ramsay, Paul. *The Patient as Person: Explorations in Medical Ethics.* New Haven, CT: Yale University Press, 2002.

NOTES

1. Michel Foucault, *The Birth of the Clinic: An Archaeology of Medical Perception*, trans. Sheridan Smith (London: Routledge, 2003), 40.
2. Georges Canguilhem, *The Normal and the Pathological*, trans. Carolyn R. Fawcett and Robert S. Cohen (New York, Zone Books, 1998).
3. René Leriche, *La philosophie de la chirurgie* (Paris: Flammarion, 1951).
4. Foucault, *Birth of the Clinic*, 46.
5. Canguilhem, *The Normal and the Pathological*, 129.
6. Michel Foucault, "The Crisis of Medicine or the Crisis of Antimedicine?", trans. Edgar C. Knowlton et al., *Foucault Studies* 1 (2004): 5–19.
7. Ivan Illitch, *Medical Nemesis; The Expropriation of Health* (London: Random House, 1976).
8. Van Rensselaer Potter, " Bioethics, the Science of Survival", *Perspectives in Biology and Medicine* 14 (1970): 127–53; Van Rensselaer Potter, *Bioethics: A Bridge to the Future* (Englewood Cliffs, NJ: Prentice-Hall, 1971).
9. Paul Ramsay, *The Patient as Person: Explorations in Medical Ethics* (New Haven, CT: Yale University Press, 2002).
10. Van Rensselaer Potter, " Aldo Leopold's Land Ethics Revisited: Two Kinds of Bioethics", *Perspectives in Biology and Medicine* 30 (1987): 157–69.
11. Aldo Leopold, *A Sand County Almanac* (New York: Oxford University Press, 2001).

TWO

Between (Bio)-Politics and (Bio)-Ethics: What Life?[1]

Luca Paltrinieri

ANECDOTES

On the 30 July 2003, a group of immigrants escaped from an Italian detention centre, having set fire to a few mattresses. The principal state broadcaster of Italy, RAI, announced the news through its main news program, specifying that the containers housing the immigrants were furnished with "air conditioning, telephones, and television". This detail, to a certain extent, should have reassured the viewer that the state had done its homework in ensuring absolutely impeccable living conditions to individuals who are simultaneously considered unworthy of both citizenship and any form of political rights. *Bios*—the ancient principle which makes a human out of a man or a woman, or existence in a political sense—is denied, while we provide these nonhumans with not only the means with which to survive but also amenities that we would consider to be comfortable.

Take the relation between the shepherd and his sheep as a microcosm of the relationship between the city-state and its citizens.[2] More often than not, over the course of history, the modern state has been a bad shepherd, failing to secure the well-being of the flock. Nevertheless, the sheep remain part of the flock, the population remains citizens of the city-state. Today, in Italy and in Europe, with the proliferation of these spaces of exception—the short-term detention centres—we witness a mirror image in which the state appears to be doing all that a good shepherd should do to ensure the prosperity of his subjects, without actually having recognized any of the rights of citizenship of a whole section of the flock, a section which is subject to the certainty of expulsion, or rather to the complete cancellation of political existence. The means with which to survive are granted, and more, but on the condition that the migrants do not demand to be taken into consideration politically. This paradoxical condition makes possible the creation of a new *lumpenproletariat*, deprived of elementary civic rights yet supposed to live and work on national soil, subject to a whole series of extortions most notably on wages, working conditions, the risk of being stopped by the police and so on. If, in the sense of the physical conditions of their existence, the intangibility of the "life" of the new *lumpenproletariat* is in perfect symphony with the triumph of the "right to a life" proclaimed by the new humanitarian doctrine, the life of the *lumpenproletariat* also becomes the mirrored and symmetrical condition of the reduction of their lives to the unbearable.[3]

On the 23 September 1990, a law was passed unanimously by the European nations which required passengers in an automobile to wear a seatbelt. The law was almost universally accepted as a necessity and so the mandatory wearing of the seatbelt quickly became a part of daily life. However, almost nobody noticed the irony in a measure aimed at protecting the users from their own cars, and from their own relationship to driving. Furthermore, nobody noticed the transformation taking place, which in this case meant that the state was no longer the go-between for individual conflicts as the classic version of Hobbesien contractualism would have it; to protect us from violence, from the egoism or malice of *the rest*. It had henceforth become the absolute protector of our material lives and the safeguard of our well-being, even and above all when these are threatened by our own irresponsibility. Without doubt, this is not the first time we have witnessed this phenomenon; the attention that individuals pay to their own health and integrity has been one of the central political debates for thousands of years. Nevertheless, the imperative to ensure one's own immunity, in the case of the law obliging passengers to wear a seatbelt, took the form of a law definitively and primarily aimed at saving us from ourselves. This law is of course destined to create a habit, to change the way we drive, or rather to change the norm. In this way, wearing a seatbelt has become a sort of reflex that represents the lifestyle of the vast majority of motorists, those educated on the risks of their own existence. Our right to live is thereby transformed into a kind of obligation, and transgressors of this obligation—those who don't take enough care of themselves, of their health, of their immunity—are faced with a certain penalty and universal condemnation.

On the 15 June 2012, my mother was hospitalized after a devastating vascular-cerebral accident. Modern resuscitation techniques and the life support machine kept her heart beating, but it quickly became apparent that she had no chance of surviving for a long time. At this exact moment, she was no more, and yet no less, than an organism which we knew with certainty would switch off its brain—a dead woman whose heart was still beating. In effect, after 1968 cardiac arrest no longer meant certain death, and thereafter only the proclamation of "brain death" could be considered as the legal and medical term for dying, even in the case of the circulation of blood.[4] Cephalic electrical activity has become the subject of a series of observations carried out day by day, waiting for it to cease. The declaration of death, for centuries left to the discretion of the family or a doctor, is nowadays the object of scientific protocol: a wait of a certain number of hours after the electrical activity of the brain disappears, the meeting of a committee of experts and doctors, posthumous observations and tests and then another few hours of waiting before informing the family. Today, the assessment of death looks increasingly like an experimental procedure of scientific order.[5] Millions of people since 1968 have been thrust into the realm of the "living dead"—if only for a few hours. This teaches us about the reality of current biopolitics and the specific nature of the relation between *zoe* and *bios*, which has progressively established itself since the implication of biological life in modern politics. The argument is an old one. In the world of ancient Greece, *zoe*—the simple fact of being alive, as is a plant—was overshadowed by the "higher" principle of liberty, *bios* (life qualified by politics).[6] Through the use of straight talk—or *parrhesia*—one would have therefore been able to put at risk one's own existence in public political action; this courage was the principle of a political life.[7] For modern politics, life itself is taken to mean biological existence, and has thus become the most precious of commodities for biopower, to the point that a whole series of extremely costly measures are undertaken in the hope of prolonging our "biological" life. This biological life—which we mistake for the *zoe* of the ancients—has become more important for contemporary governmentality than the ability to participate politically (*bios*). So, the power to put people to death—the force on which ancient sovereign power is founded—is diminishing, giving way to a power which must absolutely ensure the survival of its subjects, and this shift is solidified by a sort of a disqualification of death itself. As a limitation on biopower, "death is power's limit, the moment that escapes it".[8] Death constantly finds itself more and more removed from the world of common experience and quickly becomes a scholarly domain.[9]

Elsewhere, when thinking about the end of life, and even when considering the definition of death, we make the error of looking at it as if it were a "bioethical" question, as if it were only an issue of medical ethics. To the contrary, these debates clearly show that both life and death are political concepts. Not so much because they unearth the timeless struggle to determine the power of the sovereign on bare life, as Agamben would suggest, but rather because they underline the articulation between a scientifically conceptual space, and a political space which is no longer solely defined by sovereign decision.[10]

ON THE RELATIONSHIP BETWEEN THESE ANECDOTES

In all three cases, we see a sort of charitable, benevolent power at work, where the preservation of life and its immunization against all danger is both logical and fundamental. In this sense, as Foucault would suggest, contemporary biopolitics is actually the heir to an ancient pastoral power, finding its origins in the Christian Occident and exerting itself on living beings.[11] At the same time, in all three cases, we can see exactly how the action of protecting life on a biological level at all costs, ends by contradicting the existing system of law and deforming it in some way. Whether it be the conflict between the right to existence and the right to citizenship, or the law designed to create a habit and to transform normal behaviour, or finally the interminable debates taking place in every arena (from television to the Senate)[12] on the legal limits to life and death and on the right to euthanasia; it is always the balance between the conservation of life as the absolute value and the right of the individual to define their own limits to life that is brought into question.

But there the analogy stops because this battle, with the protection of human life as the absolute value on one side and the rights of individuals on the other, does not take the same form in all three cases. In the first case, we see the extent to which sovereign power, here the nation-state, structures the differential inclusion of migrants into its boundaries; through the construction of lines of division that are in accordance with the interests of the international division of labour of migrant populations, and also by using national symbols and inciting patriotism.[13] In the next two cases, we see a "biopolitical delegate" at work, where the forms of our observations on the body, on health or on vital functions are changing and taking two different directions. In the case of the law on wearing a seatbelt, the government of life—or biopolitcs—insists on informing us of our responsibilities, notably where our aptitude to master our habits and where the calculation of gains and losses is concerned, despite us already being capable of self-control and rationality. In the case of the definition of death, the decision about the end to life is transferred to a set of medical and social authorities, and thus the government consults a "myriad of social agents"—doctors, nurses, social workers—before taking the essential decisions. So, the power to define life and death relies on a fragile system of standards, discourses and practices, relevant both to medical expertise and to the capacity to play with legal boundaries.[14]

This situation can be seen as the point of arrival of long process. According to Foucault, at the end of the eighteenth-century life enters not only into the calculations of power/knowledge but, above all else, where a more and more clear division appears between the political and scientific domains. Science is thereafter designed to provide an "external" guide to political decisions. This separation—splitting the "magma" of the ancient art of governing in an effort to create a new relationship between political and scientific institutions[15]—is precisely that which should be examined today as the hidden biopolitical principle of bioethics, in an effort to ease the supposed conflict between politics and neutral scientific knowledge, and in order that we might liberate ourselves from the idea that politics is capable of completely absorbing the scientific domain. This is why it is less a matter of wondering about the limits that politicians should impose on scientists—as far as human life is concerned and in the name of "ethical" principles—than shedding light on the network of possibilities underpinning this new power/knowledge on life.

The issue is all the more complicated since biopolitics of the neoliberal age cannot be defined as a linear and definitive transformation of power. Instead it refers to an open debate, a complex of questions and ways of thinking about investing in our present, ways of using the present to innovate and extend the technologies of power. Even at the culminating moment of the creation of the Welfare-State and of social politics, biopolitics was not an unequivocal strategy for the state's regulation of human life but instead a network of many authorities, both political and scientific, as is shown by the different forms of power/knowledge seen in the three cited examples: the persistence of the state's sovereign power at the heart of biopower; the regulation of behaviour by discipline; the emergence of scientific standards on life and on death. These three forms are of course successive, historically speaking, but they come together in our present to form complex structures.[16] To work on the current concept of *biopolitics* means to reconfigure the blend of the archaic and the current, the combination that forms our present and leaves it irreducible to any re-updating by an epistemic frame or by the politics of the past.

BACK TO THE CONCEPT OF BIOPOLITICS

It is clear that as a result, the biopolitics that was put into motion during the eighteenth and nineteenth centuries can no longer be used to account for contemporary politics on life.[17] In effect, the modern understanding of life is completely different, a new conception of the links between power and knowledge, another dynamic of subjectivization underpinning contemporary biopolitics, and it is only in relation to this new biopolitical configuration that we can define our relationship to the history of biopower. The concern of the present, our concern, is not only to understand biopower itself but also to understand the forces that oppose it, as well as the link between these forces and the resistances seen in the past. Consequently, the possibility of practicing intrinsically political—and performative—critique should here be embedded into the work of historians.

Elsewhere, even in Foucault's work, the use of the word "biopolitics" is less of an "invention" of his own than a transformation of the sense of the word which accurately reveals its relation to the present, and its need to write "a history of the present". By the 1970s, in effect, the terms "biopolitics" and "biopower"—far from being Foucauldian neologisms—already had a long history which Foucault knew well. During the 1960s in France, E. Morin used the term frequently, and in 1968 Andre Birre created the *Cahiers de la biopolitique* (*Journal of Biopolitics*).[18] Furthermore, the conference on "Biology and Biopolitics" which took place in Paris in January 1975—promoted by the International Association of Political Sciences—brought together a string of writers from Albert Somit to Thomas Thorson, all of whom had already used this term "biopolitics" over the course of their research.[19] For these writers in particular, it was necessary to rely on the research on ethology, neurobiology and physiology in order to show the role of the physiology of the body on political conduct: the fundamental aspect could be found in the complex relation between the biological variables which determine physiological characteristics, and political behaviour. Wilson's sociobiology was able to add the final branch of biology to the human and social sciences with a procedure which raised the biological nature of man to a norm of moral behaviour: biological concepts and research were therefore called into action to explain, to predict and to prescribe political conduct.[20] Bringing political behaviour back to scientific and biological data on human life meant that human nature ceased to be the problem that, since Hobbes, modern politics traditionally overcame with the social contract: human nature would become the genetic origin of politics and at the same time, implicitly, its model.

If, for this school of thought, biopolitics means the expulsion of the history of politics through biological life becoming an unalterable element at the base of political action, for Foucault, life itself ceased to be a natural element in the strictest sense; there was no longer a total opposition between history and nature. One must not seek, says Foucault, "brute and definite biological facts which, from the depths of 'nature', impose themselves on history".[21] Simply by refusing the great overviews in which "biological and historical are consecutive to one another, as in the evolutionism of the first sociologists",[22] Foucault describes modern biopolitics as if it were an adventure in "biohistory", and at the same time he "rehistoricizes" the attempt to grasp an eternal human nature. The research into the conditions of the historical and political possibilities of biopolitics—notably its development within the liberal governmentality from the eighteenth to the twentieth centuries—was a logical progression: here Foucault's work normally raises the question of "the interference between our present and what we know of our past history",[23] the question that genealogy also strives to raise.

The uncertainty created by this Foucauldian interpretation—a doubt which has become the guarantee of the propagation of the word and of the extraordinary proliferation of its analysis—derives less from the systematic confusion maintained by Foucault between biopolitics and biopower, than from a way of reading his work that connects his analysis with the inner works of the classical blueprints of political philosophy (those of Arendt, for example).[24] Generally, this interpretation does not take into account the variations and transformations of our concept of life, and also fails to acknowledge that Foucauldian analysis is founded on a rigorous "periodization" of our concepts, our rationalities and the *modus operandi* of government.

The first interpretative line, which is largely inspired by Agamben's works and insists on the idea that modern biopolitics stems from the technical and economic objectification of human multiplicity, leads to the description of a biopolitical domination of bare life, which, in modernity, finds itself completely objectified by the dominant politico-scientific complex. The biological life in which biopower is thoroughly invested is therefore nothing more than a secularized political concept—that is to say, a secularized form of bare life originally produced by sovereign power. Such an interpretation systematically erases the pattern of power relations which define our political modernity, and with it the relationship between the government of human conduct and the practices of subjectivation. In the end, and in what amounts to a subtle combination of both denunciation and fascination, the description of the ever perfect and self-sustaining mechanisms of power creates a perfect break between the power/knowledge of the governors and the agency of those who are governed. Consequently, the relational nature of power is denied; power is seen solely as the ability to govern others, and the subjects who are governed are deprived of all power to react.

In other words, biopolitical ambivalence would here consist in thinking that the investment in biological life made by the knowledge and power of modern times coincides with the reduction of political life to "bare life", which is by nature incapable of resistance.[25] Therefore, the dehistorization of life and also of biopower leads to powerlessness. So, the only possible form of resistance should reside either in "desertion" or in an idleness (*désœuvrement*) that is supposed to render useless the political-theological machine which encircles life in its death grip.[26] Can we be sure that this hopeless setting does not derive from this description itself, or from the choice to consider biopolitics solely as a program of the integral management and objectification of life? Do we not thereby fall into the trap of describing the calculations of government as genuine descriptions, or ideological aberrations, rather than as programs of action which rely on resistance, bodies and experiences?

The second interpretative line consists in taking human life itself to be the home of resistance. In biopower's total investment in life, there is the possibility of inversion, created by the fact that biopower can only be effective if individuals voluntarily instigate and incorporate it into their lives. According to Negri and Hardt, this second approach is advantageous, given that it distinguishes between biopower and biopolitics: "Biopower stands above society, transcendent, as a sovereign authority and imposes its order. Biopolitical production, in contrast, is immanent to society and creates social relationships and forms through collaborative forms of labor".[27]

Through a systematic denaturalization and de-essentialization of our political concepts,[28] Judith Revel also emphasizes the fundamental asymmetry between biopower and biopolitics. If biopower operates through the de-singularization of individuals by disciplining them and through the improvised creation of great homogenous groups of living beings (populations), biopolitics ought to be able to indicate "not only another power in symetrical opposition to the first one, not only a counterpower, but an essential asymmetry where we can find the creative capacity of life, a new politics of resistance".[29] This hypothesis is a seductive one, above all in that it shows an internal conflict within the government of life, constantly at work in biopolitical production. In my view, however, it seems to avoid giving a definition of life, or rather it evades the task of including resistance in our concept of life. In effect, we do not know how to define this life which could be the site of a biopolitical counterpower: is it life in an ontological, biological or existential sense? From this point onward the proliferation of the word "biopolitics"—the politics of life—involves a risk of inflation: the word biopolitics henceforth refers to *all* possible politics involving a creative power over life, including in antiquity.

Over the last few years, the innumerable junctions between these two interpretative branches have given way to the most implausible interpretations. Some refer to Greek *Politiea* to show that politics no longer troubles itself with the question of ethical good or the ideal community in which the individual finds felicity, but instead "with the biological attitude of the individual towards the integration of the state's constitution".[30] Others prefer to begin on the horizon of biological life in an effort to define the perspectives of resistance which permit this same life to escape from biopolitical domination. Therefore, the preliminary reduction of biopower to a kind of domination over a bare life—animalized or biologized—and next the exaltation of this form of life as an enterprise of emancipation, ends by unearthing the exact biological horizon that Foucauldian critique rejects.

WHAT LIFE?

To counter this ultrabiologization of politics, thoroughly invested in the relations of power in their entirety, one must insist on the fact that the very life that is in play within biopower is not exclusively biological, but instead an object of economic, technical, political and even existential definitions.[31] Incidentally, Foucault himself had clearly shown that the beginnings of biopolitics did not coincide with the reduction of life to a nondescript "biological sequence", but in fact with the development of a dynamic of interests and desires which found their voice in the market, as they did in public debates about the limits of governmental practice.[32] It speaks of the displacement of governamentality itself (provoked by the consideration of biological life as an object of power), which today both targets and rests on a whole selection of relations between individuals, even between men and women, particularly in the case of the relative choices concerning a way of life. Nutrition, procreation, clothing, hygiene, but also religion, rights, opinions, relations between gender and sexuality[33]—they all became the subject of politics in the writings of the enlightened civil servants of the eighteenth century, and this is precisely the reason why even liberal governmentality is beginning to be characterized by an "action on an action", or an action exercised on the actions of others. If there is to be one element which defines the solidarity and continuity between liberalism and neoliberalism, it is neither the supremacy of the market nor the promotion of the individual auto-entrepreneur, but instead it is clearly the idea that "nothing is political, everything can be politicized, everything may become political".[34] This principle of the indefinite extension of politics towards all the horizons of existence is particularly ambiguous in that it encompasses the integral governmentalization of existence as well as the outside possibility of resistance. Incidentally, this principle would not be in favour of a biological life which must be emancipated from the "politics of life", but instead of a life which becomes entirely political.[35] From this "political life", it is clear that one must seek alternatives to contemporary biopower, starting with a redefinition of politics which in Foucauldian terminology is most certainly the "interplay" or the conflict between many "arts" of government,[36] but only to the extent that this conflict and this game are those which are "born with resistance to governmentality, the first uprising, the first confrontation".[37] The possibility of political life rests on the affirmation that nothing is political in advance, that it does not suffice to "politicize life" to the point that it includes politics, but instead that life should *become* politics, especially when its definition is the object of a fight between the pluralities of our ways of life.

In other words, resistance does not resemble a sort of *property* that intrinsically characterizes communities, a plebe or a "biological life". Rather, it resembles that which is born out of the fact that humans voluntarily say "no" as an act of disobedience, a confrontation, an insurrection.[38] Only such a decisive will—namely, the will not to be governed in a certain way (which today is *this* manner)—can break the circular argument of the protection of life versus the negotiations of rights. Therefore, if nothing is political, it is a question of looking for the politics at the heart of our practices and representations.

The life of the immigrants in the detention centre is not a bare one, abandoned to the decisions of the rulers: it is occupied by the incessant battles for circulatory liberty; it is political from the moment that it reveals the trace of a colonial history which Europe forgets and conceals by imposing the paradigm of the regulation of the streams of immigration, and of an optimal management of our well-being. The value of my mother's life while on her deathbed is not determined by medical norm, but by her story. This norm is just a part of the game of the underground battles entered into by medicine, written law, families, nurses and carers. Clearly, there are many definitions of life, which still hide behind the definition of life as a political object, and today the role of critique consists precisely of unearthing these definitions and retracing their historical origins. Admittedly, philosophers can appear sceptical about or even contemptuous of all these definitions of life, which certainly do not belong to the great philosophical tradition, or of scientific objectification, which we would be able to study in the setting of the history of science, but simply of the fact that living beings produce their own definitions of life, and it is only from this point that we can begin to realize, to have some idea of what constitutes life today.

For such a life might be seen clearly and not be systematically cast on the reaction to the more powerful entity of biopower, one must first cease to oppose biopower and biopolitics, an opposition which itself refers to the clash between the objectification/domination of a scientific-political complex and the subjectivity/resistance of primitive life. Rather, I propose to consider that, by nature, the double meaning of the word "power" ("to be authorized to", and "to be capable of") clearly shows us that the possibility to act on life politically is linked on the one hand to a technical structure, and on the other to a battle for the legitimization and authorization of all that politics concerns. In this sense, I want to underline the fact that the *political* battle for the definition of what life is or is not is *already* within the horizon of biopower. If politics is nothing but the conflict between the many arts of governing and of self-governing, biopolitics—the capacity and the authority with which to self-govern as human beings—is a perspective whose dissonant tracks must be unearthed from within the line of development of biopower itself, by accentuating the forms of "political life" that are constantly underevaluated or deleted by the paralysing descriptions of the domination of biopower.

BIBLIOGRAPHY

"A Definition of Irreversible Coma: The Report of the Ad Hoc Committee". *Journal of the American Medical Association* 205 (1968): 337–40.

Agamben, Giorgio. *Homo Sacer: Sovereign Power and Bare Life*. Translated by Daniel Heller-Roazen. Stanford, CA: Stanford University Press, 1998.

———. *State of Exception*. Translated by Kevin Attell. Chicago: Chicago University Press, 2005.

———. *The Open: Man and Animal*. Translated by Kevin Attell. Stanford, CA: Stanford University Press, 2004.

Amato, Pierandrea. *Antigone et Platon. La biopolitique dans la pensée antique*. Paris: Mimesis, 2008.

Ariès, Philippe. *The Hour of Our Death*. New York: Oxford University Press, 1981.

Blum, Carol. *Strength in Numbers: Population, Reproduction and Power in Eighteenth-Century France*. Baltimore: Johns Hopkins University Press, 2002.

Butler, Judith. *Frames of War: When Is Life Grievable*. London: Verso, 2009.

Caldwell, Lynton K., "Biopolitics: Science, Ethics and Public Policy". *Yale Review* 54 (1964): 1–16.

Cutro, Antonella. *Technique et vie. Biopolitique et philosophie du bios dans la pensée de Michel Foucault*. Paris: Harmattan, 2011.

Esposito, Roberto. *Bios: Biopolitics and Philosophy*. Translated by Timothy C. Campbell. Minneapolis: University of Minnesota Press, 2008.

Foucault, Michel. *Dits et Ecrits*. Paris: Gallimard, 1994.

———. *The History of Sexuality: An Introduction*. Vol. 1. Translated by Robert Hurley. New York: Vintage Books, 1978.

———. "*Omnes et singulatim*: Towards a Criticism of Political Reason". In *The Tanner Lectures on Human Values*, vol. 2, edited by Sterling Mc. Murrin. Salt Lake City: University of Utah Press, 1981.

———. *Security, Territory and Population*. Translated by Graham Burchell. London: Palgrave Macmillan, 2007.

———. *The Birth of Biopolitics: Lectures at Collège de France. 1978–1979*. Translated by Graham Burchell. New York: Palgrave Macmillan, 2008.

———. *The Courage of Truth: The Government of Self and Others II*. Translated by Graham Burchell. New York: Palgrave Macmillan, 2011.

Hardt, Micheal, and Negri, Antonio. *Multitude: War and Democracy in the Age of Empire*. New York: Penguin Books, 2005.

Hennette-Vauchez, Stéphanie, and Graciela Nowenstein. "Dire la mort et faire mourir. Tensions autour de la mort encéphalique et la fin de vie en France". *Sociétés contemporaines* 75 (2009): 37–57.

Hess, Charlotte, and Luca Paltrinieri. "Orbis Tertius #1". *Chimères* 70 (2009): 55–73.

Liesen, Laurette T., and Mary Barbara Walsh. "The Competing Meanings of 'Biopolitics' in Political Science". *Politics and Life Sciences* 31 (2012): 2–15.

Memmi, Dominique. *Faire vivre et laisser mourir. Le gouvernement contemporain de la vie et de la mort*. Paris: La Découverte, 2003.

Paltrinieri, Luca. "Quantifier la qualité. Les théories du capital humain entre démographie, économie, education". *Raisons Politiques* 52 (2013): 89–108.

Revel, Judith. "Identity, Nature, Life. Three Biopolitical Deconstructions". *Theory, Culture and Society* 26 (2009): 45–54.

Revel, Judith. *Foucault. Une pensée du discontinu*. Paris: Les Mille et une Nuits, 2010.

Rose, Nikolas. *The Politics of the Life Itself: Biomedicine, Power and Subjectivity in the Twenty-first Century*. Princeton, NJ: Princeton University Press, 2007.

Sahlins, Marshall. *The Uses and Abuses of Biology: An Anthropological Critique of Sociobiology*. Ann Arbor: University of Michigan Press, 1976.

Sassen, Saskia. *Territory, Authority, Rights: From Medieval to Global Assemblages*. Princeton, NJ: Princeton University Press, 2006.

Somit, Alfred, and Steven A. Peterson, "Review Article: Biopolitics after Three Decades—A Balance Sheet". *British Journal of Political Science* 28 (1998): 559–71.
Thorson, Thomas L. *Biopolitics*. New York: Holt McDougal, 1970.
Veronique, Guienne. "Faire ou laisser mourir, autopsie d'une controverse entre sénateurs". *Droit et sociétés* 74 (2010): 107–27.
Wilson, Edward Osborne. *Sociobiology: The New Synthesis*. Oxford: Oxford University Press, 1975.

NOTES

1. Translated by Asher Korner and adapted by Luca Paltrinieri, from the French version "L'equivoque biopolitique", *Chimères* 74 (2010): 153–66.
2. See Michel Foucault, "*Omnes et singulatim*: Towards a Criticism of Political Reason", in *The Tanner Lectures on Human Values*, vol. 2, ed. Sterling Mc. Murrin (Salt Lake City: University of Utah Press, 1981), 223–54.
3. Giorgio Agamben, *Homo Sacer: Sovereign Power and Bare Life*, trans. Daniel Heller-Roazen (Stanford, CA: Stanford University Press, 1998).
4. Agamben, *Homo Sacer*, 92–94. See also "A Definition of Irreversible Coma: The Report of the Ad Hoc Committee", *Journal of the American Medical Association* 205 (1968): 337–40.
5. Stéphanie Hennette-Vauchez and Graciela Nowenstein, "Dire la mort et faire mourir. Tensions autour de la mort encéphalique et la fin de vie en France", *Sociétés contemporaines* 75 (2009): 37–57.
6. Michel Foucault, *The History of Sexuality: An Introduction*, vol. 1, trans. Robert Hurley (New York: Vintage Books, 1978), 135–45.
7. Michel Foucault, *The Courage of Truth: The Government of Self and Others II*, trans. Graham Burchell (New York: Palgrave Macmillan, 2011), 142–53.
8. Foucault, *The History of Sexuality*, vol. 1, 138.
9. Philippe Ariès, *The Hour of Our Death* (New York: Oxford University Press, 1981).
10. See Nikolas Rose, *The Politics of the Life Itself: Biomedicine, Power and Subjectivity in the Twenty-first Century* (Princeton, NJ: Princeton University Press, 2007), 9–40; see also Dominique Memmi, *Faire vivre et laisser mourir. Le gouvernement contemporain de la vie et de la mort* (Paris: La Découverte, 2003).
11. See Foucault, "*Omnes et singulatim*: Towards a Criticism of Political Reason"; see also Michel Foucault, *Security, Territory and Population*, trans. Graham Burchell (London: Palgrave Macmillan, 2007), 145–208.
12. Véronique Guienne, "Faire ou laisser mourir, autopsie d'une controverse entre sénateurs", *Droit et sociétés* 74 (2010): 107–27.
13. Saskia Sassen, *Territory, Authority, Rights: From Medieval to Global Assemblages* (Princeton, NJ: Princeton University Press, 2006).
14. Hennette-Vauchez and Nowenstein, "Dire la mort"; Nikolas Rose, *The Politics of Life Itself*, 131–54.
15. Foucault, *Security, Territory, and Population*, 289.
16. Foucault, *Security, Territory, and Population*, 22–23.
17. Luca Paltrinieri, "Quantifier la qualité. Les théories du capital humain entre démographie, économie, education", *Raisons Politiques* 52 (2013): 89–108.
18. See Antonella Cutro, *Technique et vie. Biopolitique et philosophie du bios dans la pensée de Michel Foucault* (Paris: Harmattan, 2011), chap. 5; Laurette T. Liesen and Mary Barbara Walsh, "The Competing Meanings of 'Biopolitics' in Political Science", *Politics and Life Sciences* 31 (2012): 2–15.
19. According to Roberto Esposito (*Bios: Biopolitics and Philosophy*, trans. Timothy C. Campbell [Minneapolis: University of Minnesota Press, 2008], 21), the author who first used the term "biopolitics" in the current sense was Lynton K. Caldwell in the article "Biopolitics: Science, Ethics and Public Policy", *The Yale Review* 54 (1964): 1–16. See

also Thomas L. Thorson, *Biopolitics* (New York: Holt McDougal, 1970), and Alfred Somit and Steven A. Peterson, "Review Article: Biopolitics after Three Decades—A Balance Sheet", *British Journal of Political Science* 28 (1998): 559–71.

20. Edward Osborne Wilson, *Sociobiology: The New Synthesis* (Oxford: Oxford University Press, 1975). See also the critical reading of Marshall Sahlins, *The Uses and Abuses of Biology: An Anthropological Critique of Sociobiology* (Ann Arbor: University of Michigan Press, 1976).

21. Michel Foucault, "Bio-histoire et bio-politique", in *Dits et Ecrits*, vol. 3 (Paris: Gallimard, 1994), 97.

22. Foucault, *The History of Sexuality*, vol. 1, 152.

23. Foucault, "Foucault étudie la raison d'état", in *Dits et Ecrits*, vol. 4, 40.

24. Agamben, *Homo Sacer*, 71–77, 97; Esposito, *Bios*, 177–78.

25. This conception was revised in Giorgio Agamben, *The Open: Man and Animal*, trans. Kevin Attell (Stanford, CA: Stanford University Press, 2004), and *State of Exception*, trans. Kevin Attell (Chicago: Chicago University Press, 2005).

26. Agamben, *State of Exception*, 87–88.

27. Michael Hardt and Antonio Negri, *Multitude: War and Democracy in the Age of Empire* (New York: Penguin Books, 2005), 121.

28. Judith Revel, "Identity, Nature, Life: Three Biopolitical Deconstructions", *Theory, Culture and Society* 26 (2009): 45–54.

29. Judith Revel, *Foucault. Une pensée du discontinu* (Paris: Les Mille et une Nuits, 2010), 311.

30. Pierandrea Amato, *Antigone et Platon. La biopolitique dans la pensée antique* (Paris: Mimesis, 2008).

31. Judith Butler, *Frames of War: When Is Life Grievable* (London: Verso, 2009), 21–28.

32. Michel Foucault, *The Birth of Biopolitics: Lectures at Collège de France, 1978–1979*, trans. Graham Burchell (New York: Palgrave Macmillan, 2008), 321–22.

33. Carol Blum, *Strength in Numbers: Population, Reproduction and Power in Eighteenth-Century France* (Baltimore: Johns Hopkins University Press, 2002).

34. Foucault, *Security, Territory and Population*, 505.

35. See the "philosophical militancy" of the Cynics: Foucault, *The Courage of Truth*, 283–87.

36. Foucault, *Birth of Biopolitics*, 313.

37. Foucault, *Security, Territory and Population*, 505.

38. See Charlotte Hess and Luca Paltrinieri, "Orbis Tertius #1", *Chimères* 70 (2009): 55–73.

THREE

What Is Vital?[1]

Frédéric Worms

To ask "what is vital?" is to pose a question that is both self-evident—even urgent (for how could anyone live without knowing that which is vital?)—and paradoxical—even manifoldly paradoxical (for who inquires explicitly about that which is vital?). My aim, here, is precisely to show what the urgency of this question consists in, or rather, in what way the question is *itself vital*—not only in general terms but also today, in a "moment" which, in my view, it defines in part. However, it would be impossible to do so without acknowledging the paradoxes that emerge along the way, and without attempting not so much to erase them, but to confront them. It's only by going through those paradoxes that we'll be in a position to understand the relevance of that question, and only subsequently to answer it. The course of my argument will follow quite naturally from such a claim. I'll begin by formulating the three main paradoxes implicit in the question. I will the address them one by one.

THE THREE PARADOXES OF THE VITAL

The first and the most important paradox, which will recur throughout the chapter, and which it will be necessary to maintain until the end, is the one I will designate as the paradox *of the minimum and the extensive.* When we ask "what is vital?" we can think of a twofold answer. We can begin by acknowledging what I would call a *minimum,* a minimum that would be *vital*—that is, something *in the absence of which* one couldn't live. It corresponds to what we normally call a "need": bread, water, air. That being said, not only does the minimum in question take the form of a *list,* but the list itself also seems entirely open. It would find itself immediately extended by an additional "but also". Thus, we would say that what's vital is not simply this particular minimum, or the minimum of this particular need, *but also* something else. Such a list, then, would include things like air, water or bread, but also perhaps what one calls "tastes", as well as "principles", and thus remain constitutively indeterminate. It would include something that, in any event, would exceed the sphere of strictly vital *necessities*, and intimate, or at least appear to intimate, the *superfluous*—something that would indicate that the question of the superfluous, or of *luxury*, isn't just a *political* question, but, rather, an *ontological* one, which concerns our very existence or being, and certainly one of the primary ontological questions.

Perhaps then, the list wouldn't be as indefinite as we had initially thought. It may well be the case that, beyond needs, we find not fluctuating and arbitrary tastes, but that, behind those "tastes", we also find the fact of *holding*, or the capacity to *hold* something (apparently superfluous) as vital. It may well be that this *capacity*—namely, the capacity to attribute vital *importance* to something, is itself vital (this time in the sense of acutely needed). Similarly, it may be the case (as I'll show later on) that *principles* (like justice) aren't the subsequent effects of "needs", but actually precede them, and that without these principles the concept of need itself would not be self-evident. Yet, independently of how, for the time being, we handle this essential issue, the first question or paradox, that of the minimum and of the extensive, remains in place: is it possible to reduce the vital to a fixed minimum, without also immediately extending it beyond that minimum?

Beyond the minimum, should we understand the reference to the "vital" figuratively, as a "metaphor", which would also imply the existence of a "literal" meaning—namely, that of need? Or should we consider (while relying, for instance, on François Zourabichvili's analysis[2]) that the meaning of this expression exceeds the very opposition between the literal and the figurative, and is always both literal *and* figurative? In any event, the first paradox would consist in the following: one cannot hold to a minimum without having recourse to the extensive, yet one *also* has to preserve the extensive within the minimum, in other words, within the range of "that without which" or "in the absence of which" life is impossible, and which defines the minimum as such. That first paradox concentrates all the stakes, including ethical and political, regarding what's vital.

But this is where the "philosopher" steps in. He steps in to address and perhaps throw in another question, another paradox, which, by virtue of being a philosopher, he is tempted to regard as even more important than the first. "What!" he or she will exclaim (in a deeply Socratic manner). "How can you name *that which* is vital, consider which things are vital, without having first defined *what* vitalness is, what it means to be vital, and finally what life itself is?" He will perhaps add, "Shouldn't the vital be something not only external, but also internal to life? not a condition, but a principle, or at least a force, an 'impulse' [*élan*] (to borrow a concept from Bergson)? How can you substitute a vital, albeit 'extensive', minimum for a vital, albeit minimal, momentum (as Bergson himself did)?"

Now the aim of this chapter is precisely to work through this *reversal* and to claim it as such. In other words, its aim is to claim that the question "what is actually vital?"—the question regarding which "things" can be considered vital—doesn't actually follow from, but in fact *precedes*, the question "what is life?"—even if (and this is a crucial point), when we consider it *in all its extension*, the former question can only guide us to the latter! In other words, not only does this second paradox (of condition and principle, of exteriority and interiority) not come first, but it also presupposes the first, without which it has no solution, and to which it eventually leads.

Before I return to that paradox, let me mention the third and final one, which is as necessary and urgent as the other two. This time, the paradox takes the form of a new objection to the very question "what is vital?" and can be referred to as the paradox of the positive and the negative. Indeed, and by virtue of a fundamental ambivalence inherent to a number of languages, to ask "what is vital?" is also to ask about what is *mortal*.[3] Here, no doubt, a risk presents itself, for doesn't the paradox in question imply that we ask about life from not only an external but also a negative perspective, from the point of view—that is, of its other, its threat, disappearance or loss? Here, once again, it's a question of asserting not so much the "negativity" of life as its *polarity*, and recognizing that it is irreducible, but only on the condition that we consider it *in all its extension*!

The risk, the danger even, in that respect, is not to contemplate life from the point of view of that which threatens it, or from the point of view of death, but to restrict this threat, and death itself, to a minimum as it were. In actual fact, and quite spontaneously, we tend to extend the meaning of death to include something beyond the point at which life ceases. We need only consider a certain linguistic expression on which, following Adorno and Lyotard, Derrida reflects, and which consists in recognizing certain dangers or risks as being *worse than death*.[4] If there is something worse than death, or even more *lethal/mortal* [*mortel*] than death itself, it's also because the vital is not reducible to mere life, or rather because there are variations within the vital, whether of an extensive or intensive kind, which endow it with an ambiguity, between a minimum and a maximum, similar to the ambiguity inherent to the notion of *survie* (or survival), and which Derrida thinks through in its very ambivalence.

With this third and final paradox, we circle back to the first. Having said that, the first also requires the other two. Everything happens as if we could give full meaning to the question of the vital by drawing on not only the categories of the minimum and the extensive but also the various degrees (including ethical and political) of the *negative*, and by simultaneously revisiting the question "what is life?" (including in its metaphysical dimension). If we need to begin from the point of view of the "minimum" and the "extensive", it is only and ultimately with a view to recovering the two other aspects.

THE MINIMUM AND THE EXTENSIVE: CONTINUATION, INDIVIDUATION, RELATION

I now need to demonstrate why we can think of the vital only as both minimal and extensive, and then to measure its extremities. To do so, I'll start from a provisional definition, not of life itself, but rather of the vital—a definition that might well presuppose that of life itself, but which, initially at least, will have to suffice, *no matter what* definition of life we end up adopting with a view to completing it. Besides, one can add that this definition of the vital is also minimal in that it appears to be required by the objective, scientific, *biological* approach, which is itself the most minimal and rigorous approach. This means that the minimum is also, as such, inseparable from *science*, which accounts for it in all its extension; it is inseparable from a biology that finds itself, as such, positioned at the centre of our entire philosophical "moment" (unless it's the "vital" which, through its multiple scientific dimensions, finds itself at the very heart of our present).

The provisional definition of the vital, the fundamental stakes of which I'll return to shortly, is the following: I call vital *that without which the continuation of a living being's life, in the context of its relation to the environment and to other living beings, is no longer possible.* To begin with, what matters here is the expression "that without which", which indicates the conditions that need to be identified and investigated. Having said that, one might want to object that my definition is circular (regardless of whether the circularity be vicious), and that I'm attempting to define the vital by means of the living [*le vivant*], and of life. In my view, however, the three key aspects of continuation, individuation and relation allow for, and at the same time overcome, the apparent generality and circularity of my definition.

Let me begin with continuation. The "vital" couldn't possibly be the abstract condition without which life in general wouldn't be "possible", but the concrete condition without which life couldn't *go on.* It is impossible to think life otherwise than as a continuation and without the minimal finality or teleology it presupposes, which is not only compatible with scientific naturalization but also inseparable from it. It is not actually even a matter of "preservation", for that notion evokes the representation of risk and the necessity to protect oneself from it; it is a question of a kind of "continuation", in the absence of which neither the philosopher nor the scientist, neither Darwin nor Spinoza, can even begin to talk of life. To naturalize the study of life doesn't only mean to reduce life to a material substance; it also means to examine the objective conditions—indeed physical and chemical—under which it *remains* in its singularity, and which already presuppose its emergence.

To live, therefore, is not a thing but an act, one that's *extensive,* if only from a strictly temporal point of view, and beyond the possible "givenness" of an "object" which would be the living being. It is impossible to think "that which is vital" without this initial dimension. Let us, therefore, and following Beckett, "go on".[5]

But it is also impossible to think life without talking about actual *living beings*—that is, living beings that are more or less individual. This is not simply a linguistic or narrative constraint, a constraint that would be merely descriptive or "biographical" in a narrow sense. Here, too, the constraint is inherent to *biology* itself. If there is *something vital*—that is to say, continuation and interruption—it's always for *this* or *that* particular life, *this* or *that* living being, which is always *individuated,* if only through a process of observation, manipulation, knowledge or action, but also in itself, through its separation from the environment, or through its own totality and history. Canguilhem and Simondon both showed this in relation to the individual organism as such.[6] But it also holds true of "life", which isn't a merely abstract concept, but always a concrete and individual history, which unfolds on our planet, emerged from an origin, and is threatened with extinction (as we're beginning to realize). There is thus never anything vital *in general.* We might lack a model to think the "individual" outside that of the living being—a being that's temporal and polarized, with its birth, its "life" (as we say, in an extremely evocative manner!) and the latter's interruption!

But the most important and yet less frequently emphasized point is, without doubt, the fact that this individuation cannot in itself be thought in the absence of a dual relationship, not only between the living being and its environment but also *between different living beings* themselves. Certainly, genuine individuation starts precisely from within relationships—not the interspecific and generic, but rather *intraspecific* relations, which take place within the boundaries of a same species. Here, we need to go straight *to the extremes,* since it is there that we discover the meaning of the extension of the vital, or of the "that without which". In doing so, we're still within the boundaries of the vital or the minimum, since we haven't left the horizon of that without which life is neither thinkable nor possible. And yet, we've reached its greatest extension, since we enter the domain of mental and moral relations.

This, in fact, is what *ethology* teaches us: individualizing relationships, it claims, are vital for certain species, and for human beings in particular, where it takes the form of aggression, but also, and especially, of *attachment*, as Bowlby and Harlow, independently of each other, discovered in the 1950s, and as Lorenz was developing his own studies on intraspecific aggression as a factor of individuation. Thus, it is necessary to turn biology away from the study of the physical and chemical conditions of life, and the study of organisms (physiology, but also medicine), and towards the study of the relational behaviours of living beings—behaviours that still belong to life as such, and constitute the object of ethology. This reorientation of biology is necessary to grasp how, through these relational behaviors, one accesses culture, language and the symbolic, which are both distinct and inseparable from them.

To be sure, there are differences of degree within relationality, as in continuation and individuation. But the criterion for determining what's vital doesn't change. To understand how even the most individual relations are vital we need to look at the effect their absence, their loss or their destruction would have on life itself. It's precisely the fact that the latter can render the continuation of life impossible that led Bowlby to define attachment not as a construction (in the rigid, Freudian sense), but as a *need*. Thus, even though there are always thresholds and ruptures between a difference of degree of the extension of the vital and another, the vital always unfolds within the minimum, or within "that without which" life isn't possible. To verify this claim, though, we have no choice but to refer to the *polarity* that defines the vital in opposition to the mortal/lethal, yet on the condition that we recognize it in all its extension, and according to the levels I've just distinguished.

THE VITAL AND THE MORTAL: URGENCY, DISEASE, FRATRICIDE

There is nothing accidental, therefore, about the ambivalence of the vital. It is not a matter of "shifting" to a negativity that would suppose a prior positivity, and which we could define independently. On the contrary, one understands that what is vital is also intrinsically mortal, so much so that one cannot grasp the vital without at the same time grasping that which is mortal, in all its extension and precision. The scope of this polarity in the understanding of life is therefore both theoretical and existential, or, perhaps better said, epistemological and ontological. In that respect, we need to recall, and even extend, what, in a famous page from *Birth of the Clinic* (which Georges Canguilhem once described as "admirable, touching"),[7] Foucault strikingly called not vitalism, but rather "mortalism" [*mortalisme*]. This is what Foucault wrote about Bichat: "death was the only possibility of giving life a positive truth".

The irreducibility of the living to the mechanical or the chemical is secondary only in relation to the fundamental link between life and death. Vitalism appears against the background of this "mortalism".[8] Certainly, for Foucault and Canguilhem, as for Bichat or Claude Bernard before them, it is less a question of allowing death in general to illuminate life, than of allowing pathologies or diseases to illuminate norms. This point is fundamental. But it's possible and indeed necessary to emphasize not merely one, but two other aspects, the *beneath* and the *beyond* of disease, respectively, to account for the whole *extension* of the polarity between the mortal and the vital.

The first aspect actually takes us to a level beneath or prior to what I've hitherto referred to as the minimum, or as the sphere of needs. In fact, what the mortal [*le mortel*] forces us to consider, in a precise and rigorous manner, is not so much need, or the minimum, as the situation of emergency [*urgence*].[9] It is the situation of emergency (and not need), I want to claim, that's vital and/or mortal. Emergency is a *spatio-temporal* concept: it signals what needs to be done *here* and *now*, or *this, immediately; this* organ or *this* wound, requiring this specific such treatment, and with no time to waste. Currently, nothing holds greater importance than the fact of having a precise, exacting, extensive, but also limited concept of emergency, and this (as I've already outlined elsewhere) at the level of not only individuals but also ethical and political life.[10]

Yet the paradox of emergency consists in the fact that it lies both beneath and *beyond* need. As such, it already alludes to the other process of overcoming what's at stake here. In fact, if emergency, by virtue of its proximity to death (one must act, *now*), signals a life beneath that of need, it is also, and at the same time, already beyond it in that it appeals not only to a need (air, water, bread) but also to *care*—that is, to a relation to the other. If the concept of need can make us believe in a secluded autonomy of the living being, the concept of emergency returns us immediately to a relationship *between* living beings, and thus to the dimension of care. Certainly, one can imagine emergency situations in which one can act alone, but how can we define emergency, if not as a cry for help? Thus, the minimum, even in its first extreme manifestation, from the very start, and from within, connects us to its other extreme manifestation—namely, relationality!

This is the point at which we need to transport ourselves as it were into the extreme of this extreme, into the opposite end of the "mortal" or "mortalism", which has oriented my analysis thus far, and into what we could thus call the *other extreme of human mortality.* From this point of view, actually, it's not a matter of addressing death in general, or even of addressing the death of the other, albeit that of the family member or the loved one; it won't be a matter of loss or mourning (and even less of the death of the self). It won't even be a matter of looking at death in the hands of another, or as murder. Rather, it's a matter of moving beyond death in those senses and straight into that which crushes the ultimate dimension of the vital—namely, the killing *that follows from relationality itself*. At stake here are killings that we could call "intimate": parricide, infanticide, fratricide and perhaps also suicide.

I want to make the following, possibly radical claim: it is through this particular kind of killing (and more generally through what I would term *violation*,[11] which presupposes relationality) that ethics makes its entrance into life (and into the world). It is not from a general prohibition to kill that one arrives at the specific prohibition to kill one's brother. Rather, and on the contrary, it is by starting from the experience of the brother's murder that we can infer the prohibition of murder in general, and perhaps even the totality of prohibitions and moral obligations. To be sure, I would need more time and space even to outline the demonstration of such a claim, and would need to bring in ethics as well as politics, religion and the foundational narratives of all human communities. But let me simply remark that if the individualizing relationship is integral to life, then the killing that originates in it and shatters it does so in a twofold manner: it kills both the object (the killed brother) and the subject (the murderous brother). Prior even to disclosing the vulnerability of the one and the culpability of the other, or this dual abyss of morality, the individuating relationship reveals the ethical and hitherto hidden nature of relationality, in the absence of which this act of killing wouldn't be particularly significant.

Again, it's impossible for me to dwell on this important issue. Let me simply point out the following aspect, which I consider crucial: if the idea of vital emergency refers to a moral obligation, the reverse also holds true. It's only the imperative not to kill, to cause or even allow death to happen (when something can be done, and someone treated) that makes possible not only the imperative but perhaps even the very *concept* of "vital" emergency. How, otherwise, could we interpret suffering as a call? We could always refer the dying person to the answer that, according to Rousseau, a prince once gave to a beggar who was imploring him to help him survive: "I don't see the need for that!"[12] Ultimately, there is nothing "natural" about emergency—that is to say, about either care or assistance, insofar as they are not inscribed as such in the suffering body. Crucially, though, they don't emerge from a purely moral obligation which would be, as such, disconnected from life: they originate in a moral generalization of the ethical awareness arising in the face of the most *vital* killings, those which destroy not only the body, or even the individual, but also the vital and moral relationships themselves.

We're now in a better position to understand how only the mortal [*le mortel*] in its most extreme manifestations can shed light on what's vital: not just through need, but through emergency; not just through the relation, but through its violation; not just through love (or attachment, which also explains the feeling of, and the taste for, being alive, as well as the taste for things that matter) but also through justice (and the principles associated with it). Let me also note, in passing, how, between the two extremes of emergency and killing, there is the experience of disease, which I began by evoking in relation to Canguilhem and Foucault. We could also, and finally, mention the pathology of relations that we call madness, and which reveals the extent to which relations are constitutive of human lives.

I now arrive at the third and final paradox announced at the beginning of the chapter. It will allow us not only to consider, one more time, the meaning of the question we're concerned with, but also to try to answer it. If "that which is vital" is both minimal and extensive, positive and negative, does it also teach us "what it is" to be vital, or what life itself is? In other words, does our problem have a dimension that is ethical and political, as well as metaphysical? It is to this last series of questions that I now turn.

THE VITAL AND LIFE: ETHICS, POLITICS, METAPHYSICS

I would like to formulate a claim concerning ethics, or rather bioethics, which will also allow me to confront the metaphysical consequences of what I've been arguing. The claim is the following: the two extremes of the vital (namely, corporeal emergency on the one hand, and individual relations on the other) can finally be reconciled within a "bioethics"—in which the stakes of the present philosophical moment somehow manage to cluster together—while avoiding the major risk, which always threatens to return, of keeping those two extremities apart, and always in the name of a supposed essence of life! Let me call the attitude that consists in the essentialization of organic life, and in the desire for its "continuation" at all costs, bioethical minimalism. Bioethical maximalism, on the other hand, designates the attitude that transforms life into a purely moral and in itself absolute value.

Curiously enough, it ends up producing the same result. Everything happens as if the two extremes, when separated from each other, ended up denaturalizing the vital itself: curiously, "bare life" and "sacred life" converge. It's against this double essentialization, and without leaving the domain of life, that we reach here a tense (and often broken) equilibrium between the two poles of the minimum and the extensive, of continuation, individuation and relationality. There is no genuine corporeal life without individuating relationships, however minimal: the hand we shake, the voice we recognize, the gaze we respond to. But there is also no moral life without the embodied conditions required for its continuation, without the most tangible and equitable care, the attention paid to people's needs and well-being, and even to their viability and vitality. Equally, it is perhaps when embodied life is open to relational life that there can be an experience of "life" that is not only ethical but also metaphysical.

But before I reach my conclusion on this point, let me venture another, more strictly political observation. In this instance too, it can't be a question of either essentializing or neglecting the vital. What works, in this context, is to combine the minimum and the extensive, and to do so in a twofold direction—that of what I would call the minimum *for all*, but also that of the extensive for *each* individual! In truth, both are maximal demands: that of survival for humanity as a whole, but also that of a relational individuation and creation aimed at each human being. Consequently, what is *socially* vital is never just the help provided in case of emergency, but also the justice accompanying it; equally, what's vital is not justice, but also relief and care in the various dimensions I've tried—all too quickly, no doubt—to describe. This observation also has the benefit of allowing us to insist on that which places the question of the "vital" at the heart of the "contemporary moment" in philosophy.

It is the result of a singular convergence of scientific, moral and political questions, and the philosophical problems they pose. In no way is it a matter of a simplistic reduction, for instance of morality to a genetic fact, of politics to power over the living; on the contrary, what I've just highlighted is that biology itself doesn't naturalize "life" except extensively, and by spanning physical chemistry, ethology and even culture (including internal levels and ruptures). Similarly, bioethics and biopolitics don't really deal with life unless they consider it in all its dimensions—not only corporeal but also individual and relational—and at the cost of generating tensions and ruptures in relation to every concrete and necessarily tense and difficult question internal or immanent to the relations between living beings. Our moment is certainly the moment of the vital in its most material and urgent dimension, which includes the dimension of relationality and of justice.

Could this provide us with a clue of a metaphysical kind, one that would allow us to return to the question of life itself from the question "what is vital?" In any event, it seems to me that the greatest philosophies of life have always been those that don't derive what's most vital for us from an abstract principle defined as "life", but those that do exactly the opposite. Bergson, for instance, arrived at the problem of life in *Creative Evolution* only by starting from the following two extreme manifestations of life: on the one hand, the constraints of need or action (which for human beings are expressed as space), and, on the other hand, the primitive character of time, individuation and creation (which relies on duration).

Moreover, nothing provides better access to the meaning of life, according to Bergson's *The Two Sources of Morality and Religion*, than our dual moral experience, the closure that attests to a force closed upon itself, and the openness that is rooted in an *élan* that can always be reenacted. But both dimensions of the moral experience possess, above all else, their own moral criterion: mutual exclusion and war on one side, and inclusive openness and peace on the other. Both experiences allude to two aspects constitutive of life: its limit, as translated in the quest for mere survival, and its repetition, through which it aspires to creativity and to a better life.

No doubt, we can wonder whether it's still possible to go beyond this double experience and arrive at a metaphysical principle, whether within ourselves or within things. Are we still entitled to think (possibly by appealing to Freud, beyond Bergson) that what's at stake here is not only that which is vital for us but also an ultimate polarity, if not within life itself, then at least between life and "death"? In any event, minimally, and by taking this time a completely different path, we might want to insist again on the tension, constitutive of "life" itself, between the minimum and the extensive, continuity and creativity, resistance to destruction and the constantly individuating relationship between living beings. In what follows, I'll consider two further extreme symptoms of this tension by answering two final objections.

The first objection is the following: since we can sacrifice ourselves, our own life, how could life be seen as the ultimate value? Didn't the greatest moral philosophies always insist on this possibility, on this twisting free? My answer is the following: yes, we can sacrifice our lives, but the sacrifice is *still* made in the name of values related to life, of vital values; it's made for the sake not of destruction, but of creation; not to add to the death of bodies the murder of relationships, but, on the contrary, because the traumatic nature of the situation (which doubtlessly defines the tragic, and which constitutes to a certain extent a criterion for sacrifice, not all sacrificing being "moral" as such) reveals the vital character of both!

The second objection is the following: are the greatest joys really relational and vital? Don't morality or justice, but also science or art, the beautiful or the veridical exceed the relational and the vital? To this I respond that not only the relational joys themselves (birth, love, friendship) but also the joys which tear themselves away from relations are in fact the products of life (in the same way that Winnicott shows the symbolic, culture, the "place where we live" to be made possible by the very relations between living beings); moreover, they, too, are necessary for life, and shine over it. I don't know whether joy indicates the metaphysical sense of life, but what I do know, and without doubt, in a way that's general, minimal as well as maximal, is that joy is vital.

BIBLIOGRAPHY

Adorno, Theodor W. *Negative Dialectics*. Translated by E. B. Ashton. London: Routledge, 1973.

Beckett, Samuel. *Three Novels: Molloy, Malone Dies, The Unnamable*. London: Grove Press, 1989.

Canguilhem, Georges. "A New Concept in Pathology: Error". In *The Normal and the Pathological*. Translated by Carolyn R. Fawcett and R. S. Cohen. New York: Zone Books, 1989.
———. *The Knowledge of Life*. Translated by Stefanos Geroulanos and Daniela Ginsburg. New York: Fordham University Press, 1992.
Derrida, Jacques. *The Work of Mourning*. Translated by Pascale-Anne Brault and Michael Naas. Chicago: University of Chicago Press, 2001.
Foucault, Michel. *The Birth of the Clinic: An Archeology of Medical Perception*. Translated by A. M. Sheridan, London: Routledge, 2003.
Simondon, Gilbert. *L'individuation psychique et collective*. Paris: Aubier, 2007.
———. *L'individuation à la lumière des notions de forme et d'information*. Paris: Jérôme Millon, 1995.
Worms, Frédéric. "Critique de l'exception". *Esprit* 5 (2007):150–53.
———. "La négativité dans les relations morales: perte, rupture, violation". Forthcoming.
———. "Penser la violation". *Esprit* 2 (2000): 66–81.
———. "Pour aller au-delà de l'urgence". *Esprit* 12 (2006): 177–81.
———. "Quand les relations deviennent-elles morales?" In *Philosophes en liberté. Positions et arguments*. Edited by Francis Wolff. Paris Ellipses, 2001, 67–84.
———. *La vie qui unit et qui sépare*. Paris: Payot et Rivages, 2013.
Zourabichvili, François. *La littéralité et autres essais sur l'art*. Paris: Presses universitaires de France, 2011.

NOTES

1. Translated by Claudia Alexandra Manta and revised by Miguel de Beistegui.

2. François Zourabichvili, *La littéralité et autres essais sur l'art* (Paris: Presses universitaires de France, 2011).

3. In English, although in a way that is less common than in French, "mortal" can refer to the being who can die, but also to that which can cause death, as when we say "a mortal blow".

4. Cf. Theodor W. Adorno, *Negative Dialectics*, trans. Edward B. Ashton (London: Routledge, 1973), 368–73. Derrida returns to this question particularly in his tribute to Lyotard, included in *The Work of Mourning* (trans. Pascale-Anne Brault and Michael Naas [Chicago: University of Chicago Press, 2001]).

5. Samuel Beckett, *Three Novels: Molloy, Malone Dies, The Unnamable* (London: Grove Press, 1955), 404.

6. Cf. Gilbert Simondon, *L'individuation à la lumière des notions de forme et d'information* (Paris: Jérôme Millon, 1995) and *L'individuation psychique et collective* (Paris: Aubier, 2007). Cf. Georges Canguilhem, *The Knowledge of Life*, trans. Stefanos Geroulanos and Daniela Ginsburg (New York: Fordham University Press, 1992).

7. Georges Canguilhem, "A New Concept in Pathology: Error", in *The Normal and the Pathological*, trans. Carolyn R. Fawcett and R. S. Cohen (New York: Zone Books, 1989), 285.

8. Michel Foucault, *The Birth of the Clinic: An Archeology of Medical Perception*, trans. Alan M. Sheridan (London: Routledge, 2003), 145.

9. The French word *urgence* can be translated as both *urgency* and *emergency*, in that it indicates the purely temporal sense of urgency, as well as the dimension of care. In the context of the hospital, *les urgences* is the department known in the United Kingdom as "Accident and Emergency" (A&E). Translator's note.

10. See Frédéric Worms, "Pour aller au-delà de l'urgence", *Esprit* 12 (2006): 177–81 and "Critique de l'exception", *Esprit* 5 (2007): 150–53.

11. I develop this point elsewhere. See Frédéric Worms, "Penser la violation", *Esprit* 2 (2000), 66–81; "La négativité dans les relations morales: perte, rupture, violation" (forthcoming); "Quand les relations deviennent-elles morales?", *Philosophes en liberté.*

Positions et arguments, ed. Francis Wolff (Paris: Ellipses, 2001), 67–84; *La vie qui unit et qui sépare* (Paris: Payot et Rivages, 2013).

12. Jean-Jacques Rousseau, *Emile: Or, On Education*, trans. Allan Bloom (New York: Basic Books, 1979).

Part II

Norms and Normality

FOUR

The Return of Vitalism: Canguilhem, Bergson and the Project of a Biophilosophy

Charles T. Wolfe and Andy Wong

Tout ce que j'ai écrit était vitaliste, du moins je l'espère.

—Gilles Deleuze[1]

BIOPHILOSOPHY AND VITALISM

French biophilosophy in the 1950s–1960s means at least three names—Georges Canguilhem, Raymond Ruyer and Gilbert Simondon; for reasons of space, we focus on Canguilhem in this chapter. Biophilosophy during its relatively brief tenure was a project distinctively different from Anglophone "philosophy of biology".[2] Notably, it does not present philosophy as coming second in relation to a foundational or normative status of scientific practice. In this context we frequently encounter reference to Life, the thinking of Life, the meaning (*sens*) of Life, and, of course, the idea of a philosophy of Life, along with a focus on vitalism. Here, the influence of Bergson is nonnegligible.

These thinkers blend the historical and the normative when dealing with vitalistic themes in the conceptual foundations of the life sciences ("biology", natural history, medicine, etc.). This appears vividly in Canguilhem, who presents himself at least to all but close readers as a scholar, with a "*thèse d'État*" on the origins of reflex physiology, and yet declares quite bluntly that he *is* a vitalist. Canguilhem often refers to vitalism in his work, going as far as describing himself as one in the foreword of the previously mentioned work on reflex action ("Il nous importe peu d'être ou tenu pour vitaliste"); he presents the book itself as a "defense of vitalist biology".[3] Additionally, some years earlier, he had devoted an article to the topic: "Aspects du vitalisme" (originally lectures at the Collège de Philosophie in Paris, in 1946–1947).[4] Here, Canguilhem asserts from the outset that when the philosopher inquires into biological life, she has little to expect or gain from "a biology fascinated by the prestige of the physico-chemical sciences, a biology reduced or reducing itself to the role of a satellite of these sciences".[5] In other words, the philosopher in this position is almost inexorably led to a vitalist *positionnement*.

We are not the first to note that there is an unusual combination here of the historical and the normative, or the scholarly and the speculative. In a little-known but interesting book entitled *La notion d'organisation dans l'histoire de la biologie* (1978), which is marred by frequent polemical outbursts (these also contribute to rendering it interesting), Joseph Schiller targeted the historian of the life sciences Jacques Roger, Foucault and Canguilhem as anti-Cartesians who attempted a "vitalist" revision of the history of science, so as to deemphasize the key role of Descartes in particular and the mechanistic "paradigm" in general. Schiller opposes "good" history of science, which he understands as being in agreement with what the scientists say, and thereby mechanistic, from Descartes to Bernard and beyond, to "bad" history of science, which obeys certain philosophical imperatives, in this case vitalistic ones.

As it turns out, Canguilhem explicitly reflects on the dual nature of vitalism as both historical object and conceptual stance, thus mirroring Schiller's critique but also becoming a moving, self-aware target. Our main focus will be this dual nature of vitalism, as presented "by" but also "in" Canguilhem—that is, according to both his analyses and his own philosophical performance. The historical side is unique to him, whereas the conceptual argument for thinking Life on particular terms—with the primacy of activity, and the devalorization of the paradigm of the machine—bears a strong Bergsonian imprint.

Canguilhem shared the idea with Bergson of pursuing a biological philosophy in which life and knowledge would be united.[6] In fact, there is a continuity from Bergson to Canguilhem in their shared emphasis on the "meaning" (*sens*) of life. But it is also "the meaning of life" that distinguishes Canguilhem from Bergson. For Bergson, metaphysics and science constitute two different types of knowledge of reality, and they have developed respectively with their particular approaches—namely, intuition and intelligence. Intuition is a method to enter into the duration of life, while intelligence is an approach to the science of matter. In metaphysics the meaning of life is found by intuition, while in science it is found by intelligence. For Canguilhem, the meaning of life is a "counter-intuitive" knowledge which has nothing to do with the constitution of knowledge by means of philosophical intuition.[7] The point is not to attribute the meaning of life to metaphysics rather than to science. Instead, in Canguilhem's words, "Philosophy should not begin at the place where science terminates, because science in its own manner is already a philosophy".[8]

Canguilhem carried on the Bergsonian objective in his project of a biophilosophy: "philosophy should create a new perspective, faced with the vital fact".[9] The project aims to introduce a philosophical, "vital perspective" on life, instead of the reductionist project of reducing life to matter. Since the tradition of biological philosophy had been totally neglected from Descartes to Sartre,[10] Bergson's philosophy of life, with *L'évolution créatrice* in 1907, was a challenge to the French philosophical tradition.[11] In fact, "the oblivion of life" in French philosophy existed for a long time. First, it can be traced back to the rationalism of Cartesian mechanism and its "mistrust and hostility" towards life (regardless of whether this is a fair assessment of Cartesian physiology), by assimilating living beings to mechanical and material objects.[12] The nature of life is not granted any metaphysical originality when life is completely analysed by reason and in matter. Second, existentialism eliminated the biological aspect of human life from the notion of existence. Life is defined there in a rational form, as a pure "existence" attributed to the condition of human beings alone, not to all living beings. In Descartes, "in the philosophies of Alain, Brunschvicg and Sartre, life is not recognized as a proper metaphysical object".[13] Canguilhem makes use of the same opposition in a later essay on "environment" in biology, where he opposes the restrictive, Cartesian view of animal motion (this time as presented in behaviorism) to a richer understanding of motion and perception, in Gestalt theory and von Uexküll's ethology.[14] Canguilhem's desire to present a unified picture of life and knowledge sometimes has surprisingly humanist overtones, as when he opposes Life to technology and the various forms of the "mechanization of life", and speaks of human biology and medicine as belonging to an "anthropology"; by extension, "medical vitalism" is the expression of "a distrust, shall we say an instinctive one, of the power of technique over life".[15]

In his desire to "roll back" some of this denial of a metaphysics of life, Canguilhem is then a vitalist, indeed, a self-proclaimed one. But what sort of vitalist was he, and what role did Bergson play in this evolution in his thought?[16]

HISTORICAL EPISTEMOLOGY OF LIFE OR METAPHYSICAL VITALISM?

Unlike Bergson, Canguilhem does not begin with a metaphysical assertion of the uniqueness of Life. He often acts as the historical epistemologist, seeking to defuse some of the reductionist challenges to vitalism by problematizing it as a historical object. Yet at the same time, even if he wears the mask of the scholar, looking at the construction of a concept (say, the cell theory), Canguilhem the philosopher asks highly "motivated" questions of science, in a manner which probably owes a great deal to Bachelard, precisely in the context of a historical epistemology: "A philosophy that asks science for clarifications of concepts cannot remain uninterested in the construction of this very science"; "Truth is not constituted in a history of truth but in a history of science, in the experience of science"; "the pursuit of truth is the effect of a choice which does not exclude its opposite".[17] The history of science has to study possible conceptual developments rather than just invalidate the past (the error of "presentism"). What this entails for vitalism is that it has a specifically *philosophical* place, whether it is scientifically "validated" or "refuted", and apart from its status as a scientific "construction".

In this sense, as Canguilhem suggests, *vitalism is not like geocentrism or phlogiston: it is not refutable in quite the same way.*[18] Vitalism is generally considered to have been "refuted" twice. First, according to a celebrated scientific tale, with Friedrich Wöhler's synthesis of urea in 1828, which showed that organic substances can be produced out of inorganic compounds, thus rendering the claim that the chemistry of the living body is categorically distinct from that of inanimate bodies, invalid. Second, a century later, this time because of physics, in early twentieth-century Vienna Circle arguments against Hans Driesch and Bergson, in the name of the causal closure of the space-time world (given the causal closure that physics reveals in the world, how could there be an immaterial vital impulse force, entelechy or *élan vital* which causes events in this world without itself being caused?).[19] In both cases, a form of vitalism may be refuted, but not what Canguilhem has in mind. The undead character of vitalism appears in the first case, with Wöhler's synthesis of urea, when people start to describe the purported refutation as a "chemical legend" (including because the synthesis was actually only performed by Berthelot later on, and chemists like Berzelius continued to speak of vital forces afterwards[20]); in the second case, substantival vitalism is refuted, not what we might call *explanatory* or *heuristic* vitalism.

So not only is vitalism a unique kind of historical object, but, much more metaphysically, it is *Life itself* which dictates a certain kind of attitude on the part of the inquirer. There is something about Life that places the knower in a special relation to it. Indeed Canguilhem frequently makes an overtly metaphysical, ahistorical claim that the living animal is necessarily a knower, so that conversely, the nature of Life itself forces the knower to approach it in a certain way (with echoes of the beginning of Aristotle's *Metaphysics*). In the 1966 essay "Le concept et la vie", one of Canguilhem's most difficult and ambitious pieces, which deliberately indulges in high metaphysics, he begins by reflecting on Aristotle, declaring that the thinker is interested in Life insofar as it is "the form and potential of the living".[21] Foucault emphasizes the same point in his fine essay on Canguilhem: "To form concepts is a way of living, not of killing life".[22] This is also true at the historical level, for Canguilhem describes Life itself as transcending the oppositions familiar to historians of science: "The opposition between Mechanism and Vitalism, or Preformationism and Epigenesis, is transcended by life itself, extending itself [*se prolongeant*] as a theory of life" ("Aspects du vitalisme", 85). Of course, if vitalism and mechanism are simply the two poles of the activity of Life and its interpretation, why should vitalism be any better than mechanism? We will not attempt to answer this question now, as we mainly want to emphasize that Canguilhem is operating with an extremely robust, one might even say overdetermined concept of Life.

ORGANOLOGY, MECHANISM AND LIFE

Reflecting on the "situation of biological philosophy in France" in 1947,[23] Canguilhem speaks of "the value of Bergsonian philosophy" in its contribution to French philosophy:

> for understanding the true relationship of organism and that of mechanism, for being a biological philosophy of machinism, treating machines as the organs of life, and laying down the foundations of a general organology.[24]

This "biological philosophy of machinism" aims to integrate machine and organism, biology and technology. In other words, this biological philosophy or organology should be more fundamental than the world of machines or technology itself, again in a "push-back" against the sovereignty of Cartesian mechanism, in which the organism was understood on the model of the automaton, explaining its structure and function "on the basis of the structure and function of an already-constructed machine".[25] The ideal type of the machine has become a proof of the self-sufficiency of mechanism. By contrast, Canguilhem reverses the priority of machine over organism in mechanism: "biological organization must necessarily precede the existence and meaning of mechanical constructions".[26]

Bergson criticizes mechanism in *Creative Evolution* for its construction of an artificial system in which life is treated as no different from inert matter. But life is also a sort of mechanics, in its technical character, which is irreducible to anything else. There is "a mechanics of transformation" in organic activity that cannot be mathematically developed as a mechanism articulated with the theme of geometrical and spatialized modes of thought.[27] Canguilhem writes:

> The philosophy of *Creative Evolution* appears to us like the most clairvoyant (if not totally successful) attempt to complete the explanation of

> the mechanisms, including the mechanisms of life—which belongs to science—with a comprehension of the construction of machines taken as cultural rather than just physical facts, which requires the reinsertion of mechanisms in living organization as a necessary condition of anteriority.[28]

For Canguilhem, all of mechanics essentially retains a vital origin which is irreducible to any rational forms. This vital dimension is, additionally, part of a broader embeddedness of mechanisms in cultural and organic activity (here Canguilhem echoes much of contemporary cognitive archaeology): the machine is "a fact of culture".[29] If mechanical invention is derived from the origin of life, the machine is a cultural exemplar which is inseparable from the organization of life.[30]

This is where the Bergsonian motif of an "organology" comes in: the project "to inscribe the mechanical within the organic",[31] "to return mechanism to its place in life and for life" and "to reinsert the history of mechanism into the history of life".[32] The organization of matter is attributed to an act of the *élan vital* insofar as the act of intelligence on matter also belongs to an organic activity. Recall Canguilhem's Bergson-nourished attitude towards the Cartesian notion of animal-machines: he views them, on the one hand, as inadequate representations of organisms, but, on the other hand, as the ruse of reason (!), as a form of skill, referring back to the original term μηχανή. As such, he considers that mechanistic representations are subsumable once again under the category of Life and its productions—that is, as modalities of the organic world.[33]

What troubles Bergson in mechanism is a "mechanistic idea of matter" that treats life as inert matter in a closed and artificial system. To this repetitive picture of matter he opposes its transformation by the *élan vital* in organization, generating the distinction between organized matter and unorganized matter. The *élan vital* distinguishes Bergsonian from classical vitalism.[34] The latter is nothing other than "a sort of label affixed to our ignorance" about the irreducibility of life, while mechanism "invites us to ignore that ignorance".[35] In fact, the debate between vitalism and mechanism ultimately becomes a question of the compatibility between life and knowledge. Mechanism is a rationalistic idea according to which "there exists a fundamental conflict between knowledge and life, such that their reciprocal aversion can lead only to the destruction of life by knowledge or to the derision of knowledge by life".[36] By contrast, for Bergson and Canguilhem, there is an interaction between life and knowledge.

FORMS OF VITALISM: BERGSON AND CANGUILHEM

Canguilhem's identification of life with knowledge can be viewed as echoing Bergson's project of constituting knowledge in the reciprocal dynamic between "theory of life" and "theory of knowledge", but in Bergson this reciprocity does not lead towards the concept of life, since there is a break that cannot be reconciled by the intellect between "life explained" and "life experienced":[37] this marks an important difference between the two thinkers, and their two forms of vitalism. Conceptual knowledge, which is dedicated "to thinking matter", has a "natural inability to comprehend life".[38] The concept is incompatible with life, and, by extension, a philosophy of the concept is not united with a philosophy of life. Canguilhem comments:

> It is evident that a philosophy of life conceived in this way cannot be a philosophy of the concept, since the genesis of living forms is not a completed development nor a n integral derivation and therefore a replica. [39]

Bergson sees concepts as tools employed by life in its relation to the environment. But these conceptual tools—unlike Canguilhem's more "Aristotelian" emphasis on a concept (life continuity, discussed here)—are incapable of exploring the durational movement of life because it is molded by intelligence on an immobile matter. Nevertheless, intuition can renew the concept once the intellect attends to the creation of life without returning to the natural intellectual habit of generalization. The *intuitive* concept is vitalized by intuition, rendering it susceptible to duration, and thereby articulating the meaning of life in relation to the experience of duration.[40] Intuition is always prior to the concept and generates it as such.

The intuition of life contradicts the pure intellectual practice of generalization of concepts: this is Bergsonian "intuitionism".[41] However, Bergson also attributes the origin and value of general ideas to "the fundamental requirements of life" that determine the vital significance of the faculty of generalization.[42] As Canguilhem notes, Bergson grants "the essential resemblances or objective generalities which are inherent in reality itself".[43] These natural general ideas are generated from all types of organisms, and are distinct from the artificial ones fabricated by the human mind:

> every living being, perhaps even every organ, every tissue of a living being generalizes, I mean classifies, since it knows how to gather, in its environment, from the most widely differing substances or objects, the parts or elements which can satisfy this or that of its needs; the rest it disregards. Therefore it isolates the characteristic which interests it, going straight to a common property; in other words, it classifies, and consequently abstracts and generalizes.[44]

Generalization is a biological function of the organism. All living beings generalize in their living world: their generalizing activities are biological in essence.

In Canguilhem's view, the linkage of conceptual knowledge to life could be alternatively developed with a Bergsonian inspiration, since the conceptual activity of a living being in its living context is generated from the act of biological generalization. Bergson's vitalism is less concept-friendly, more suspicious of the transformation of life that occurs in and through the intellect. In contrast, for Canguilhem "life is concept".[45] This is so in at least three ways: in his focus on conceptual activity, on biological knowledge ("vitalism ultimately means the recognition of life as an original realm of phenomena, and thus the recognition of the specificity of biological knowledge"[46]), and on vitalism as a kind of *fundamental existential attitude*, "immanent in living beings":

> We can thus suggest that vitalism translates a permanent exigency [*exigence*] of life in the living, the self-identity of life immanent to the living. This explains one of the characteristics that mechanist biologists and rationalist philosophers criticize in vitalism: its nebulousness, its vagueness. If vitalism is above all an exigency, it is normal that it would have some trouble formulating itself in terms of determinations.[47]

He also calls it an "ethics", again using the term "exigence": "That vitalism may be an exigency rather than a method and a morality rather than a theory".[48] For Canguilhem, the concept is an emanation or outgrowth of a more fundamental vital activity, neither abstract generalization nor mere "determination"—more an ethics than a theory. The living being is a "center of reference" radiating activity outwards, including conceptual activity:

> The living is precisely a center of reference. It is not because I am thinking, it is not because I am a subject in a transcendental sense; it is because I am alive that I must look to life for the reference of life.[49]

Canguilhem also refers (with a hint of irony) to Hegel's imprudent leap away from Kant—from a deliberately regulative, projective vision of life (organism) to a "rational metaphysics", or, in Canguilhem's terms, an explicit identification of concept and life: "Hegel accepted what Kant refused to accept. In the *Phenomenology* as well as in the Jena *Real-philosophie*. . . . Concept and Life are identified with each other".[50] And yet, at times Canguilhem appears to side with Aristotle (whom he enthusiastically describes as the first to understand concept and life together) and Hegel rather than with this regulative, non-metaphysical view. We should really distinguish between a Bergsonian, intuition-based vitalism—which, as we have seen, Canguilhem credits for awakening French rationalism and existentialism from their combined dogmatic slumbers—a more regulative, heuristic vitalism in which Life is more of a construct, and third, a "conceptualist" vitalism, in which concept and life are one, or at least unified.

Canguilhem's identification of concept and life is indeed Bergsonian-tinted; yet it is not just more intellectualist ("concept-friendly") but also more naturalistic in the sense that it does not seek to be more fundamental than the conceptual activity of biological science itself (although this naturalism has to be taken with a grain of salt, as we discuss in closing). Take the example of genetics: it is "an anti-Bergsonian science" because it clings to "the belief in the stability of the structures produced by generation" in contrast to the Bergsonian forms of living being produced by the *élan vital*. Life is *élan* and the biological heredity in the formation of living forms is "the transmission of the *élan*": to overcome the obstacle of matter which divides, diversifies, disperses and multiplies the *élan* of life for the individualization of living forms. Finally, the obstacle to the *élan* is the *élan* itself when the *élan* becomes a limit for its self-overcoming in its transmission.[51]

By contrast, the account given by genetics for "the formation of living forms by material presence" is "information". It explains how the biological function of heredity is compared to a transmission of information in which the sense of life is identified as "a *logos*, inscribed, conserved and transmitted" in living matter. Canguilhem allows for "an *a priori* that is properly material and no longer just formal":[52] this is far removed from the Bergsonian hostility to matter as the obstacle to the *élan vital*.

CONCLUDING REMARKS

Like Bergson, Canguilhem is not a substantival vitalist asserting the existence of special vital forces. Instead, he focuses on the relation between concept and life, additionally bringing out an "existential", constructive dimension of vitalism as a requirement or demand (*exigence*) expressed by living beings.[53] Canguilhem is furthest from Bergson in the way he historicizes the issue. Recall his suggestion that vitalism is not like (the theory of) phlogiston or geocentrism. This can be the case for two different reasons:

- it's not like phlogiston because it's *true* and thus one's ontology needs to include it (an ontological claim, which can be explicated in Aristotelian, Hegelian, Bergsonian or even Drieschian ways)
- it's not like phlogiston because it has this *heuristic value*, or explanatory power (a heuristic claim: living phenomena need to be approached in a certain way in order to be understood)[54]

In fact, it's not entirely clear where Canguilhem falls in this divide. However, his comments on vitalism as an "orientation" (what we have called an attitude) tend towards the latter interpretation. Indeed, it is clear that as an *épistémologue* he is careful to distinguish his claims from the more inflated ones of substantival vitalism. We seem to be far from a metaphysics of Life, then. Yet the concept-and-life side of Canguilhem ("life is concept"), and also the side in which historico-scientific formations like "mechanism" express an aspect of a deeper level called "Life", point to a different moral of the story—not a safe piece of historical epistemology.

If Canguilhem's vitalism is not an ontological commitment to the existence of vital forces, and at times explicitly recognizes the irreducibility of historico-instrumental forms of grasping "life science", how can it be a metaphysics? What is "this vitalist confidence in the spontaneity of life"?[55] In a very real sense one cannot distinguish between a historical claim and a philosophical claim in Canguilhem's "history of vitalism" or "vitalism". To put it in the form of a slogan (which concludes "Le concept et la vie"), "Contemporary biology, read in a certain way, is somehow a philosophy of life".[56] But the strongest claim of all is that Life itself is a positing of norms. Canguilhem's recurring Nietzschean point, that what it is to be alive rather than a crystal or mineral is to be capable of error, or, conversely, that life could be the result of an error, must be understood in support of his more general claim that norms are derived from vital activity itself. A vital error is something like an anomaly, which is why the history of biological thought always includes the problem of monsters: "If life has any meaning, we have to admit the possibility of a loss of meaning, of aberrations and misdeals".[57] Hence, as Canguilhem often says, there are no monstrous crystals, or monstrous machines.

Canguilhem's revisionary project to put the life sciences at center stage in the history of science overall (traditionally dominated by the hard sciences) is bound up with strong ontological commitments, and a certain conceptual vagueness to boot. Namely, his project must amount to a claim regarding the specificity of its object, but it is not easy to make out exactly which claim he wants to make:

- Life itself as an object is ontologically unique, including in its anomalousness.
- Living entities are meaningful and meaning-producing entities and thus have to be understood as such (this covers both the existential and the Goldsteinian aspects of his claim).

Of course, both of these can be coherently regrouped under the heading of a medical vitalism, in which "the problem of the specificity of disease and the threshold it marks among natural beings" marks a kind of challenge to the integration of the objects of the life sciences within the sciences in general; "the possibility of disease, death, monstrosity, anomaly and error" (Foucault).[58] Yet Canguilhem's vagueness appears—for example, when he denies that vitalism is a metaphysics, and then adds immediately afterwards that it is "the recognition of the originality of the fact of life [*le fait vital*]".[59]

We shall close with a brief consideration of a particularly difficult passage in "Aspects du vitalisme", where Canguilhem rejects substantival vitalism more clearly than anywhere else:

> the classical vitalist accepts the insertion of the living organism into a physical milieu to whose laws it constitutes an exception. Therein lies, in our opinion, the philosophically inexcusable fault. There cannot be an empire within an empire without there being no longer any empire, neither as container nor as contents. There can be only one philosophy of empire, that which refuses any division: imperialism. The imperialism of physicists or chemists is thus perfectly logical, pushing to its limit the expansion of logic or the logic of expansion. One cannot defend the originality of the biological phenomenon, and consequently the originality of biology, by demarcating within the physico-chemical territory—that is, within the milieu of inertia, of externally determined movements—enclaves of indetermination, zones of dissidence, or foyers of heresy. If one is to assert the originality of the biological, this must be in terms of the originality of one realm over the whole of experience, and not over islets of experience. In the end, classical vitalism sins, paradoxically, only in its excessive modesty, in its reluctance to universalize its conception of experience.[60]

"Classical" vitalism as described here is substantival vitalism. And Canguilhem's diagnosis of an "inexcusable philosophical mistake" is clear enough. But what should we make then of his defense of the "originality of the biology"—that is, the autonomy of biology, as a "reign over the totality of experience"? What looks at first glance like metaphysical holism might instead be an "attitudinal" conception—that is, a *point of view* on experience.

Unlike the "classical vitalist", Canguilhem insists, using Spinoza's phrase, that we are *not* an *imperium in imperio*! That is, the laws of the physical world fully apply to all living beings, humans included, without exceptions. So all problems would appear to be solved; yet this statement creates new problems. Granted, to the standard question, "how can one be a vitalist and reject any *imperium in imperio*?", we can answer on Canguilhem's behalf that one can be a constructivist or heuristic vitalist, but what do we do with the talk of "Life itself"? Similarly, if we grant that the "Aristotelian" dimension in his vitalism—the stress on how Life itself creates a certain attitude on the part of the knower—is not to be confused with an appeal to substantival vital forces, we are left with the rather opaque invocation in the earlier quotation of "experience". This may sound mysterious, unless we recall both Canguilhem's Bergsonian background and his conceptually oriented nuance.

BIBLIOGRAPHY

Ansell-Pearson, Keith. "Bergson's Encounter with Biology: Thinking Life". *Angelaki* 10, no. 2 (2005): 59–72.

Bergson, Henri. *The Creative Mind: An Introduction to Metaphysics*. Translated by Mabelle L. Andison. New York: Citadel Press, 1946.

Bergson, Henri. *Creative Evolution*. Translated by Arthur Mitchell. New York: Dover, 1998.

Bergson, Henri. *An Introduction to Metaphysics*. Translated by Thomas E. Hulme. New York: Palgrave Macmillan, 2007.

Bianco, Giuseppe. "At the origins of Georges Canguilhem's 'Vitalism.' Against the Anthropology of Irritation". In *Vitalism and the Scientific Image in Post-Enlightenment Life Science, 1800–2010*. Edited by Sébastien Normandin and Charles T. Wolfe, 243–67. Dordrecht: Springer, 2013.

Canguilhem, Georges. "Note sur la situation faite en France à la philosophie biologique". *Revue de métaphysique et de morale* 52, no. 3 (1947): 322–32.

———. *La formation du concept de réflexe aux XVIIe et XVIIIe siècles*. Paris: Vrin, 1977.

———. *Études d'histoire et de philosophie des sciences*. Paris: Vrin, 1968.

———. "De la science et de la contre-science". In *Hommage à Jean Hyppolite*, 173–80. Paris: Presses Universitaires de France, 1971.

———. "Vie". In *Encyclopaedia Universalis*, t. 16, 746–69. Paris: Encyclopaedia Universalis, 1973 (first published 1968).

———. "Le problème des régulations dans l'organisme et la société". In *Écrits sur la médecine*, 101–25. Paris: Éditions du Seuil, 2002.

———. "Commentaire au troisième chapitre de *L'évolution créatrice*". In *Annales bergsoniennes III: Bergson et la science*. Edited by Frédéric Worms, 113–60. Paris: Presses universitaires de France, 2007.

———. *Knowledge of Life*. Edited by Paola Marrati and Todd Meyers. Translated by Stefanos Geroulanos & Daniela Ginsburg. New York: Fordham University Press, 2008.

Deleuze, Gilles. *Pourparlers, 1972–1990*. Paris: Minuit, 1990.

Foucault, Michel. " La vie: l'expérience et la science". *Revue de métaphysique et de morale* 90, no. 1 (1985): 3–14.

Frank, Phillip. *The Law of Causality and Its Limits* (1932). Edited and translated by M. Neurath and R. S. Cohen. Dordrecht: Kluwer, 1998.

Gayon, Jean. "La philosophie et la biologie". In *Encyclopédie philosophique universelle*, t. IV: *Le Discours philosophique*, edited by Jean-François. Mattéi, 2152–71. Paris: Presses Universitaires de France, 1998.

———. "The Concept of Individuality in Canguilhem's Philosophy of Biology". *Journal of the History of Biology* 31 (1998): 305–25.

———. "Bergson's Spiritualist Metaphysics and the Sciences". In *Continental Philosophy of Science*. Edited by Gary Gutting, 43–58. Oxford: Blackwell, 2005.

Le Blanc, Guillaume. "From Matter to Materiality according to Canguilhem". *Graduate Faculty Philosophy Journal* 22, no. 1 (2000): 255–70.

———. *Canguilhem et la vie humaine*. Paris: Presses Universitaires de France, 2002.

———. "Le problème de la création: Bergson et Canguilhem". In *Annales bergsoniennes II: Bergson, Deleuze, la phenomenology*. Edited by Frédéric Worms, 489–506. Paris: Presses universitaires de France, 2004.

Lecourt, Dominique. *Marxism and Epistemology: Bachelard, Canguilhem and Foucault*. Translated by B. Brewster (originally published 1972). London: NLB, 1975.

———. "La question de l'individu d'après G. Canguilhem". In *Georges Canguilhem. Philosophe, historien des sciences*, 262–70. Paris: Albin Michel, 1993.

McKie, Douglas. "Wöhler's 'Synthetic' Urea and the Rejection of Vitalism: A Chemical Legend". *Nature* 153 (1944): 608–10.

Ramberg, Peter J. "The Death of Vitalism and the Birth of Organic Chemistry: Wöhler's Urea Synthesis and the Disciplinary Identity of Organic Chemistry". *Ambix* 47, 3 (2000): 170–95.
Schiller, Joseph. *La notion d'organisation dans l'histoire de la biologie*. Paris: Maloine, 1978.
Schmidgen, Henning. "Thinking technological and biological beings: Gilbert Simondon's philosophy of machines". *Revista do Departamento de Psicologia UFF* 17, no. 2 (2005): 11–18.
Wolfe, Charles T. "From Substantival to Functional Vitalism and Beyond, or from Stahlian Animas to Canguilhemian Attitudes". *Eidos* 14 (2011): 212–35.
———. "Was Canguilhem a Biochauvinist? Goldstein, Canguilhem and the Project of Biophilosophy". In *Medicine and Society: New Continental Perspectives*. Edited by Darian Meacham. Dordrecht: Springer, forthcoming.
Wolsky, Alexander A., and Issekutz Wolsky. "Bergson's Vitalism in the Light of Modern Biology". In *The Crisis in Modernism: Bergson and the vitalist controversy*, edited by Frederick Burwick and Paul Douglass, 153–70. Cambridge: Cambridge University Press, 1992.

NOTES

1. Gilles Deleuze, *Pourparlers, 1972–1990* (Paris: Minuit, 1990), 196.

2. No less a resource than Wikipedia informs us that "biophilosophy" is not guilty of the reductionist excesses of analytically oriented "philosophy of biology". We shall ignore these kinds of valuative definitions as our aim is to understand what kind of claim was being made by Canguilhem in this context, with particular reference to the notion of vitalism. On the emergence of philosophy of biology as a discipline in contrast to "biophilosophy", see Jean Gayon, "La philosophie et la biologie", in *Encyclopédie philosophique universelle*, ed. Jean-François Mattéi, t. IV (Paris: Presses universitaires de France, 1998), 2152–71.

3. Canguilhem, *La formation du concept de réflexe aux XVIIe et XVIIIe siècles* (Paris: Vrin, 1977), 1 (translations are ours unless otherwise indicated).

4. Georges Canguilhem, "Aspects du vitalisme", in *Knowledge of Life*, ed. Paola Marrati and Todd Meyers, trans. Stefanos Geroulanos and Daniela Ginsburg (New York: Fordham University Press, 2008), 59–74.

5. Canguilhem, "Aspects of Vitalism", 59.

6. For instance, for Keith Ansell-Pearson, although Canguilhem is definitely not a Bergsonian philosopher, his works share the common Bergsonian idea that "the project of knowledge is one with thinking life", "Bergson's Encounter with Biology: Thinking Life", *Angelaki* 10, no. 2 (2005): 69. Also, Jean Gayon notes that "Bergson's reflection on biological facts closely resembles Canguilhem's notion of normativity of the living [being]". "Bergson's Spiritualist Metaphysics and the Sciences", in Gary Gutting, ed., *Continental Philosophy of Science* (Oxford: Blackwell, 2005), 57. These biological facts are facts "intended" by nature, and therefore norms are posited by the living being. For Guillaume le Blanc, the common theme connecting Bergson and Canguilhem is the affirmation of life's creative character. "Le problème de la création: Bergson et Canguilhem", in Frédéric Worms, ed., *Annales bergsoniennes II: Bergson, Deleuze, la phenomenologie* (Paris: Presses universitaires de France, 2004), 489–506.

7. As Jean Gayon writes, "Such a conception of 'meaning' or 'concept' inscribed in matter is of course counterintuitive". "The Concept of Individuality in Canguilhem's Philosophy of Biology", *Journal of the History of Biology* 31 (1998): 323.

8. Canguilhem, "Commentaire au troisième chapitre de *L'évolution créatrice*", in *Annales bergsoniennes III: Bergson et la science*, edited by Frédéric Worms (Paris: Presses Universitaires de France, 2007), 135.

9. Canguilhem, "Commentaire", 120. See Henri Bergson, *Creative Evolution*, trans. Arthur Mitchell (New York: Dover, 1998), 194–99.

10. Canguilhem, "Note sur la situation faite en France à la philosophie biologique", *Revue de métaphysique et de morale* 52, no. 3 (1947): 324.

11. Canguilhem, "Note sur la situation", 323. Canguilhem emphasized the importance of Bergson's 1907 *Creative Evolution* at the beginning of this article. Bergson and Ruyer had provided "an interpretation of the fundamental biological phenomena, such as embryonic development, starting from psychological models, to trace back the Platonic theory of ideas in the relations of individuals to species, to revive in reinventing these relations, not to dust off them, the Aristotelian concepts of power and form".

12. Canguilhem, "Note sur la situation", 324, 326.

13. Canguilhem, "Note sur la situation", 324. See Giuseppe Bianco, "At the origins of Canguilhem's 'vitalism'. Against the anthropology of rrritation", in Sébastian Normandin and Charles T. Wolfe, eds., *Vitalism and the Scientific Image in Post-Enlightenment Life Science* (Dordrecht: Springer, 2013), 243–67. Another core influence on Canguilhem here besides Bergson and Alain is Kurt Goldstein. For more on this relation, see Charles Wolfe, "Was Canguilhem a biochauvinist? Goldstein, Canguilhem and the project of biophilosophy", in Darian Meacham, ed., *Medicine and Society: New Continental Perspectives* (Dordrecht: Springer, forthcoming).

14. Canguilhem, "The Living and Its Milieu", in Canguilhem, *The Knowledge of Life*, 110.

15. Canguilhem, "Aspects of Vitalism", 62.

16. We do not seek to provide an exhaustive biographical reconstruction of Canguilhem's thought here, including for reasons of space; see the discussion in Bianco, "At the origins of Canguilhem's 'vitalism'".

17. Respectively Canguilhem, "Aspects of Vitalism", 80, and "De la science et de la contre-science", in *Hommage à Jean Hyppolite* (Paris: Presses Universitaires de France, 1971), 173–80.

18. Canguilhem, "Aspects du vitalisme", 80.

19. See Phillip Frank, *The Law of Causality and Its Limits* [1932], ed. and trans. M. Neurath and R. S. Cohen (Dordrecht: Kluwer, 1998), especially chapter 4; M. Wolsky and A. A. Wolsky, "Bergson's Vitalism in the Light of Modern Biology", in *The Crisis in Modernism*, Frederick Burwick and Paul Douglass, eds. *Bergson and the Vitalist Controversy* (Cambridge: Cambridge University Press, 1992).

20. Douglas McKie, "Wöhler's 'Synthetic' Urea and the Rejection of Vitalism: A Chemical Legend", *Nature* 153 (1944): 608–10; Peter Ramberg, "The Death of Vitalism and the Birth of Organic Chemistry", *Ambix* 47, no. 3 (2000): 170–95.

21. Georges Canguilhem, "Le concept et la vie", in *Études d'histoire et de philosophie des sciences* (Paris: Vrin, 1968), 335.

22. Michel Foucault, "La vie: l'expérience et la science", *Revue de métaphysique et de morale* 90, no. 1 (1985), 13.

23. Canguilhem, "Note sur la situation".

24. Canguilhem, "Note sur la situation", 332. Guillaume Le Blanc gives a detailed account of Canguilhem's appropriation of Bergsonian organology in *Canguilhem et la vie humaine* (Paris: Presses Universitaires de France, 2002), 188–205.

25. Canguilhem, "Machine and Organism", in *Knowledge of Life*, 76.

26. Canguilhem, "Machine and Organism", 91.

27. Bergson, *Creative Evolution*, 31–32.

28. Canguilhem, "Note sur la situation", 332.

29. Canguilhem, "Machine and Organism", 92, 93.

30. Le Blanc, *Canguilhem*, 201.

31. Canguilhem, "Machine and Organism", 96.

32. Canguilhem, "Note sur la situation", 332. See Bergson, *Creative Evolution*, 139–40, 161.

33. Canguilhem is thus closer to a biologistic, anti-computational position (like Ruyer's) than to a philosophy of technology (like Simondon's); see Henning Schmidgen, "Thinking technological and biological beings: Simondon's philosophy of machines", *Revista do Departamento de Psicologia UFF* 17, no. 2 (2005), 12.

34. Although Bergson is not a vitalist in a traditional sense, his philosophy of life is vitalistic in inspiration. Notably, Bergson's vitalism is contingent upon the *élan vital* not being reified as a thing or a substance, as its reality is a tendency acting on matter that is tied up with the temporal experience of duration: Bergsonian vitalism is a durational vitalism (see Bergson, *Creative Evolution*, 96). For a distinction between substantival, functional and existential forms of vitalism, see Charles Wolfe, "From Substantival to Functional Vitalism and Beyond, or from Stahlian Animas to Canguilhemian Attitudes", *Eidos* 14 (2011): 212–35.
35. Bergson, *Creative Evolution*, 42.
36. Canguilhem, *Knowledge*, xvii.
37. Bergson, *Creative Evolution*, xiii, 10.
38. Bergson, *Creative Evolution*, ix, 165.
39. Canguilhem, "Le concept et la vie", 348.
40. Bergson, *An Introduction to Metaphysics*, trans. Thomas E. Hulme (New York: Palgrave Macmillan, 2007), 40.
41. Canguilhem, "Le concept et la vie", 339.
42. Bergson, *The Creative Mind: An Introduction to Metaphysics*, trans. Mabelle L. Andison (New York: Citadel Press, 1946), 61.
43. Canguilhem, "Le concept et la vie", 348; Bergson, *The Creative Mind*, 56.
44. Bergson, *The Creative Mind*, 54.
45. Canguilhem, "Le concept et la vie", 364.
46. Canguilhem, *La formation du concept de réflexe*, 113.
47. Canguilhem, "Aspects of Vitalism", 82.
48. Canguilhem, "Aspects of Vitalism", 73. The word "exigence" seems important here: Canguilhem uses it seven times in the article.
49. Canguilhem, "Le concept et la vie", 352.
50. Canguilhem, " Le concept et la vie", 345.
51. Canguilhem, " Le concept et la vie", 339, 353.
52. Canguilhem, " Le concept et la vie", 339, 362.
53. Canguilhem, "Aspects du vitalisme", 86.
54. One of Canguilhem's students, Dominique Lecourt, confuses this heuristic dimension with ontological vitalism: "The assertion of . . . 'vitalism' as an intellectual requirement which aims to acknowledge the originality of Life, entirely retains its significance today, when the combination of a type of biochemical materialism and a type of mathematical formalism tend to deny this originality of Life, the better to neuronalize thought" ("La question de l'individu d'après Canguilhem", in *Georges Canguilhem. Philosophe, historien des sciences*, ed. Etienne Balibar et al. [Paris: Albin Michel, 1993], 269). In contrast, in his earlier work Lecourt had judged Canguilhem's vitalist tendencies severely—they were the part of his thought that the Marxist interpreter should not keep (*Marxism and Epistemology: Bachelard, Canguilhem and Foucault*, trans. B. Brewster [1972; London: NLB, 1975]).
55. Canguilhem, "Aspects du vitalisme", 63.
56. Canguilhem, "Le concept et la vie", 364.
57. Canguilhem, "Le concept et la vie", 364.
58. Foucault, "La vie: l'expérience et la science", 11.
59. Canguilhem, "Le normal et le pathologique", 156.
60. Canguilhem, "Aspects du vitalisme", 70.

FIVE

Life and Objective Norms: Canguilhem in the Context of Contemporary Meta-ethics

Sander Werkhoven

This chapter explores the possibility that the growing philosophical focus on life results from a deep dissatisfaction with the general tendency to regard all values and norms as nothing other than subjective preferences or contingent cultural constructs. The turn to life could be viewed, at least in part, as an attempt to ground normativity in a more objective and naturalistic stratum than human subjectivity or cultural contingency. This development looks *prima facie* warranted: there seems to be an important sense in which living beings, in virtue of being alive, have certain things that are good and beneficial to them and other things that are not. In this respect, life seems to gives rise to a type of normative differentiation that is significantly more objective than values originating from individual preference or particular cultural dynamics. Our guiding hypothesis shall therefore be that the concept of life can indeed help ground certain normative claims and do so in an objective way.

The purpose of this chapter is to understand the nature of this life-based normativity—if there is indeed any to be identified—and to analyse its meta-ethical status. The writings of Georges Canguilhem will play a central role in this pursuit, as he has made the most explicit and sustained argument in favour of the idea that life gives rise to normative distinctions. The first sections of this chapter discuss Canguilhem's ideas on life's normativity in detail, while subsequent sections aim to formulate a meta-ethical theory based on his views. In the final sections of this chapter recent work from moral realists Peter Railton, Philippa Foot and

Christine Korsgaard shall be utilized to further support the proposed meta-ethical theory. The chapter concludes, first, that there is a normative domain that can be considered objective and grounded in life, and, second, that our concept of life must accommodate the fact that life generates normative distinctions.

THE NORMATIVITY OF HEALTH AND PATHOLOGY

In *The Normal and the Pathological* Canguilhem's primary objective is to argue against a supposedly value-neutral account of pathology and health in terms of deviations from statistical normality, and, more generally, against an understanding of living organisms as being *only* comprised of physical and chemical facts.[1] The latter claim follows from the former: the idea that living beings are norm-generating—or that life is "normative", as Canguilhem prefers to say—and that life is therefore in some sense irreducible to the level of chemistry and physics, which follows from his reflections on the nature of pathology and health.[2] His thoughts on pathology and health shall thus also be our starting point.

The conception of health and pathology Canguilhem opposes is the conception formulated by Claude Bernard, who thought that health and disease were quantifiable deviations from a normal mode of organic functioning. In Canguilhem's words, Bernard thought that "disease is the exaggerated or diminished expression of normal functioning".[3] Like many medical theorists after him, Canguilhem felt that statistical normality of physical measurements and functional outcomes could not be the defining feature of health, just as a deviation from normality does not constitute, in and of itself, a pathological condition, and medical interventions should not be blindly directed at the re-establishment of normality. What gives functional outcomes and physical measurements their meaning, and what renders them truly indicative of health and pathology, is what Canguilhem calls the "biological value" of these measurements, "value" here intended in a distinctly normative sense: "what distinguishes the physiological from the pathological is not a physico-chemical objective reality but a biological value".[4] Perhaps the best way to understand Canguilhem here is by means of concrete examples, like the glucose levels in our blood and blood pressure in a state of rest. What renders glucose concentration outside the scope of 4.0–9.0 mmol/l and systolic blood pressures in rest higher than 160 mmHg potentially pathological, is not the deviation from statistical normality *as such*, but the *value* these measures have for the organism as whole. Various physical states might follow from hypoglycaemia and high blood pressure, such as sweating, lethargy, fatigue, aneurisms, a heart attack or even death. Considered physico-chemically, these potential consequences are simply different states, mere material reconfigurations, just as the measurements belonging to them would be mere numerical deviations. For the living organism, on the other hand, these alterations and different states are clearly not mere alterations and reconfigurations. For living organisms, there is a clear normative difference between these various states: some states are worse for it than others. The *value* of one state compared to another is what renders certain measures indicative of pathology and others of health. The point Canguilhem wants to drive home—and about which he seems to be right—is that certain processes or scenarios are simply bad for an organism, like being in a coma or dying, while others are positive from the perspective of the organism, like being well nourished and being secure in existence. Similarly, certain external objects, environments or happenings can be good for organisms, like sunshine and rain are for a plant—while others are bad, like heavy frost or an overdose of pesticide. Life distinguishes between what is positive and negative at this rudimentary level of normative distinctions, in the same way, as Canguilhem points out, that life distinguishes "food from excrement".[5] Life itself establishes these differentiations, norms and thresholds; norms don't follow from statistical calculus but precisely from the self-regulating and self-preserving makeup of living bodies. In Canguilhem's words, "It is life itself and not medical judgement which makes the biological normal a concept of value and not a concept of statistical reality".[6]

The norms and thresholds generated by life are therefore not just quantitative demarcations and variations: the norms constituted by living beings are what Canguilhem calls "vital norms".[7] Vital norms, I presume, are to be distinguished from social norms or subjectively held norms. The norms set by life are "vital" norms and the values created by life "biological" values because they maintain and promote the life and health of individual organisms rather than any other ends or purposes. The recognition that for all living beings there are normative differentiations between different states of being, external substances and environmental conditions, that life discriminates between what is compatible with it and what is not, that it constitutes vital norms on the basis of its own constitution and regulatory processes, drives Canguilhem to the conclusion that "life is polarity and thereby even an unconscious position of value; in short, life is in fact a normative activity".[8] And precisely this characteristic of life is lost once it is analysed purely at a physico-chemical level. Health and pathology cannot be predicated over mere chemical assemblages or physical constellations, whereas a differentiation between health and pathology certainly does pertain to all beings that live. Canguilhem insists that it is a *defining* feature of life that "for-it" the world is polarized into positives and negatives, structured around vital norms and coloured with values rooted in self-sustaining biological processes. Canguilhem, then, endorses and systematizes the thought Nietzsche once jotted down in a notebook—namely, that "'alive': that means already *valuing*",[9] and that "valuations lie in all functions of the organic being".[10] Canguilhem writes in full agreement that "there is no life whatsoever without norms of life"[11] and that "even for an amoeba, living means preference and exclusion"[12] So, according to Canguilhem, and to be slightly more precise, establishing norms is an *essential* property of life: there can be no life without there also being norms of life and objects of value.

The alternative account of health that Canguilhem develops is one based precisely on the ability to *diverge* from normality, averages and constants. So the ability to cope with abnormal glucose levels and blood pressures, for instance, without incurring significant tissue damage or dying, would typify a state of health. Organisms fixed to some normal state or set of constants, unable to deviate from this state, incapable of temporarily transcending their vital norms, unable to recover from environmental infractions—that is, organisms that are *dependent* on being normal—are precisely those that suffer from pathology. "What characterizes health", Canguilhem writes, "is the possibility of transcending the norm, which defines the momentary normal, the possibility of tolerating infractions of the habitual norm and instituting new norms in new situations".[13] Canguilhem's conception of health is therefore one characterized by flexibility, malleability, adaptability and the power to recover from illness—that is, what he calls "the margin of tolerance"[14] and the level of "biological luxury".[15] "Healthy life", he writes in a later essay, "is a life of flexion, suppleness, almost softness".[16] Pathology, on the other hand, is precisely the kind of process that decreases the "margins of tolerance" and levels of "biological luxury". In a sense, Canguilhem inverts Bernard's account: being fixed to normality is pathological, while the ability to be abnormal is precisely what characterizes health.

OBJECTIVITY OF LIFE-BASED VALUES

Canguilhem's reflections on the normativity of health and pathology, in combination with the recognition that life stands at the basis of certain norms and values, has profoundly interesting meta-ethical implications. It opens the possibility for a theory of normativity which is quite compelling, as I hope to demonstrate. To be clear, in meta-ethics the principal concern is not with what is good, bad, valuable, healthy or pathological. In meta-ethics, the second-order question is raised regarding the status of any first-order normative claims—for example, whether normative claims ever be true and on what grounds, whether things are valuable independent of valuing subjects, whether normative claims are reducible to nonnormative claims, and so on. Canguilhem's ideas about the norm- and value-generating character of life are relevant to these discussions because they open up the possibility for a theory in which life-based values are genuinely objective, without, however, existing in some abstruse way in the external world independently of beings for which things can have value. Canguilhem's account of vital norms and biological values can provide the first steps towards an intuitively plausible middle way between meta-ethical subjectivism and objectivism, an opposition that continues to divide much of the meta-ethical literature. To appreciate how, I shall briefly sketch the most important positions on both sides of this divide.

On one side of the meta-ethical spectrum there are subjectivists and sceptics, who, mostly following Hume, think that values and normative judgements are nothing but subjective or emotional responses to value-neutral things and occurrences. Error-theorists claim that we make a fundamental error each time we make a normative judgement and when we call something good or bad. They claim that objectively—that is, according to the way the world is in itself, or, as some prefer to say, the way the world is *anyway* (independent, in any case, of our own valuating practices)—things are neither valuable nor invaluable. Emotivists and other noncognitivists deny that such a categorical error is committed in normative judgements. Noncognitivists endorse the semantic theory that normative statements do not set a requirement on the objective world to start with: normative claims only express our emotive and subjective responses to this objective world. Quasi-realists, finally, are emotivists but of a less sceptical strand, as they claim that normative claims can be true or false, even though what makes things valuable are ultimately our emotions and subjective responses; subjective responses that we project onto a value-less world. Regardless of these internal differences, subjectivists and sceptics share the conviction that there are no emotion and subject-independent normative facts.

On the other side there are moral realists and objectivists, who think that there are facts about values and normative matters and that the truth or falsity of normative claims obtains independently of our own theorizing, emotions and subjective responses to the world. Nonnatural realists think that these normative facts cannot be reduced to value-neutral natural facts. They argue the world contains normative truths irreducible to truths about the natural world. According to the naturalist realists, on the other hand, there are facts about normative matters, but these facts supervene on, or are grounded in, some combination of natural facts. The shared conviction among normative realists, however, is that normative facts have a truth and existence independent of valuing subjects or any of our projective tendencies.

Canguilhem's claims about the normativity of life can be developed as a middle position between meta-ethical subjectivism and objectivism. If we endorse his claim that living beings, by virtue of being alive, constitute vital norms, polarize their environment in positives and negatives, differentiate between what is compatible and incompatible, and in doing so establish values and norms, then clearly these vital norms and biological values are not *merely* subjective and emotional responses projected onto an objective and value-neutral world. Rather than emotions or any other noncognitive attitudes, it seems to me that the *structural* properties of organisms determine what is conducive to individual living beings and what is not. Structural properties of organisms are precisely the material and functionally interrelated systems that make up of living bodies and the features that underlie the vital norms. Structural properties, I imagine, are features of organisms like the metabolic system, the composition of outer membranes and skin, reproductive systems and so on; properties that determine what the precise vital norms of the organism are. Life-based norms and values, on this view, are not the projections of desiring and affective subjects—as noncognitivists, error-theorists and quasi-realists think—since the kind of mental processes required for such projections cannot be attributed to most forms of life. The norms and values that Canguilhem identifies express a specific type of *relation*: a relation between objective properties of organisms, like the digestive system of a cow, and objective properties of the surrounding world, like grasses that can be digested by a cow—without, it seems, invoking an image of projection of essentially mental responses.

On the other hand, the vital norms and biological values that Canguilhem points to are also not independent facts in the world out there, to be accessed with some uniquely human faculty of moral intuition, as certain nonnatural realists have claimed. The values and norms under consideration are also not normative properties that supervene on—or are grounded in—the physical makeup of the external world considered in isolation, as moral realists of a naturalist bent tend to think. Canguilhem's views oppose the thought that normative facts emerge out of physical objects or processes themselves, independent of the presence of living organisms for which these objects or processes would have value. Canguilhem's thoughts would do justice to the anti-realist intuition that in the life-independent world things are neither valuable nor invaluable, while submitting to the realists that there are facts about what is good and bad that exist independently of our attitudes, emotions and ways of knowing the world—although those facts *are* of course dependent on the presence of life and the particular structural features of individual life forms.

Placing Canguilhem's position on the traditional meta-ethical axes, we could claim that life-based normativity is *realist* in the sense that these norms and values factually obtain, and judgements about them can be true or false. They are *objective* in the sense that facts obtain independently of our subjective faculties, emotions and ways of knowing the world. Yet, vital norms and biological values are *antirealist* in the sense that there would be no such normativity if there were no living beings. With some hesitation we could even say they are *subjective* in the sense that they are generated by—or arise only in the presence of—living beings. Instead of these traditional labels, then, I return to the suggestion made earlier that it is more appropriate to say that vital norms and biological values express a certain *relation* between life and the physical world, a relation that despite its relationality carries a significant level of objectivity. Working out the idea further that biological values and vital norms are relational and yet objective requires that we move beyond Canguilhem's writings and onto more recent literature.

Before doing so, however, it has to be emphasized that the meta-ethical status outlined above only applies to norms and values grounded in life. If a certain society or human subjects value things or activities wholly independent from the norms arising out of our own organic constitution, then those norms and values remain subjective and ungrounded. Even though it makes perfect sense to claim that nothing has value in a universe without any life, this does not imply that all values are objectively grounded in structural features of life; only the norms set by the objective features of organisms are objectively grounded. That is, only objects and processes that promote the self-sustaining and self-regulatory processes and that are capable of improving the health of organisms are objective and grounded in life. Norms that do not follow from our organic constitution or that are only loosely connected to vital norms, like norms belonging to some particular dining etiquette, are not objectively grounded in life—while norms about which foods are good for us and which are not, clearly are grounded in this way. The meta-ethical position defended here, therefore requires a fundamental split in the totality of norms and values: some are objectively grounded while others remain essentially subjective.

The question of whether typically *moral* statements can be grounded in life is a complex question and one I shall not try to answer here. Moral statements like "killing is wrong" are of a different kind than statements about what is naturally good for an organism given its internal constitution. They are nonetheless not completely disconnected: it is not hard to imagine that in a society of beings for whom it is bad to be killed, moral principles like "killing is wrong" arise and are treated with utmost seriousness. It might even be possible to ground moral statements of the kind "killing is wrong" on the basis of what is nonmorally good and bad for individual human beings, such that statements of this kind are also in some way objective and indirectly grounded in life.[17] This would not even be a possibility, however, if what is nonmorally good or bad for individual human beings is itself not grounded and objective. The concern of this chapter is therefore only to isolate a normative domain and argue for its groundedness and objectivity. Whether this domain could serve as the basis for claims about what is morally right and wrong I shall leave as an open question. In either case, however, the meta-ethical position outlined earlier and further detailed next only pertains to norms and values that are life-based, which might provide a basis for nonrelational moral statements, but certainly does not apply to all normative concepts or our entire normative vocabulary.

RELATIONALITY AND OBJECTIVITY

We followed Canguilhem in his conviction that there are normative differences between an organism's various states and that something about life underpins these normative differentiations. I argued that life-based norms and values are neither objective in an absolute sense nor subjective in an emotivist sense, but rather expressive of a kind of relation. The task is now to explain the nature of this relation more clearly: first, by specifying which two items actually stand in a relation to each other; and second, by showing how this relation can itself be viewed as something objective.

We agreed with Canguilhem that for a living being there is a difference in value between the possible states it can be in. This implies that norms can express a relation internal to a living organism—that is, between the organism as a whole on the one hand, and the state of its parts on the other. For example, it is bad for the human being as a whole when its legs are broken or when its glucose levels are too high, not for the legs or the blood levels themselves. Hence, normative relations obtain between the organism as a whole and the state of its parts. But this is not the only normative relation we mentioned. It was also said that external objects or happenings could be good for living beings, like a certain amount of rain is good for plants and certain types of grasses are good for cows. In that case values do not express a relation internal to an organism but a relation between the organism as a whole and features of the external world. Grass is good for a cow not because of the relationship between a cow's stomach and properties of grass, but rather because of the relation between the cow as a whole—including its digestive system—and features of the external world. Life-based normativity, I suggest, expresses either a relation between the organism as a whole and its parts or a relation between the organism as a whole and external objects and events.

Having determined the *relata* of the relation we can now turn to the proclaimed objectivity of the relation. The abstract claim that relational concepts can be objective is itself nothing controversial. When two things stand side by side, for instance, the notions "standing right of the other" and "standing left of the other" are clearly relational in a three-place way, but the relationality of spatial position certainly does not undermine its objectivity. The question, then, is whether it can be defended that life-based values are objective in a similar way.

Peter Railton has most explicitly defended the idea that values are relational and yet objective.[18] Nonmoral goodness, Railton claims, is always goodness in relation to some being for which things are good or bad. He writes that "although there is no such thing as absolute goodness—that which is good in and of itself, irrespective of what or whom it might be good *for* or the good *of*—there may be *relational* goodness"[19] For relational goodness to be objective, however, we require a distinction between *relationality* and *relativity*. Railton seems to have such a distinction in mind but does not substantiate it any further. But if all values were nothing but subjective preferences projected onto the objective world, then they would also be relational in some sense, *viz.* relational to some valuing subject. I would suggest, then, that what distinguishes relationality from relativity, a relativity incommensurable with objectivity, is that the *relatum* for which things have value—the living being as a whole—is itself an objective matter, whereas subjective preference *qua relatum* is clearly not. Relativity, I propose, is a relation in which one of the *relata* is a subjective or cultural preference, while relationality implies that both *relata* are objective items. Consider one particular key passage from Railton which would illustrate this idea, dovetailing perfectly with Canguilhem's thought and the way I have been trying to extend it into a meta-ethical position:

> In a naturalistic spirit, we might think of goodness as akin to nutritiveness. All organisms require nutrition, but not the same nutrients. Which nutrients a given organism or type of organism requires will depend upon its nature. . . . There is, then, no such thing as an *absolute* nutrient, that is, something that would be nutritious for all possible organisms. There is only *relational* nutritiveness: substance *S* is a nutrient for organisms of type *T*.[20]

What renders nutritiveness relational rather than relative, I suggest, is that the digestive system of a given organism is itself an objective matter. Had the nutritiveness of substances been dependent on whatever subjects or cultures decide to call nutritive, nutritiveness would have been merely relative and subjective. The fact that it is not a subjective preference but an objective thing that determines what is nutritious—*viz.* the organism as a whole including its digestive system—renders nutritiveness relational rather than relative.

In addition to the objectivity of both *relata*, there is also an epistemic argument to make for the objectivity of life-based values. The more we know about the makeup of some living being, the better we can determine which conditions, substances and environments are compatible with it and good for the organism as a whole. On the basis of this idea, Railton builds an account argument that with "unqualified cognitive and imaginative power, full factual and nomological information about physical and psychological constitution, capacities, circumstances, history, and so on", it can be determined what is in someone's—and by extension some organism's—"objective interests".[21] The same can be argued about all vital norms and biological values, it seems to me: the more information we have about the particular physical constitution and regulatory processes of organisms, the better and more objectively we can determine the relevant norms and values. Moreover, little or false information about the structural features of organisms could result in us being wrong or misguided about what is valuable for some organism, including ourselves: we could be wrong about what is poisonous or nutritious—or, indeed, what is too high a glucose level and what is not. The fact that the objective features of organisms determine what is valuable and what is not, that we could be wrong about an organism's vital norms and biological values, and that our knowledge of them is capable of improvement, is precisely what renders life-based normativity objective.

This account of life-based normativity, developed on the basis of Canguilhem's and Railton's assertions, is one in which normativity supervenes on objective facts about living beings and the world. And this supervenience is global: it is impossible for there to be two worlds identical in material constitution whereby the valuative relations between organisms and their surrounding world or their own states are different. Although Canguilhem pertinently resisted a reductionist account of life's normativity, it seems to me that life-based values are reducible in this *weak* sense, the reduction base consisting of the structural properties of organisms considered as a whole in combination with properties of the material world. To be clear, this weak form of reducibility of life-based values is an ontological rather than semantic reduction. The claim is that objective values are ontologically dependent on the material constitution of life and the external world, not that the meaning of valuative terms reduce to natural terms. The domain of normativity claimed to be objective here, in short, supervenes on structural features of living beings and the material world.

THE EXPLANATORY CHALLENGE

Canguilhem's reflections on pathology and health, as well as the objective meta-ethical status of life-based norms and values, raise one further question we must confront: *viz.* what makes living beings such that they do indeed give rise to normative differentiations? Why can't it be claimed that certain things are objectively good for mountains, riverbanks or artefacts like tables and chairs in a way that would depend on the material constitution of such natural items? Canguilhem seems right about the claim that life is unique in standing in a valuative relation to occurrences in the world and its own physical states, but what gives life this special status? In this section I shall discuss three possible answers to this explanatory challenge.

Canguilhem's own answer to this question shows that in our analysis of the relationality and objectivity of life-based values we have moved away from an important aspect of his thought. Canguilhem thinks that life is normative—that is, norm-generating—because living beings constitute a "centre of reference" around which meaning and value are organized. "To live", he writes in a famous essay, "is to radiate; it is to organise the milieu from and around a centre of reference".[22] That is to say, Canguilhem identifies life itself with a sphere of subjectivity that gives rise to and imposes values and meanings on a milieu. Alain Badiou writes in an illuminating paper on this theme that, for Canguilhem, "living is always in some way a pre-subjective aptitude".[23] Canguilhem's pairing of life with subjectivity (or, indeed, some kind of pre-subjectivity) becomes evident when he writes things like "we think that there is nothing in science that has not first appeared in the consciousness, and that . . . it is particularly the sick man's point of view which forms the basis of truth".[24] And similarly, "that the life of the living being, were it that of an amoeba, recognizes the categories of health and disease only on the level of *experience*, which is primarily a test in the affective sense of the word, and not on the level of science".[25] If the difference between health and pathology is ultimately to be found in qualitative subjective experience, and if life's normativity indeed arises out of a subjectivity or pre-subjectivity inherent to the complete manifold of life, then our recourse to life will have been in vain. Once life itself is characterized by an ineliminable and decisive element of subjectivity then normative differentiations turn out to be grounded in subjectivity and affective responses after all, just as meta-ethical subjectivists think—an affectivity that has only been extended outside the sphere of human subjectivity. The turn to life was precisely intended to avoid the relativity and ungroundedness of subjective experience and affective responses, and seemed to give an objective basis for a domain of values and norms grounded in structural features of organisms. Canguilhem's idea that life itself is permeated with subjectivity and that experience ultimately grounds all valuative differentiations is therefore one I shall reject.

A different kind of answer to the question of how normativity arises with life is provided by Philippa Foot.[26] She also believes strongly that normativity is not to be reduced to human emotional responses in the way that Humeans think, and implicitly concurs with Canguilhem's view that all living beings have objects or states that are naturally good for them. She also thinks this "natural" or "primary" goodness is restricted to the domain of life alone, effectively excluding the possibility that inanimate objects like rivers or storms have things that are naturally good for them.[27] The reason she thinks life stands apart from the nonliving is not by reference to an element of subjectivity, but on the basis of the Aristotelian idea that the goodness for living beings depends on what she calls the "life-form" of the species of which living organisms are a member: "it is the particular life-form of a species of plant or animal that determines how an individual plant or animal should be".[28] The life-forms of plants and nonhuman animals have self-maintenance and reproduction as their end, she continues roughly in line with Aristotle, while the life-form of human life has happiness as its natural end.[29] So natural goodness can be attributed to all of life, because life participates in forms that provide a certain standard for how individual organisms should be.

This approach is more compelling than Canguilhem's as Foot effectively builds the normativity of life into the concept of life-form, by definition excluding nonliving entities. It still remains unclear what makes life-forms such that normativity is conditioned by it, however, other than that it re-asserts that life indeed has this property. Aristotle made no distinction between life and inanimate being in a way that would restrict goodness only to life. Aristotle thought that all substances had a final cause—all being is teleological for Aristotle—and it is not clear how Foot could substantiate a distinction between life and nonlife on the basis of a metaphysically weighty sense of "life-form" alone. Nevertheless, it could be argued that she does not have to answer *why* life-forms condition normativity in a way that inanimate being does not, as all she defends is the thesis *that* life-forms do so and nonlife-forms do not. In response to the question of how living beings give rise to norms, her answer is ultimately a replication of Canguilhem's core claim: *viz.* that it is just an *essential* property of life that it does so.

A recent paper by Korsgaard provides further conceptual tools to understand how life gives rise to normative differentiations in a way that inanimate being does not.[30] The first thing to note about the paper is that Korsgaard argues that *all* goods are always goods *for* a kind of entity for whom things can be good or bad. The kind of things that she believes goods are good for, following another Aristotelian principle, are functional systems. In her own words, "If something is a functional system, the properties that enable it to perform its function well are the properties that make it a good one, and the conditions that tend to promote and protect those properties are good for it. So functional systems, by their very nature, have a good".[31] But inanimate things like knives and rivers can also be viewed as functional systems, such that things can be good or bad for them as well. Here Korsgaard argues, convincingly, that the functionality of inanimate objects is always functional in relation to a living being, including the goods that follow from this functionality. So sharpening is good for a knife because it enables the knife to cut well, but the functionality of the knife and the goodness of its being sharpened obtain only in relation to a living being for which a knife is a cutting-thing and for which cutting has some value to start with—that is, for human beings. The functionality of inanimate things gives rise to a kind of normativity that is ultimately normative *only* by reference to life. Korsgaard's functionalist account, then, in contrast to Foot's account centred on the notion of "life-form", gives *some* further explanation why living beings have objective goods and nonliving beings do not: living functional systems have goods that refer to themselves rather than to anything outside of them.

The notion that Korsgaard introduces to further substantiate the self-referential functionalistic account of life-based normativity is that of "having a sake".[32] The way in which things are good or bad for plants, animals and human beings is different from the way things are good or bad for things like knifes and vacuum cleaners. In the context of a living being, things are good or bad for *their own* sake. What is good for knives and vacuum cleaners is obviously not good for the sake of knives and vacuum cleaners, but rather for our sake insofar these things fulfil a function for us. Korsgaard argues that all living beings have "a sake for which" certain states, conditions and happenings are good or bad. In this sense she comes extremely close to Canguilhem's position, were it not for the fact that she does not identify "having a sake" with occupying a distinctly *subjective* and *experiential* reference point.

Again, we may question what gives living beings "a sake", such that things are good or bad in relation to them. Just as we saw in Canguilhem and Foot, her answer seems to be that this just is what it means to be living: "having a sake" is *essential* to the concept of life. Korsgaard also builds normativity into the concept of life and is justified in doing so without having to further explain *why* life is so. The one advantage her account has over Canguilhem's and Foot's accounts is that she can further substantiate the difference between living beings and nonliving beings on the basis of life's self-referential functionality and the fact that living beings have a sake. But ultimately she too has built normativity into the concept of life itself.

CONCLUSION

This chapter considered the possibility that the motivation behind the growing philosophical focus on life results from an attempt to ground normativity in an objective and naturalistic stratum. I followed Canguilhem in his conviction that for living beings their own physical states and their environmental conditions have a certain value. I subsequently argued that these values carry a significant level of objectivity. What is objectively valuable is what promotes the self-sustaining and self-regulating processes of some organism, or, in short, the health of an individual organism. The meta-ethical status of these values is one that falls between absolutism and subjectivism, I argued, and derives its objectivity from the fact that life-based values and norms are relations between two *relata* that are themselves objective matters. Finally, three answers were considered in response to the question how life gives rise to these objective norms and values. While the authors considered in this chapter all built normativity into their concept of life, Korsgaard's account of life's self-referential functionalistic nature and the idea that all living beings have a "sake for which" things are normatively discriminated provided some further conceptual tools to articulate why only living beings have this norm-generating capacity. In sum, while the idea that life grounds vital norms and biological values opens a compelling theory of values, it demands a conception of life that supports and explains the norm-establishing nature of life. Continued philosophical interest in the nature of life will therefore remain of vital importance.

BIBLIOGRAPHY

Badiou, Alain. "Is There a Theory of the Subject in Georges Canguilhem?" *Economy and Society* 27, no. 2 (1998): 225–33.

Boorse, Christopher. "Health as a Theoretical Concept". *Philosophy of Science* 44, no. 4 (1977): 542–73.

Canguilhem, Georges. *The Normal and the Pathological*. Translated by Carolyn R. Fawcett and Robert S. Cohen. New York: Zone Books, 1991.

———. *Knowledge of Life*. Edited by Paola Marrati and Todd Meyers. Translated by Stefanos Geroulanos and Daniela Ginsburg. New York: Fordham University Press, 2008.

Foot, Philippa. *Natural Goodness*. Oxford: Oxford University Press, 2001.

Korsgaard, Christine. "On Having a Good". Paper presented at the Royal Institute of Philosophy Lecture, London, March 2012. Accessed 15 April 2013. www.people.fas.harvard.edu/~korsgaar/CMK.HG.pdf.

Nietzsche, Friedrich. *Friedrich Nietzsche: Sämtliche Werke. Kritische Studienausgabe in 15 Bänden, Band 11*. Edited by Giorgio Colli and Mazzino Montinari. Berlin: de Gruyter, 1980.

Railton, Peter. *Facts, Values, and Norms*. Cambridge: Cambridge University Press, 2003.

NOTES

1. Georges Canguilhem, *The Normal and the Pathological*, trans. Carolyn R. Fawcett and Robert S. Cohen (New York: Zone Books, 1991).

2. When Canguilhem's states that life is normative he means nothing other than that life generates norms. He writes unambiguously that "normative, in the fullest sense of the word, is that which establishes norms" (Canguilhem, *The Normal and the Pathological*, 126–27). This is slightly confusing, as these expressions do not necessarily mean the same thing. Something can be an origin of norms and establish norms, without being something normative itself. Pleasure and pain, for instance, could be understood to give rise to certain norms but are therefore not automatically something normative themselves. To avoid the confusion, I shall avoid the expression that life *is* normative as much as possible and simply state that life generates or establishes norms.

3. Canguilhem, *The Normal and the Pathological*, 75. A similar view to Bernard's is nowadays defended by Christopher Boorse. See especially Christopher Boorse, "Health as a Theoretical Concept", *Philosophy of Science* 44, no. 4 (1977): 542–73.

4. Canguilhem, *The Normal and the Pathological*, 220. Similarly, "A simple summary of quantities, without biological value, a simple fact or system of physical and chemical facts . . . cannot be called health or normal or physiological. Normal and pathological have no meaning on a scale where the biological object is reduced to colloidal equilibria and ionized solutions" (Canguilhem, *The Normal and the Pathological*, 110).

5. Canguilhem, *The Normal and the Pathological*, 220. This does not apply to organisms that feed off their own faeces of course.

6. Canguilhem, *The Normal and the Pathological*, 131.

7. Canguilhem, *The Normal and the Pathological*, 127.

8. Canguilhem, *The Normal and the Pathological*, 126.

9. Friedrich Nietzsche, *Friedrich Nietzsche: Sämtliche Werke. Kritische Studienausgabe in 15 Bänden*, ed. Giorgio Colli and Mazzino Montinari (Berlin: de Gruyter, 1980), 11.25 [433] [1884].

10. Nietzsche, *Sämtliche Werke*, 11.26 [72] [1884].

11. Canguilhem, *The Normal and the Pathological*, 228.

12. Canguilhem, *The Normal and the Pathological*, 136.

13. Canguilhem, *The Normal and the Pathological*, 126.

14. Canguilhem, *The Normal and the Pathological*, 197.

15. Canguilhem, *The Normal and the Pathological*, 199.

16. Georges Canguilhem, *Knowledge of Life*, ed. Paoloa Marrati and Todd Meyers, trans. Stefanos Geroulanos and Daniela Ginsburg (New York: Fordham University Press, 2008), 113.

17. Peter Railton has gone a long way in this direction. See especially the second part of the paper "Moral Realism", included in Peter Railton, *Facts, Values, and Norms* (Cambridge: University Press, 2003), 17–42.

18. Railton, *Facts, Values, and Norms*, 3–42.

19. Railton, *Facts, Values, and Norms*, 48.

20. Railton, *Facts, Values, and Norms*, 48.

21. Railton, *Facts, Values, and Norms*, 11.

22. Canguilhem, *Knowledge of Life*, 114.

23. Alain Badiou, "Is There a Theory of the Subject in Georges Canguilhem?", *Economy and Society* 27 (1998): 225–33.

24. Canguilhem, *The Normal and the Pathological*, 93.

25. Canguilhem, *The Normal and the Pathological*, 198. Emphasis added.

26. Philippa Foot, *Natural Goodness* (Oxford: Oxford University Press, 2001).

27. Foot, *Natural Goodness*, 26.

28. Foot, *Natural Goodness*, 32.

29. Foot, *Natural Goodness*, 66–81.

30. Christine Korsgaard, "On Having a Good" (paper presented as The Royal Institute of Philosophy Lecture, London, March 2012). Accessed on 15 April 2013, www.people. fas.harvard.edu/~korsgaar/CMK.HG.pdf.

31. Korsgaard, "On Having a Good", 20.

32. Korsgaard, "On Having a Good", 42.

SIX

Critical and Political Stakes of a Philosophy of Norms

Part I: Towards a Critical Philosophy of Norms

Julien Pieron

When he sketches his first formulations of the biopolitical hypothesis,[1] Foucault insists that we need to give up, or at least complicate, the secular conception of power as a simple agency of repression or prohibition,[2] and conceive instead of it as the production, intensification or management of the object of its exercise—in this instance, life in both its individual and its collective dimensions.[3] In his lecture of 17 March 1976, Foucault points out that the element that allows for the joining of discipline and regulation, the element that circulates and brings together these two sides of biopower as production, is nothing more than the *norm*:

> The norm is something that can be applied to both a body one wishes to discipline and a population one wishes to regularize. The normalizing society is therefore not, under these conditions, a sort of generalized disciplinary society whose disciplinary institutions have swarmed and finally taken over everything. . . . The normalizing society is a society in which the norm of discipline and the norm of regulation intersect along an orthogonal articulation.[4]

To question the possibilities of keeping or creating a critical space in the age of biopower is to inquire about the conditions, the resources, but also maybe the limits, of what we will call a *critical philosophy of norms*. This chapter aims to show how a set of contemporary studies, which cannot be reduced to mere commentary, builds an original theory of norms and normativity in the wake of Canguilhem and Foucault. Starting with a reference to the *Normal and the Pathological*, we will try to reconstruct some of the main characteristics of this philosophical paradigm.

ASPECTS OF A VITALISM

In the general introduction to his 1943 *Essay on Some Problems Concerning the Normal and the Pathological*, Canguilhem points out that he studied medicine not in order to try his hand at a scientific discipline in general, but in order to be able to pose two problems: one being the relationship between science and technology and the other the question of norms and of the normal.[5] Those two problems might seem distinct, but are in fact intimately linked. In that same introduction, Canguilhem sketches a definition of medicine or therapeutics: "Medicine seemed to us and still seems to us like a technique or art at the crossroads of several sciences, rather than, strictly speaking, like one science".[6]

Further specifying its object, he asserts that the essence of medicine resides in its clinical and therapeutic nature, as *"a technique of establishing or restoring the normal which cannot be reduced entirely and simply to a single form of knowledge"*.[7] Why is it so difficult to reduce this technique called medicine to a form of knowledge? Because the normal that has to be found or restored is not a fact that can be established by science, but a *value* that has to be instituted—or re-instituted.[8] The normal is not a fact, but a value. This, we believe, is the first core intuition of Canguilhem's work. The second core intuition, indicative of what we'll refer to as Canguilhem's *vitalism*, is that it's not science, but life, which posits, establishes or institutes the value in question. Life reveals itself as a normative activity, which establishes values.

Let's go back for a moment to this vitalism and recall the main features of the concept of vital normativity:

> We think that medicine exists as the art of life because the living human being himself calls certain dreaded states or behaviours pathological (hence requiring avoidance or correction) relative to the dynamic polarity of life, in the form of a negative value. We think that in doing this the living human being, in a more or less lucid way, extends a spontaneous effort, peculiar to life, to struggle against that which obstructs its preservation and development taken as norms . . . the fact that a living human being reacts to a lesion, infection, functional anarchy by means of a disease, expresses the fundamental fact that life is not indifferent to the conditions in which it is possible, that life is polarity and thereby even an unconscious position of value; in short, life is in fact a normative activity. . . . Normative, in the fullest sense of the word, is that which establishes norms. And it is in this sense that we plan to talk about biological normativity.[9]

What this quote affirms is that biological life in general, the one in which human life is rooted, is a movement or an activity that consists in discriminating between the good and the bad, establishing differences everywhere (in the milieu, in the body, in its actions), and creating poles of attraction and rejection. The norms and the normal emerge precisely from such a polarization, from the institution of cardinal points, and the establishment of limits towards which one strives (or deviates from). The *norm* cannot be identified as a pole or a value, but constitutes the field of tension and relative constraint that the dynamics of polarization open up: a certain "must", "it is convenient to", which, different from a physical law, always leaves room for mistakes.[10] The *normal* will constitute the rhythm or the mode by which life stabilizes itself following this polarization, or this submission to norms.[11] The fundamental idea of Canguilhem's essay is that the polarization and the vital normativity in question can only be grasped as such in the experience of the negative, or in what Canguilhem calls "negative values". It is only in its deviation that the norm appears as such; it is only in the noise generated by the disease that health, as *"life lived in the silence of the organs"*,[12] is finally understood:

> It is the abnormal which arouses theoretical interest in the normal. Norms are recognized as such only when they are broken. Functions are revealed only when they fail. Life rises to the consciousness and science of itself only through maladaptation, failure and pain.[13]

One last thing to point out concerning this vital normativity: the fundamental norm that regulates life—namely, to preserve and develop itself, or its own essence, is immanent to life itself. Thus it is the essence of life that reveals itself as the fundamental norm regulating the movement of polarization, the dynamics of life—an essence that needs to be understood in a nonnaturalistic way, as a habit[14] open to the event and to history.[15] With more time, we could dwell on the joint usage of the concepts of preservation and development of life. More than the question of preservation, though, it is the idea of development that is central to Canguilhem. In several instances, Canguilhem shows how the simple search for preservation is in fact the goal of a diseased life, whereas a healthy life is a life that plays and risks. In other words, the abuse of health is constitutive of health itself.[16] It lies in this risky endeavour, in which life calls itself into question, and which represents the privileged experience of illness, where life experiments with new possibilities, and this in such a way that it is impossible to say in advance whether they will constitute a condition of development.[17]

Having recalled the main lines of Canguilhemian vitalism, we must note that this vitalism is immediately qualified and complicated in several respects, first and foremost by the assertion of an essential connection between life and technique. One of Canguilhem's fundamental theses is that, in fact, technique is not something opposed to life, but more of an extension, even a fundamental dimension of it.[18] This explains why medicine might not be defined as a simple technique to restore the normal, but as a technique that *establishes* the normal, and constitutes the very dynamics of life. The insertion of technique in life does not necessarily lead to irenicism and technological optimism. To say that medicine is a technique that seeks not simply to restore normality, but also to establish that very normality, is tantamount to giving certain living beings—namely, doctors—an amazing power over others—namely, the potentially sick beings that we all are—since there is no perfect health. Since medicine and medical institutions constitute a mediation or a point of necessary passage between the living individual and its own values or vital norms, the self-introduction of vital values ceases to be a purely immanent affair, and becomes the potential issue at stake in a conflict or balance of powers. It is precisely on this point that a connection with political issues—primarily with the work of Foucault and the problem of biopolitics—emerges almost immediately.

Medicine is a technical establishment of the normal—a technique that contributes to, or extends the dynamics of the positing of norms, integral to life itself. Since the advent of positivism, this medical technique has spread in the social field through various enterprises of normalization.[19] Its ambition is to follow entirely from science, and to turn the normal, to which its own activity is attached, into something *obvious*: a pure fact, which would be neither problematic nor questionable, or which it would be useless and illegitimate to question. Yet what Canguilhem shows—and this theoretical gesture has obvious political implications—is that as a value or a norm, the normal is never found, but posited, and that this position is inherently problematic.[20] In the case of medicine, the normal is established by the doctor from various sources, but these sources are not in agreement *a priori*, and are potentially in conflict. In the introduction to part 2 of his book, Canguilhem notes the following:

> The physician usually takes the norm from his knowledge of physiology (called the science of the normal man), from his actual experience of organic functions, and from the common representation of the norm in a social milieu at a given moment. Of the three authorities, physiology carries him furthest. [21]

This remark says at least three things. First, there are three possible sources from which to draw the concept of the normal: science, the lived experience and the socio-historical context. In other words, there are three regimes of normativity (epistemological, vital, social), the coexistence of which Canguilhem problematizes in his book. Second, those three sources remain in a conflictual relationship. Third, the relationship between those three regimes of normativity can freeze at a given point and lead to a certain structural configuration. This schema of the three conflicting sources of the normal, on which the medical technique, coextensive with life, draws, further complicates Canguilhem's "vitalism": we have seen that the self-introduction of vital values is not a purely immanent affair, since medicine constitutes a mediation or a point of necessary passage between the living individual and its own values or vital norms; the question is to understand that the sources of this "mediated self-introduction" of vital values are not themselves purely vital, but situated at the conflicting crossroads of three different regimes of normativity (epistemological, vital, social), of which two seem to be exterior to life *stricto sensu*.

FOUR PARADIGMATIC TRAITS

From this brief first incursion into the work of Canguilhem, we can retain the following four traits as relevant to a paradigm for a critical philosophy of norms:

1. The idea that the establishment of norms is essential to life itself, that it takes part in the immanent movement of life. In this sense we will move away from Canguilhem's analysis and follow instead contemporary authors such as Judith Butler and Guillaume Le Blanc, who leave the meaning of the word *life* open, don't limit it to biological life and extend it to human life considered in its multiple dimensions—psychological, social and political. In that sense, the category of life can be identified with that of the subject, or subjectivity.
2. The idea of a nonnaturalistic vitalism, opened to the event and historicity, according to which the essence of life lies in its development, rather than its preservation. Understood in that way, life involves a series of risky experiments, the outcome of which is either fatal or creative, but in any event *a priori* undecidable. Diseases are in a way the ideal type of such experiments. We want to formalize this vitalism—the idea of which would deserve to be analysed for itself—in order to avoid limiting life to a certain region of Being (that of organic life, as opposed to inert matter) and to allow it instead to denote the possibility of a twisting free of, or deviation from, a given situation. Formalized in those terms, the category of vitalism only designates a postulate: that of the nondeductible possibility of a divergence, or an accidental invention, which generate unpredictable consequences and give life considerable leeway in relation to the instituted norm. Bachelard's concept of "break", Canguilhem's "error", Foucault's "chance" or Deleuze's "nonorganic vitality" all emerge from this vitalist assumption.
3. The idea according to which the immanent establishment of norms is a process that's always already mediated, exposed to heterono-

my, and thus constitutes the locus of a potential conflict or power relations. This is an idea that we develop by generalizing the concept of mediation, which our analysis of medicine as a technique rooted in life brought out. In other words, the idea that any stream of *normativity*, understood as the immanent and active position of norms, is always already traversed by an opposite stream of *normalization*, understood broadly as the passive submission to a transcendent norm, imposed from the outside, as it were. In the end, we need to see life as a normative/normalized activity, and the normativity and normalization of life as two aspects or moments of a complex dynamic, the distinction of which is not actual, or given, but only virtual. It is the task of the critical attitude to "actualize" the virtual distinction on a case by case basis, and according to the situation.

4. The idea that this unsurpassable entanglement of normativity and normalization works in a framework consisting of the play of multiple and conflictual normative systems. Canguilhem problematizes three such systems (epistemological, social and vital) but it is by no means forbidden to increase or to complicate his list. We will soon return to this question, by way of a new detour through *The Normal and the Pathological*.

In the first section of the "New Reflections on the Normal and the Pathological (1963–1966)", Canguilhem assembles some of his most fruitful insights on the issue of the mode of being and functioning of the norm. The framework in which this reflection on the norm unfolds is not primarily one of vital normativity, but rather that of the phenomenon—historically situated, and qualified by Canguilhem as a *"specifically anthropological or cultural experience"*[22]—of normalization (linguistic, medical, educational, industrial, economical). Here, we will limit ourselves to three aspects of this reflection, and relating them to later developments in the critical philosophy of norms. At stake here is a further stage in our attempt to delineate a "paradigm" for a critical philosophy of norms.

A POLEMICALLY IMPOSED REQUIREMENT ON AN EXISTENCE

The *New Reflections* offer a general definition of the norm, according to which it functions as the correction or transformation of an *other* reality, a preexisting reality, the alterity of which is primarily hostile: "A norm, or rule, is what can be used to right, to square, to straighten. To set a norm (*normer*), to normalize, is to impose a requirement on an existence, a given whose variety, disparity, with regard to the requirement, present themselves as a hostile, even more than an unknown, indeterminant".[23]

This hostility of existence, which the demand will seek to redress, reveals the norm and the concept of the normal as inherently controversial entities, which are always located in relations between forces: "A norm draws its meaning, function and value from the fact of the existence, outside itself, of what does not meet the requirement it serves. The normal is not a static or peaceful, but a dynamic and polemical concept".[24]

The correlate of this polemical nature of the norm is the historical or ontological primacy of divergence, or of infraction, which is nonetheless secondary from a logical point of view, since the abnormal is defined as the negation of the normal:

> It is not just the exception which proves the rule as rule, it is the infraction which provides it with the occasion to be rule by making rules. In this sense the infraction is not the origin of the rule but the origin of regulation. It is in the nature of the normative that its beginning lies in its infraction. To use a Kantian expression, we would propose that the condition of the possibility of rules is but one with the condition of the possibility of the experience of rules. In a situation of irregularity, the experience of rules puts the regulatory function of rules to the test.[25]

This general framework of intelligibility of the norm, based on the idea of a primacy of the infraction or irregularity, seems to be presupposed by all the authors we are trying to gather under the banner of a "critical philosophy of norms", including Pierre Macherey, Judith Butler, Guillaume Le Blanc and Stéphane Legrand.

IMMANENT CAUSALITY AND NOMINALIST THEORY OF THE NORM

The *New Reflections* also outline a strange "ontology" of the norm, which denies it any power of coercion *sui generis*: the norm offers itself, but does not impose itself in the manner of a natural law. It has no efficiency in itself, and gathers efficiency only by being mobilized by the protagonists of the *polemos*, the dispute or the balance of forces:

> A norm offers itself as a possible mode of unifying diversity, resolving a difference, settling a disagreement. But to offer oneself is not to impose oneself. Unlike a law of nature, a norm does not necessitate its effect. That is to say, a norm has no significance as norm pure and simple.[26]

This almost hidden remark of Canguilhem found important extensions in the work of Pierre Macherey and Stéphane Legrand. The basic idea that, following his reading of Foucault, Macherey introduces, is that of an immanent causality of the norm, which occurs as norm only through the movement by which subjects are subjected to it. Macherey introduced this extremely valuable idea in order to counter a catastrophic reading of Foucault's conception of disciplinary power as inherently productive.

Thinking about power in productive (and not only repressive) terms requires accepting that the object or the subject of power is not given prior to its exercise, but is constituted and comes to existence precisely within relations of power. That being said, Macherey argues that thinking of norms in terms of a productive capacity does not necessarily lead to the catastrophic claim that subjects, insofar as they are produced by norms, are "always already trapped" by them. The subjects are already "trapped" only if the productivity of the norm presupposes that the norm is always already given, existing even before it takes effect. Yet the thinking of a productivity of norms does not necessarily imply a priority of the norm with respect to what it normalizes, and does not necessarily mean that the norm is prior to its effects. According to Macherey, a strict reading of Foucault, on the contrary, shows that it is in the very movement by which it has effects (as the subjectivation of subjects) that the norm itself becomes a norm:

> If the norm is not exterior to its field of application, this is not only because, as we have already shown, it produces it but because it produces itself in it as it produces it. Just as it cannot be said that it acts on a content existing independently of and outside the norm, so, too, the norm is not in itself independent of its action, though this takes place outside itself, in a form which is necessarily that of division and separation. It is in this sense that it is necessary to talk about the immanence of the norm, in relation to what it produces and the process by which it produces it: that which "norms" the norm is its action.[27]

Considered abstractly, and in its speculative purity, such an understanding of the power of norms does not yet release its political power. It is the work of Stéphane Legrand, and his nominalist theory of the norm as an empty signifier, mobilized by the protagonists of a balance of power, which clears political thought of this "force of norms". According to Legrand, the originality of Foucault's use of the concept of norm is that, contrary to a theoretical gesture that finds its origin and paradigmatic example in Durkheim,[28] Foucault does not see any normative or binding power in the norm as such, any force that's *sui generis*. The norm is a statement or a fragment of code, which functions as a norm only when it's inscribed in a historically determined play of forces.[29] Let me refer, by way of example, to the analysis that Legrand provides of the Order of 1 January 1766 concerning the handling of rifles.[30] The Order in question is not a collection of inherently normative realities, the "disciplinary" functions of which would be fixed once and for all: instead, it offers nothing but a set of potentialities for normative purposes, and such uses are far from being unambiguous or determined *a priori*. The Order could be used to increase efficiency in handling guns. Equally, though, and by virtue of the radical impossibility to conform to norms that their "manic exhaustivity" (*exhaustivité maniaque*) renders impossible to observe, it could be used to make soldiers peel potatoes or clean latrines.

A THWARTED SYSTEMATIC AIM: CONTRADICTION AND OVERDETERMINATION OF REGIMES OF NORMATIVITY

The last aspect of the *New Reflections on the Normal and the Pathological* we wish to emphasize is the idea that, in the world of normalization, a norm does not operate in a vacuum, but refers to a normative decision, one that is always inscribed in the horizon of a system in which norms are unified by referring to one another. Norms need to be decided, and to constitute themselves as a system, or tend to form a system:

> In both of these examples the norm is what determines the normal starting from a normative decision. As we are going to see, such a decision regarding this or that norm is understood only within the context of other norms. At a given moment the experience of normalization cannot be broken down, at least not into projects.[31]

To be sure, norms carry within themselves the aspiration or the tendency to be a system. However, this systematic unity is not given from the start; it too involves possible conflicts, and a work of composition between the various regimes of normativity:

> The logic of technology and the interests of the economy must come to terms. . . . So we see how a technological norm gradually reflects an idea of society and its hierarchy of values, how a decision to normalize assumes the representation of a possible whole of correlative, complementary or compensatory decisions. This whole must be finished in advance, finished if not closed.[32]

This idea of an aspiration on the part of norms to become systematic, the idea of a desire thwarted by the coexistence of numerous regimes of normativity, has led Stéphane Legrand to draw a series of political consequences. What is at stake here is nothing less than the reinsertion of disciplinary techniques of normalization, considered as many "micro-powers" in the much larger "systemic" framework of a given social formation. Specifying the originality of Foucault's position, and contrasting it with a Durkheimian model, Legrand asserts:

> It [Durkheim's position] comprises the collective phenomena only through the hypostasis of a normativity that exists for us only as a meaning within language, and to which it attributes a virtue of causality on reality, whereas I wish to demonstrate that we can understand the phenomena in question only by referring to the practical manner in which social groups use those codes. . . . I judge it excessively irenic, insofar as it associates the social function of norms with the objective of ensuring the homeostasis of society. Instead, I wish to demonstrate that

> they only make sense within the framework of the social struggles in which those groups enter.[33]

Foucault's "nominalism": The norm is nothing, not something that would be effective in itself; it is primarily a fragment of a code. It is to the extent that this fragment of code is mobilized in a power struggle that the norm becomes a norm, a way of regulating conducts. Foucault's "Marxism": The balance of power, even at a "micro" level, does not float in a vacuum, but is inserted in the context of broader social struggles, which go beyond the strictly local.

It is in terms of Althusser's overdetermination that Legrand develops the relationship between the local and the global. The whole point of his reading of *Discipline and Punish* is to show that the rationality of disciplinary mechanisms or micro-powers does not make sense in itself; it is intelligible only in the broader context of class struggle. If there is overdetermination, it is because each level works and conditions the other, while itself being conditioned or worked by the other levels.[34] The reinsertion of disciplinary technologies in a Marxist-Althusserian framework also allows Legrand to shatter the somewhat monolithic idea of a "disciplinary archipelago" or "power of norms", by emphasizing the contradictions and double-bind faced by any disciplinary attempt. Consider the example of factory workers: it's a matter of producing the most profitable bodies possible within the framework of capitalist exploitation, while ensuring that these bodies do not assemble themselves politically and destroy the framework in question.[35] Those two aims are distinct, and Legrand's analysis is interesting insofar as it shows that "discipline" is not in itself the magical power that reconciles them once and for all. Those tensions and contradictions between different goals involve the invention of several technologies of discipline, as well as their connection—that is to say, various regimes of normativity. It is the nominalist theory of the norm as empty signifier, as the fragment of a code mobilized by the protagonists of a temporary balance of forces, which thinks the coexistence and coordination of these regimes of normativity. This theory allows us to grasp that some fragments of code, which function as a local norm, are likely to move to a different level, to "overcode" two codes or two different regimes of normativity, and to allow them to communicate with one another. Thus, there is a connection between the physiological and the mental, the mental and the judiciary and so forth.[36] Such a movement of overcoding seems to tend towards the great division between the normal and the abnormal, in which the signifier "abnormal" means nothing more than the pure, empty form of the divergence.[37]

CONCLUSION

Before we submit the paradigm of the critical philosophy of norms to the test of a number of contemporary political problems, let us emphasize the following seven features, constitutive of that paradigm:

1. The intrinsic link uniting life (in its greatest polysemy) with the norm.
2. The formalization of the category of vitalism, reduced to the postulate of a nondeductible possibility of invention—accidental in its origin and undecidable with respect to its consequences—of a leeway in relation to the instituted norm.
3. The thesis of an original entanglement of normativity and normalization.
4. The assumption of a play and a conflicting relation between multiple regimes of normativity.
5. The general definition of the norm as a polemic demand imposed on an existence, and which presupposes the primacy of an infraction, or a divergence.
6. The negation of the idea of a binding power *intrinsic* to the norm, and the affirmation of the immanent causality of the norm, understood as an empty signifier mobilized by the protagonists of a relation of forces.
7. The idea of a systemic aim in the normative intention, always already hindered by the coexistence of multiple regimes of normativity, of which the nominalist theory of the norm allows us to think the tensions, contradictions and overdeterminations.

Can this critical philosophy of norms open up possibilities for a renewal of social criticism and political thought? Chapter 7, the second part of this topic, will deal with this problem.

BIBLIOGRAPHY

Canguilhem, Georges. *The Normal and the Pathological*. Translated by Carolyn R. Fawcett in collaboration with Robert S. Cohen. New York: Zone Books, 1991.

Foucault, Michel. *The History of Sexuality, Volume I: An Introduction*. Translated by Robert Hurley. New York: Vintage Books, 1978.

———. *"Society Must Be Defended", Lectures at the Collège de France 1975–76*. Translated by David Macey. New York: Picador, 2003.

Legrand, Stéphane. *Les normes chez Foucault*. Paris: Presses Universitaires de France, 2007.

Macherey, Pierre. "Towards a natural history of norms". In *Michel Foucault Philosopher*. Translated by T. J. Armstrong. Hemel Hempstead: Harvester Wheatsheaf, 1992.

NOTES

1. As the power to make live and let die, founded on the articulation between disciplinary technologies directed towards individual bodies, and the control and regulation mechanisms directed towards mass bodies, constituted as populations or species: Michel Foucault, *The History of Sexuality, Volume I: The Will to Know*, trans. Robert Hurley (New York: Vintage Books, 1978), 138, 139; Michel Foucault, *"Society Must Be Defended", Lectures at the Collège de France 1975–76*, trans. David Macey (New York: Picador, 2003), 240–41, 250.
2. The "juridico-discursive" conception of sovereignty understood as the power to capture and extract.
3. See Foucault, *The Will to Know*, 137, and more broadly 181–202, which fully develops the scope of such an analysis of power.
4. Foucault, *"Society Must Be Defended"*, 253.
5. Georges Canguilhem, *The Normal and the Pathological*, trans. Carolyn R. Fawcett and Robert S. Cohen (New York: Zone Books, 1991), 34.
6. Canguilhem, *The Normal and the Pathological*, 34.
7. Canguilhem, *The Normal and the Pathological*, 34.
8. This value is experienced in fact as the "re-" of repetition, which implies the crisis, the primacy of the negative.
9. Canguilhem, *The Normal and the Pathological*, 126–27.
10. Canguilhem, *The Normal and the Pathological*, 127–28.
11. Canguilhem, *The Normal and the Pathological*, 205.
12. Canguilhem, *The Normal and the Pathological*, 118.
13. Canguilhem, *The Normal and the Pathological*, 209.
14. Canguilhem, *The Normal and the Pathological*, 169, 206.
15. Canguilhem, *The Normal and the Pathological*, 198, 203–4, 221.
16. Canguilhem, *The Normal and the Pathological*, 139, 199–200.
17. Canguilhem, *The Normal and the Pathological*, 206.
18. Canguilhem, *The Normal and the Pathological*, 130: "All human technique, including that of life, is set within life, that is, within an activity of information and assimilation of material. . . . Because life is activity of information and assimilation it is the root of all technical activity".
19. Canguilhem, *The Normal and the Pathological*, 245.
20. Even in the living who are closest to "nature", the position of the normal is not a natural immemorial given, but depends on a dialogue, a confrontation (*Auseinandersetzung*) with the milieus and the species that inhabit it. See Canguilhem, *The Normal and the Pathological*, 198 and 143–44.
21. Canguilhem, *The Normal and the Pathological*, 122.
22. Canguilhem, *The Normal and the Pathological*, 244.
23. Canguilhem, *The Normal and the Pathological*, 239.
24. Canguilhem, *The Normal and the Pathological*, 239.
25. Canguilhem, *The Normal and the Pathological*, 242.
26. Canguilhem, *The Normal and the Pathological*, 240.
27. Pierre Macherey, "Towards a Natural History of Norms", in *Michel Foucault Philosopher*, trans. T. J. Armstrong (Hemel Hempstead: Harvester Wheatsheaf, 1992), 187–88.
28. Stéphane Legrand, *Les normes chez Foucault* (Paris: Presses Universitaires de France, 2007), 4.
29. Legrand, *Les normes chez Foucault*, 3.
30. Legrand, *Les normes chez Foucault*, 11–12, note 4.
31. Canguilhem, *The Normal and the Pathological*, 245–46.
32. Canguilhem, *The Normal and the Pathological*, 247.
33. Legrand, *Les normes chez Foucault*, 5–6.
34. Legrand, *Les normes chez Foucault*, 147–48.
35. Legrand, *Les normes chez Foucault*, 147, 313.

36. Legrand, *Les normes chez Foucault*, 12–13.
37. Legrand, *Les normes chez Foucault*, 144.

SEVEN

Critical and Political Stakes of a Philosophy of Norms

Part II: Theory of Norms and Social Criticism

Florence Caeymaex

We can envisage the paradigm explored in the previous chapter as an answer to the questions raised by two Foucauldian theses—namely, that of the subject as constituted through relations of power, and that of the role which, in the contemporary context, norms play in the exercise of power. Beyond Foucault, though, the vitalism defined by our critical philosophy of norms suggests that we regard human life as an immanent process of subjectification—one which is located at the conflictual intersection of a plurality of regimes of norms—and as repeatedly supported and threatened by the interweaving of normativity and normalization.

This is the reason why this sort of vitalism is likely to continue and renew a double trend of contemporary social critique: that which developed in the margins of the sociology and psychology of work, and that of social philosophy. Both have paid special attention to the modes of subjectification implied by the recent transformations of social relationships. On the one hand, an important trend within the social and psychosocial sciences has developed in the last twenty years. It has helped to shed light on a variety of phenomena of social exclusion, which it studies from a double point of view: that of their structural mechanisms (the fragmentation of the society of wages and the evolution of the working world) and that of their effects, especially at the subjective level, where it amounts to a devaluation of a number of individual and collective identities. On the other hand, social philosophy has initiated the revival of the project of critical theory by interpreting different forms of *experience* of social injustice (also called "social pathologies") as socially institutionalized denials of recognition, which affect the subjectivity of individuals. Those two ways of raising the "social question" through an essentially negative experience renew the debate around the social conditions of possibility pertaining to individual and collective modes of subjectification.

I would like to point out how, by drawing on the social sciences and modifying the theory of recognition, the critical philosophy of norms can open up possibilities for a renewal of social critique and political thought, in a world where power is necessarily exercised through forceful norms. My aim, in what follows, is not to reproduce exhaustively the arguments and the debates that underpin the books I will talk about. Instead, I will simply try to underline the prominent elements that show the heuristic and critical utility of the "paradigm" that was described in the previous chapter.

THE SUBJECT: LIVING WITHIN THE NORMS

According to this paradigm, norms are not to be viewed exclusively as systemic constraints imposed on individuals and determining their behaviour, while ensuring the homeostasis of a particular social group. The point is not to *explain* social relationships—whether through the general process of "normalization" or by appealing to a hypothetical force intrinsic to the norms themselves—but to analyse the regimes of normativity as conflictual environments in which subjectivity takes shape. In this regard, Judith Butler famously developed Foucault's initial suggestions that the subject is shaped by games of truth and power relations.[1] Reinterpreting Hegel, Althusser, and psychoanalytic theories,[2] she suggests that it is subjection that produces the subject. Whereas the making of the subject implies involvement in the norms that arrest him/her (gender norms, for example), subjectification does not simply amount to normalization, because the subject remains incomplete, "thus never *fully* constituted in subjection",[3] but is "instead . . . in the process of being produced, it is repeatedly produced (which is not the same as being produced anew again and again)".[4] This repetition is rooted in a performative use of a constraining sequence of significations[5]—that is, in a mandatory "resignification" that *can—or not, depending on the case*—generate changes in signification or give rise to new possibilities of signification:[6] such a repetition "does not consolidate that dissociated unity, the subject, but . . . proliferates effects which undermine the force of normalization".[7] Thus, far from conforming to a simple logic of reproduction (as if norms determined lifestyles in advance), the ordinary relation that human lives develop with social norms takes on the form of both an attachment—and, in that sense, the formation of the subject is always an heteronomous process—and an invention of the self, by the self, within the instituted norms, an invention that has the capacity to sustain the "revaluation" or transformation of those norms, or even the institution of new norms: "it is a practice of improvisation within a scene of constraint".[8]

Guillaume Le Blanc takes up this Butlerian idea of a subject produced within the norms through performative acts and extends it to what he calls "ordinary lives". He claims that there is a "vital necessity to repeat the standards of social normality",[9] while insisting that difference or deviation from the norms are constitutive of such a repetition. Indeed, what he calls "ordinary life" does not conform to a rule or logic of mere reproduction, but implies creativity at the very heart of this process of repetition, similar to what Michel de Certeau observed at the level of the "arts of doing", which, according to him, feature in everyday life practices.[10] This creativity is typical of the use of things, but it also belongs to the use of the "self" in its relation to others:

> thus ordinary life doesn't consist, as we might expect, in our submission to norms that have been incorporated, repeated, and are thus taken for granted. It circulates in the practices that imply a use of the self, in order to generate what the self expects from others.[11]

In other words, and to the extent that norms sustain the production of communities, living within the norms involves not only the self but also relationships with others; they work as valuations (or qualifications) that mediate the process of social recognition—regardless of whether it's actually successful for an individual subject.[12]

CRITICAL OR CLINICAL PROCESS?

Canguilhem's thesis about the primacy of infraction or irregularity, combined with that of the immanent causality of the norm, which requires its actualization or performance by the subject, thus insists on the variation, creation or even resistance inherent to the normative play.[13] The philosophy of norms allows for a type of criticism that does not aim at a blunt denunciation of the apparatuses of normalization or the encompassing power of norms.[14] Rather, it strives to shed light on the vitality of ordinary lives and highlight the process of emancipation it might contain. However, the critical philosophy of norms isn't the expression of a political optimism that is based on the promise of an emancipation, which, by virtue of the fact that there is always creation and resistance, would always be possible. Instead, this social criticism of the present calls into question the contemporary process of social devaluation and exclusion, and shows how it weakens the construction of subjectivity and, more widely, the possibilities of any emancipatory politics of subjectification.

My aim, here, is not to describe the various forms of social exclusion, but to emphasize the fact that they all pose a threat to the individuals' capacity for establishing creative relationships with norms, whether collectively or individually, and that the capacity in question is what makes life liveable, and subjectivity possible.[15] Those types of social relegation correspond to various ways of impairing the normativity of ordinary lives. This diagnosis and the need to take normativity into account requires, in turn, that we consider two ideas: we need to accept, on the one hand, that there is a vulnerability or a precariousness that's inherent to living within norms,[16] and, on the other, that the normativity proper to life never neutralizes the subjection that results from normalization, for the simple reason that normativity never works as an autonomous process, but precisely as an heteronomous one, through strong attachments to a socially established normality. The creative capacity, the ability to act within the norms, must remain connected to social normality: "The creations of an ordinary life only acquire meaning through the firm support provided by the reception process of social normativity along the various recognized forms of social normality".[17]

In terms of norms, the experience of devaluation represents the impairing, or even the destruction of the normative capacity thanks to which someone can entertain a series of existential possibilities, or can make his or her own existence singular in the realm of hegemonic or dominant norms of life. This entails a decrease of one's own power or agency. What's interesting in this perspective is that it renders intelligible the continuous process of inclusion and exclusion, and the unequal distribution of agency due to social normality, both phenomena being dominant features of our societies.

The critical significance of this philosophy of norms does not target social normality as such, but the ways in which society jeopardizes the possibilities for some to deploy a creative vitality that would allow them to be valued within and through the play of norms. It isn't simply a question of assessing the relevance or efficiency of all the forms of social support that are supposed to ensure the physical, psychic and social viability of devalued and precarious existences, which we could refer to as a "clinical process"—that is, as an assemblage of actions and social measures destined to restore this creativity. Nor is it simply a question of calling for the strengthening or transformation of those types of support. Rather, we need to take stock of the fact that they are themselves powerful conveyors of normalization. They amount to a sort of interpellation [*interpellation*] by social normality which, at the same time, plays the role of a devaluing designation. This is the problem we're faced with when we try to combine a form of social criticism with a clinical process that aims to alleviate the effects of social pathologies:

> By giving priority to the clinical process . . . to the care provided to those who find themselves in a precarious situation, or excluded, do we not risk setting aside more radical forms of social critique . . . ? Can a genuine critical potential be maintained within the clinic [*clinique*] of precariousness and exclusion? Or do we risk effacing the critical within the clinical and thus perpetuate the elementary forms of precariousness and exclusion . . . ?[18]

If the subject's agency depends on his or her capacity to signify anew and appropriate the recognized elements of social normality, we, as thinkers, need to question the specific conditions that deprive some individuals of such a capacity, and thus turn social normality into a major factor of devaluation.

THE PHILOSOPHY OF NORMS AND THE SOCIAL SCIENCES

The critical philosophy of norms extends and sheds light on the results recently established by a branch of the social sciences. The branch in question attempts to find a form of sociological objectivity in the *feeling* of social devaluation. It takes the latter as indicative of an existing social stratification, without reducing the subjective or experienced dimension to a set of objective reasons.[19] According to that approach, precariousness cannot be reduced to a mere state of economic deprivation, or material suffering; it must also be envisaged as a "positional suffering", defined in relation to the one who experiences it,[20] and as a weakening of subjective identifications. In Paugam's words, it refers to "the feeling of not being sufficiently appreciated within the social space of reference that one belongs to".[21]

In fact, if it is possible to regard the philosophy of norms as a certain interpretation of the social question as it is being redefined today, it is also important to point out that this critical philosophy of norms largely relies on the theoretical frame developed in the particular field of the social and psycho-social sciences I've mentioned. The concept of "precariousness" analysed by Le Blanc is not a universal anthropological invariant; it refers to its socially established forms, which are to be understood in relation to work and employment. The experience of precariousness, which corresponds to an undermining of the socio-personal identity, is thus linked to a deterioration of working conditions and a shrinking of the social protections associated with employment, at least for a whole stratum of contemporary societies.[22] According to the sociological tradition initiated by Simmel, devaluation is not a matter of social exclusion but of a specific type of *integration* that affects some precarious wage workers—a type that could be designated as "peripheral" and contrasted with the "central" form of integration, normally achieved through stable and satisfying jobs.[23] Considered in terms of norms, this view suggests that the ideal type of secured (vs. precarious) professional integration plays precisely the role of an authoritative normative reference through its power to both bring about social integration and distribute (i.e., differentiate) social status. Depending on the cases and contexts, its effects either support or threaten what I call here the "creativity" of ordinary lives.

FAIRE ŒUVRE AND THE PROBLEM OF RECOGNITION

By acknowledging that the institutionalized forms of work play a decisive role not only for the normativity of ordinary life but also for the very definition of social normality, we insist on the practical dimension of the life of the subject, in his/her relation to himself/herself, as well as others. While norms are operators of the process of recognition and misrecognition, and take part in the play of the relationship to oneself and to others, this process is not purely symbolic. Work nowadays plays a fundamental role among the different modalities through which one becomes human. Indeed, work is not a mere question of responding to vital necessities, but should instead be considered "one of the main ways of participating in the world and (hi)story of others".[24] It is from within the historically established ways of working—and in relation to the norms they bring into play, the work activity being equated here with a "*genre*" that regulates human activities[25]—that the "*faire œuvre*" (i.e., a finalized production that can leave behind lasting material artefacts or immaterial goods available for use by others) is made possible or impossible.[26] The normativity proper to life must be understood as a form of agency, a capability for leaving some visible imprint on reality and therefore for influencing the actions of others: "*faire œuvre*, for a single life, means experiencing that the 'self' is involved in a wider environment to which it contributes through actions that go beyond the self and create the conditions of a broader existence".[27]

The problem with devaluation is not merely the social blame or contempt suffered by individuals, but mostly the distress caused by stunted life possibilities resulting from unemployment and therefore uselessness. In other words, the problem of the devaluation borne by precariousness is not first that of misrecognizing or denying recognition to a mere subjective identity, but that of being deprived of some possibilities to act and to deploy the powers of a creative life—that is, to connect with others while transforming the common world. It means that there is something below recognition, a power-to-work (*pouvoir faire œuvre*) underlying the visibility and thus the recognition of the subject.[28]

CRITICAL AND POLITICAL STAKES

As already mentioned, the critical philosophy of norms consists neither in a philosophical understanding of the theoretical coordinates and empirical facts of the sociology of precariousness, or of the psycho-sociology of work, nor in a simple adaptation of the theories of recognition. This coupling of philosophy and the social sciences generates the idea that a form of criticism concerning social norms needs to take into account the socially and economically determined contexts in which they take place. If precariousness can be understood as the impairing of an ordinary life, it is also rooted in the increase of precarious jobs at the very heart of the labour market, negatively determined in relation to the norms still prevailing in the wage labour standards. This major discordance is neither accidental nor temporary. Rather, it seems to have turned into an almost structural aspect of contemporary governmentality. Precariousness and vulnerability aren't anthropological universalities. But they affect many individuals in environments where regulation is exerted through the power of norms. The vulnerability that characterizes anyone living within the norms makes sense for an anthropology that considers the different "historical modes of institution of human existence".[29] As such, it cannot be separated from the governmentality of the world of labour that either grants or denies access to the apparatuses that are likely to compensate this vulnerability.

It's worth repeating that what is at stake here concerning the philosophy of norms is the agency that challenges normativity—that is, the socially constructed capacities to shape one's existence into something singular, or, in short, the capacity to become the subject of one's own existence—and not a unilateral denunciation of the normalizing apparatuses, nor a dogmatic claim about ontological resources supposedly inherent to life itself. The problem of knowing how to connect such criticism to what has been termed here a "clinic" [*clinique*] or "clinical process"—which raises the questions of how to avoid turning the clinic into an instrument assigned to social normality, or how to maintain the critical potential of existences that are not in line with the dominant standards—implies that the attachment to norms, depending on the context, may expose the subject to being disqualified by social normality, or even becoming the object of a normalizing coercion, instead of opening the realm of possibilities. Are precarious and outcast existences (e.g., through the measures aimed at the unemployed) not also the most exposed to social control? If an array of social norms can help some to actualize their capabilities in various ways and, at the same time, operate as criteria to disqualify the existences that even try to conform to them, the critical issue lies in what, in the previous chapter, was referred to as the "original entanglement of normativity and normalization".

Envisaging the play of norms regardless of its integration in material circumstances would probably render those different effects indistinguishable, being at once what creates the possible for the subject and what neutralizes it. Only a combination of the philosophy of norms with an inquiry into the material conditions in which the various normative systems operate would enable its *critical* power, as a power of *discernment*. The question is about neither how to escape normalizations nor how to use the powers of vital normativity against the barriers of normalization. Rather, it's about our ability to discern and assess the combined and conflictual effects of the normative apparatuses, which can never be said to be alienating or emancipating *prima facie*. The power of norms is always exercised at the same time as a power of social inclusion and as a power of differentiation, thereby potentially excluding some forms of life. The criticism relating to norms should henceforth aim at a case-by-case assessment of the immanent effects—carrying subjection or emancipation—that are produced within and through these differentiations. This criticism begins in the inevitable normative conflicts, for instance, about the standards and features that are supposed to define, and thus to "norm", what "work" or "labour" really means[30]—its temporality, its quality, its social distribution, its value.

This type of criticism is, of course, eminently political. The issue that we are dealing with here—the devaluation of some existences—calls for discernment for another reason: it is necessary to question the processes that impede the emergence of political stages where power relationships can turn into a meaningful social conflict. The normativity (or *creativity*) manifests itself as a form of agency in a common world where norms can be represented, contested and reconfigured precisely through the social conflict that take them as objects. The fact that precariousness tends to weaken political subjectification and the formulations of common claims and strategies of struggle is, in that respect, highly significant. Indeed, the processes that impair the normativity and, as a result, the subjectification of individuals are the same as those that prevent access to the political stage, where social norms can be disputed and appear as produced through former, and perhaps forgotten, collective decisions, rather than the expression of some kind of fate or inevitable process. It is clear that, in recent history, the first conflictual "spaces" where the social norms inherent in the institution of social citizenship were contested emerged from within the world of labour—especially, though not exclusively, through trade unions.[31] The questioning of what makes political subjectifications so difficult for precarious and unemployed workers raises the broader question of the legal, institutional and socioeconomic conditions that make it possible for subjects to establish spaces of struggle. There, the polemic power of social norms can be unveiled and the power relations they support can be showed, made problematic and transformed. It is a matter of politics, a question of how to convert power relations into an appropriable social conflict—the outcome of which is the mere possibility to regain collective control over the normative differentiations that define the contours of what makes a life human.

The paradigmatic case of labour shows how the establishment of modern citizenship, particularly social citizenship, remains, as Balibar has argued, underpinned by a condition of "normality" that is tied to the very definition of "humanity":

> citizenship is historically engaged in an uninterrupted process of the extension, deepening, and adaptation of norms. The social norm must be represented to individuals or subjects in order to be able to define or delimit the rights of man (humans) and social rights, that is, to identify their (transcendental) conditions of possibility.[32]

The inclusive dynamics of social citizenship, as a producer of norms and values to which groups and individuals can get attached in various ways, has always exercised a normalizing power, potentially relegating some forms of life to the status of a one-dimensional and useless life. Balibar formulates the hypothesis that, after being organized essentially through the question of labour, the contemporary problem of access-to-rights citizenship reattaches to "anthropological differences" the decisive status they used to have in antiquity. But, unlike what happened then, these differences appear today as problematic, ambiguous and mobile, depending as they do on a play of norms with no fixed nature and free of any natural foundation—which does not mean that they are not fully tangible and powerful. One of the questions of a politics of norms would thus be: How can we turn their definition into a collective issue, allow them to be appropriated and disputed?

BIBLIOGRAPHY

Balibar, Etienne. *Equaliberty: Political Essays*. Translated by James Ingram. Durham, NC: Duke University Press, 2014.

Bourdieu, Pierre (dir.). *La misère du monde*. Paris: Seuil, 1993.

Butler, Judith. "Gender as Performance: An Interview with Judith Butler". *Radical Philosophy* 67 (1994): 32–39.

———. *The Psychic Life of Power: Theories in Subjection*. Stanford, CA: Stanford University Press, 1997.

———. *Undoing Gender*. New York: Routledge, 2004.

Butler, Judith, and Paul Rabinow. "Dialogue: Antigone, Speech, Performance, Power". In *Talk, Talk, Talk: The Cultural Life of Everyday Conversation*, edited by Shelley I. Salamensky. New York: Routledge, 2001.

Castel, Robert. *From Manual Workers to Wage Laborers: The Transformation of the Social Question* (1995). Translated and edited by Richard Boyd. New Brunswick, NJ: Transaction, 2003.

Clot, Yves. *Interpréter l'agir, un défi théorique*. Paris: Presses universitaires de France, 2011.

———. *Travail et pouvoir d'agir*. Paris: Presses universitaires de France, 2008.

de Certeau, Michel. *The Practice of Everyday Life*. Translated by Steven Rendall. Berkeley: University of California Press, 1988.
Gorz, André. *Reclaiming Work: Beyond the Wage-Based Society* (1997). Translated by Chris Turner. Malden, MA: Wiley-Blackwell, 1999.
Le Blanc, Guillaume. *L'invisibilité sociale*. Paris: Presses universitaires de France, 2009.
———. *Vies ordinaires, vies précaires*. Paris: Seuil, 2007.
Paugam, Serge. *La disqualification* sociale. Paris: Presses universitaires de France, 1991, 2002.
———. *Les formes élémentaires de la pauvreté*. Paris: Presses universitaires de France, 2005.
———. *Le salarié de la précarité. Les nouvelles formes de l'intégration professionnelle*. Paris: Presses universitaires de France, 2000.
———. *La société française et ses pauvres*. Paris: Presses universitaires de France, 1993 and 2002.
Querrien, Anne. "La Lutte contre la précarité: un mouvement de fond". *Multitudes* 25 (2006). Accessed 25 March 2014. http://www.multitudes.net/La-Lutte-sur-la-precarite-un/.

NOTES

1. Butler claims that her starting point is "the Foucauldian premise that power works in part through discourse and it works in part to produce and destabilise subjects". See Judith Butler, "Gender as Performance: An Interview with Judith Butler", *Radical Philosophy* 67 (1994): 33.

2. Judith Butler, *The Psychic Life of Power: Theories in Subjection* (Stanford, CA: Stanford University Press, 1997), 106–7.

3. Butler, *The Psychic Life of Power*, 94.

4. Butler, *The Psychic Life of Power*, 93.

5. Performativity refers to performative speech acts "that bring into being that which they name. This is the moment in which discourse becomes productive in a fairly specific way". Far from being a performance freely accomplished by the subject, "performativity" means that discourse produces the subject, that subjectivity, willingly or not, is invested with norms that have to be repeated or recited, and that subject production always implies a "trope" or a "figure" which is a creation of its own ontological effects. See Butler, "Gender as Performance", 33.

6. Butler, *The Psychic Life of Power*, 94.

7. Butler, *The Psychic Life of Power*, 93.

8. Judith Butler, *Undoing Gender* (New York: Routledge, 2004), 1.

9. Guillaume Le Blanc, *L'invisibilité sociale* (Paris: Presses universitaires de France, 2009), 184.

10. Guillaume, Le Blanc, *Vies ordinaires, vies précaires* (Paris: Seuil, 2007), 40; Michel de Certeau, *The Practice of Everyday Life*, trans. Steven Rendall (Berkeley: University of California Press, 1988).

11. Le Blanc, *Vies ordinaires, vies précaires*, 43.

12. We will return to this question. For the time being, let me simply note that, according to Butler, "the capacity for recognition and what is recognizable is given to us through various discourses that constrain the recognizable as such: what will and what will not be a recognizable person, what will and will not be recognizable gender". See Butler and Paul Rabinow, "Dialogue: Antigone, Speech, Performance, Power", in *Talk, Talk, Talk: The Cultural Life of Everyday Conversation*, ed. Shelley I. Salamensky (New York: Routledge, 2001), 41.

13. "The deviation is nothing else than the distance that the norm requires, and which allows it to grow in the very act that allows it to operate as a rule". See Le Blanc, *Vies ordinaires, vies précaires*, 39.

14. Butler argues that certain ways of living or existing are socially aberrant, considering the accepted laws: the ones who live that way still have to resort to "the dominant language and use it in a way that misuses it and contaminates it in order to allow it to mean something else or something different". Opposing Jameson's opinion that attempting to distort significations in order to mean something different "culminates in the domestication of the claim", she states, "I think that that view of power, that kind of systemic totality that eats things—I think that that's not quite right". Such is the reason why she gives priority to the analysis of situations where discourse power meets its own limits. See Butler and Rabinow, "Dialogue", 40.

15. Le Blanc, *Vies ordinaires, vies précaires*, 45.

16. As already said, subject formation goes through a repetition process; the life of the subject is "outside the self", for a subject is first dependant on his attachments to norms and therefore exposed to subordination, and second, because it is fundamentally unachieved, never fully formed, two reasons for both his/her power *and* vulnerability.

17. Le Blanc, *Vies ordinaires, vies précaires*, 127.

18. Le Blanc, *L'invisibilité sociale*, 164–65.

19. Let me mention here the works of Serge Paugam: *La disqualification sociale* (Paris: Presses universitaires de France, 1991, 2002), *La société française et ses pauvres* (Paris: Presses universitaires de France, 2002), *Le salarié de la précarité. Les nouvelles formes de l'intégration professionnelle* (Paris: Presses universitaires de France, 2000), *Les formes élémentaires de la pauvreté* (Paris: Presses universitaires de France, 2005). See also the works of Yves Clot: *Travail et pouvoir d'agir* (Paris: Presses universitaires de France, 2008) and *Interpréter l'agir, un défi théorique* (Paris: Presses universitaires de France, 2011). Paugam analyses the precarious wage labour, and draws four categories, each corresponding to four forms of social integration through work. What he calls the "secure integration" (the ideal type of all professional integration) is characterized by "the double guarantee of a material and symbolic recognition of work *and* of a social protection ensuing from employment" (Paugam, *Le salarié de la précarité*, xvii). This results in the double possibility of gaining satisfaction from work *and* of planning the future thanks to a protection against the hazards of life. The combination of those two criteria allows Paugam to identify the deviant forms of professional integration (uncertain, laborious and disqualifying) that belong to the more general category of "precariousness". These can be considered heterogeneous experiences of social devaluation, which happen when the individuals "cannot recognize themselves in their work and cannot act according to their own moral self-representation" (Paugam, *Le salarié de la précarité*, 378).

20. *La misère du monde*, ed. Pierre Bourdieu (Paris: Seuil, 1993), 17. My translation.

21. Paugam, *Le salarié de la précarité*, xvi.

22. See also Robert Castel, *From Manual Workers to Wage Laborers: The Transformation of the Social Question* (1995), trans. and ed. Richard Boyd (New Brunswick, NJ: Transaction, 2003).

23. "Simmel analyses the margin, but also that which, simultaneously, ties it to the centre and makes it a constituent part of the whole that is society". See Paugam, *Les formes élémentaires de la pauvreté*, 225.

24. Le Blanc, *L'invisibilité sociale*, 24.

25. Le Blanc, *L'invisibilité sociale*, 27.

26. As said earlier, the idea of "use" here implies the putting into action of a certain *style* of use, just as in the "arts of doing" identified by de Certeau.

27. Le Blanc, *L'invisibilité sociale*, 29–30.

28. Le Blanc, *L'invisibilité sociale*, 31.

29. Le Blanc, *Vies ordinaires, vies précaires*, 248–49.

30. Etienne Balibar notes that from the moment the problem of citizenship becomes inseparable from social justice claims, the contours of "social citizenship" are defined on the basis of social rights, these being closely linked to labour—which is what Castel's works highlights, as he claims that this social citizenship was developed

through the strengthening of wage labour. Balibar calls it a "new foundation for Equaliberty" and sees it as a new way of rendering a life human that conforms to a double logic.

31. This is one of the reasons why those who undertook to raise new social claims on the grounds of a resurgent precariousness have started to question the central role conferred to standard wage labour as the exclusive source of income and social rights. See Anne Querrien, "La Lutte contre la précarité: un mouvement de fond", *Multitudes* 25 (2006), accessed 25 March 2014, http://www.multitudes.net/La-Lutte-sur-la-precarite-un/; see also André Gorz, *Reclaiming Work: Beyond the Wage-based Society*, trans. by Chris Turner (Malden: Wiley-Blackwell, 1999).

32. Balibar, *Equaliberty: Political Essays*, trans. James Ingram (Durham, NC: Duke University Press, 2014), 123.

EIGHT

The Racial Politics of Life Itself: Goldstein, Uexküll, Canguilhem and Fanon

Robert Bernasconi

Foucault posed the question of whether the exercise of biopower can ever be separated from racism. In this chapter I ask the parallel question of whether one can employ the concepts of the normal and the pathological in contemporary society without also remaining under the sway of a certain systemic racism. To do so I develop Fanon's critical evaluation of Canguilhem's distinction between the normal and the pathological, an evaluation all the more pertinent following Canguilhem's decision, probably unknown to Fanon, to expand his use of the terms "normal" and "pathological" beyond the sphere of biology to apply them to society. I show that Canguilhem was to some degree sensitive to the issue Fanon raised and attempted to accommodate these concerns in his expanded version of *The Normal and the Pathological*, but it is not clear whether those who draw on Canguilhem are equally alert to the dangers that arise from his heavy reliance on Kurt Goldstein and Jacob Baron von Uexküll.

Michel Foucault closed the final lecture of *Society Must Be Defended* with this question: "How can one make a biopower function and exercise the rights of war, the rights of murder and the function of death without becoming racist?"[1] The immediate context is a discussion of the racism of the socialists up until the time of the Dreyfus Affair. It forms part of a more general polemic that Foucault directed against socialist states: at the end of the nineteenth century, the French socialists were racists "to the extent that . . . they did not reevaluate—or, if you like accepted as self-evident—the mechanisms of biopower that the development of society and State had been establishing since the eighteenth century".[2] Foucault thereby set a strict standard: the French socialists were to be considered racists not because of what they said explicitly about race, nor because of their actions, but because of the questions they failed to pose.

Because this question of whether the exercise of biopower or biopolitics can be divorced from racism was hurriedly posed at the end of the lecture course, one cannot be certain how Foucault himself would have answered it. Was Foucault announcing that he would subsequently demonstrate how biopolitics can free itself from racism, which is indeed the task Nikolas Rose takes up in the sixth chapter of *The Politics of Life Itself*?[3] Or was Foucault asking the question in an attempt to suggest to his audience that racism is, at least for the foreseeable future, an inevitable function of biopower? In the lecture course Foucault indicated that at very least he believed that the exercise of biopower had been racist in the past and that continued to be the case at the time when he was speaking in March 1976.

One might look for an answer to this question in Foucault's brief review of Jacques Ruffié's *De la biologie à la culture* in *Le Monde* later the same year. Foucault suggested that Ruffié had posed the question of a biopolitics that would be based not on division, conservation and hierarchy, but rather on communication and polymorphism.[4] In that case, society might not only change the ways in which it makes its evaluations but also dispense with all evaluations of that kind. As Foucault observed, the myth of the superiority of racial purity, which had been an instrument of biopower, gave way in Ruffié to the scientific truth that often racial mixing is biologically useful because it provides the basis for biological adaptation.[5] But if that is Foucault's answer to the question of how biopolitics might escape racism, then he would be reverting to a narrow conception of racism. Ruffié was following the limited strategy promoted by Ashley Montagu and UNESCO in its 1950 "Statement on Race". It fails to do justice either to Sartre and Fanon's ideas of systemic racism, which incorporated a recognition of the way in which past racisms become incorporated within institutions and indeed the very materiality of society, or to Foucault's own influential conception of a biopolitical racism.[6] The biopolitical racism that Foucault helped scholars to locate in contexts where the narrow biological notion of race was absent cannot be addressed by an argument, such as Ruffié's, which is directed against a narrow biological notion of race. In other words, Foucault's remarks on Ruffié's book do not provide a definitive answer to the question Foucault posed at the end of *Society Must Be Defended*.

One can read Roberto Esposito's book *Bios: Biopolitics and Philosophy* as offering a sustained answer to Foucault's question about biopower's ties to racism, even if he is more concerned with addressing the question of whether the concept of life can be freed from its Nazi resonances than he is with the mechanisms of biopower as such. For four chapters Esposito explains that the Nazis committed their genocidal crimes in the name of *life* with such resilience that the reader is led to believe that there is a clear proximity of biopolitics to racism. Then in the final chapter a remarkable reversal takes place. In just a few pages he attempts to free a concept of life—and thus of biopower—from racism by examining the contributions of Gilbert Simondon, Gilles Deleuze and Georges Canguilhem.[7] Nevertheless, even if we identify what precisely was so dangerous in the Nazi concept of *life* and establish another conception of *life* uncontaminated by the history of National Socialism, there would still be a need to investigate the role of *life* in the histories of colonialisms, genocides, segregation and slavery. These were largely ignored by both Foucault and Esposito. The concept of life has to answer the same set of charges that after 1945 were directed against the concepts of race and the *Volk*. And there is also the philosophical question—arguably the question that has dominated the discussion of philosophical strategies over the last sixty years—as to whether and how a concept is freed from a heritage that contaminates it.

Nor should we forget that one of the key concepts in the debate about the overcoming of metaphysics was at its very outset the concept of life. It was an issue raised by Heidegger in the context of the Nazi appropriation of Nietzsche's philosophy. Heidegger had always been suspicious of the concept of life following his early exposure to Heinrich Rickert's critique of the philosophy of life, but the decisive moment came when Heidegger abandoned the question of the Being of life, which he was still posing in the mid-1930s, in order to locate life within his account of the history of Western metaphysics. He understood that for Western metaphysics races were the units of life, which constituted a decisive association between National Socialism and the history of Western metaphysics.[8] This is a measure of how deeply entwined biopolitics is with racism.

The question of whether and how biopolitics can be disentangled from racial politics, whether or not of the kind some Nazis called *Rassenpolitik*, will remain in the background of the closely related question that I explore in the remainder of this chapter: how can one apply the concepts of the normal and the pathological to an account of contemporary society without remaining under the sway of the systemic racism that organizes that society and determines what counts as normal and what does not? The question was already raised by Frantz Fanon in "The Black Man and Psychopathology", the sixth chapter of *Black Skins, White Masks*, where he argued that the psychopathology of Freud and Adler had been based exclusively on studies of whites and so could not be understood as presenting a reliable understanding of blacks.[9] Early in the chapter he announced that "a normal child brought up in a normal family will become a normal adult". To explain what he meant here by "the normal" he referred in a footnote to "the extremely instructive work of G. Canguilhem *The Normal and the Pathological*, though it focuses solely on the biological".[10] Fanon's reference to Canguilhem[11] is followed by this enigmatic comment: "Let me just add that in the psychological field the abnormal is he who demands, appeals, and begs". I interpret that to mean here that those who refuse or are unable to adapt to the environment as it is given are abnormal.

Fanon's understanding of the normal and abnormal is clarified in the course of his argument. He makes two claims. The first is that "a normal black child, having grown up with a normal family, will become abnormal at the slightest contact with the white world".[12] The second is that "Very often the black man who becomes abnormal has never come into contact with Whites".[13] The resolution to the apparent tension between these two claims lies in a broader view: "any neurosis, any abnormal behavior or affective erethism in an Antillean is the result of his cultural situation".[14] More broadly, because of the disproportionate distribution of power between the races within modern society, a black family has a different relation to the nation or to society from that of a white family, and in a society controlled by whites it is the white perspective that determines what is and what is not abnormal. Fanon's focus is not on the individual in isolation but on the relation of the individual to the environment, which in a racist society is different for blacks and whites to the point where we are forced to say that they live in different but reciprocally related environments, divided between the oppressors and the oppressed. And if the environment is that of anti-black racism and one is black, then the expectation would have to be that one would be judged abnormal in the sense that one is unable to adapt to this environment successfully. Fanon made the point more simply when, in "Racism and Culture", he wrote, "The racist in a culture with racism is therefore normal".[15]

So far as I am aware, Canguilhem never commented on Fanon's discussion, but, in a manuscript he left unpublished, he referred to an admission by Theodule Ribot in 1870 that could be used to support Fanon's claim about the inability of the psychopathology of his day to assist blacks. Ribot, in the context of a discussion of "inferior races", wrote that "psychology, instead of being the science of psychical phenomena, has simply made man, adult, civilized, and white, its object". To be sure, in paraphrasing Ribot's comment, Canguilhem omits all mention of race and reflects on the possibility that psychology had "a hidden assumption that the respectable, cultivated male's mode of thought was somehow valid and normal".[16]

In fact, what Canguilhem read in Ribot was a comment on race: "We might show that ordinary psychology, in restricting itself to man, has not even included the whole of mankind; that it has taken no heed of the inferior races (black and yellow), that it has contented itself with affirming that the human faculties are identical in nature and various only in degree, as if the difference of degree might not sometimes be such as to be equivalent to a difference of nature".[17] It is as if Canguilhem was unwilling to think about the role of race in the history of biology and psychology. This is not the only instance of it. Even when discussing Maupertuis's *Venus physique* or the account of cell theory that Gobineau described in his *De la vie individuelle,* Canguilhem neglected to mention race.[18] Elsewhere discussions of human races were left undeveloped.[19] This was perhaps out of a misguided attempt to follow the UNESCO strategy of hoping that racism would go away if one stopped thinking about race.[20]

Nevertheless, Canguilhem and Fanon are clearly in agreement at a more general level: "Try as one will, a plurality of norms is comprehensible only as a hierarchy. Norms can coexist on a footing of equality only if drained of the normative intention that called them into existence as codified, normative decisions embodied in institutions, customs, dogmas, rites and laws. A norm cannot be normative without being militant, intolerant".[21] This, he explained, is why tolerance is the value in the name of which one becomes intolerant, just as relativity is the value in the name of which one becomes absolute. Canguilhem did not develop the point further in that place, but in the context of a critique of Soviet theories of heredity, he made what was for him a rare reference to racism. His concern was that the Soviet conviction that the milieu has a determining action legitimated the cult of "man's unlimited action on himself".[22]

Still thinking of the Soviets, he continued: "One can understand how it is that genetics could be charged with all the sins of racism and slavery, and Mendel presented as the head of a retrograde, capitalist, and idealist biology". Canguilhem knew that Soviet scientists like Trofim Lysenko rejected the Mendelian theories of hybridity in part because it seemed to present obstacles to the desire to dominate nature by breeding new kinds of human beings: "The Mendelian theory of heredity, by establishing the spontaneous character of mutations, tends to damp human—and specifically Soviet—ambitions for the total domination of nature and to limit the possibility of intentionally altering human beings". Canguilhem chose to neglect the role that the Mendelian theory played in legitimating the establishment of the one-drop rule in the United States, as well as in the Nazi racism of prominent figures like Eugen Fischer. In this way Canguilhem conformed to the general tendencies on the part of the postwar generation of historians of biology to ignore the theories used to justify segregation, sterilization and extermination because they were held to be so obviously illegitimate. But is this not another way of failing to address the question of biopower and racism on which Foucault insisted?[23]

Canguilhem was from early on concerned with the relation of theories of the organism to theories regarding the organization of society, and sometimes he warned against drawing parallels between the two.[24] His decision to expand his discussion of the normal and the pathological from the realm of biology to embrace society as a whole was already anticipated in 1947 in his review of Friedmann's *International Society: The*

Emergence of the Human Problem of Automation.[25] The key theoretical issue for Canguilhem in this context became that of the relation of the human being to his or her environment and it is because of their contributions to that topic that two figures dominate *The Normal and the Pathological*: Kurt Goldstein and Jakob Baron von Uexküll. The importance of Kurt Goldstein's *Der Aufbau des Organismus* for Canguilhem's study has been ably demonstrated.[26] By contrast, the importance of Uexküll for the expanded version of *The Normal and the Pathological* has not received the same attention. Canguilhem's name belongs alongside those of Ernst Cassirer, Martin Heidegger, Maurice Merleau-Ponty, Arnold Gehlen and Gilles Deleuze, as philosophers who acknowledge the significance of Uexküll's concept of the environment.[27] For Canguilhem the reversal of the traditional account of the relation of the organism and the milieu already took place in Uexküll. Whereas for Buffon and Lamarck time and circumstances gradually constitute the living by a "formation by deformation", for Uexküll, according to Canguilhem, "time and favorable circumstances are relative to living beings".[28] In 1946, in "Aspects of Vitalism", Canguilhem paired Goldstein and Uexküll as two prophets of work still to be done: "from an authentically biological point of view a general theory of the milieu of man as technician and scientist—one like Jakob von Uexküll's theory for the animal and Kurt Goldstein for the sick—remains to be elaborated".[29] But what united Uexküll and Goldstein most particularly was their sense that technology had disturbed the relation between the human being and the environment. That is to say, it had separated peoples from their respective *Umwelten* and they agreed that this called for decisive—one might say, drastic—action.

In another essay from 1946 Canguilhem remarked on the relation of vitalist biology to totalitarian political philosophy, but he did not follow it up.[30] He could well have been thinking of Uexküll who was a self-conscious philosophical advocate of biopolitics in his book *Staatsbiologie. Anatomie—Physiologie—Pathologie des Staates*, where the pathology of the state, although named last, seemed of primary importance. Although a reviewer of the 1920 edition simply dismissed *Staatsbiologie* as "a false concept" on the grounds that the state is not a living being (*Lebewesen*), nor an end in itself (*Selbstzweck*),[31] Uexküll was undeterred. The 1920 edition is largely pessimistic with its focus on the inevitable death of the state.[32] By the time of the 1933 edition Uexküll had come to see in Hitler and the Nazi movement a means for administering the *Staatsmedizin* needed by the German people.[33]

To the assertion of 1920 that the decline of the European states is only a matter of time, the 1933 text now added this sentence: "This danger is exorcized for Germany through Hitler and his movement".[34] Further additions included an account of the dangers of race mixing drawing on Mendel and a theory about the cleft between the different *Umwelten* of different peoples (a theory which was taken up by Walter Gross in his racial politics), and these changes seem to have been derived by an increased concern with the impact of technology.[35] One cannot argue that these ideas were isolated from Uexküll's more theoretical works. The idea of the state as an organism is integrated into both editions of his *Theoretical Biology*.[36] Uexküll aligned himself with the Nazi worldview: his 1913 collection *Bausteine* was dedicated to the notorious racist Houston Stewart Chamberlain,[37] and in 1928 he edited a selection from Chamberlain's works with a long section on the concept of life.[38] Nor was the potential of Uexhüll's ideas lost on those seeking to develop a Nazi philosophy. Ernst Krieck, one of the leading Nazi philosophers, began his chapter on "Politische Biologie" in his major work *Leben als Prinzip der Weltanschauung und Problem der Wissenschaft* by announcing that Uexküll had overcome the materialism and mechanism of Haeckelism and Darwinism and assigned to biology the task of forming a new worldview for the "*rassisch-volkish-politischen Revolution der Nationalsozialismus*".[39] And Oswald Spengler took Nietzsche's idea of life as a struggle of the will-to-power and, as part of a larger argument, applied it to Uexküll's idea of the environment as set out in his *Theoretical Biology*.[40] The degree to which Goldstein drew on Uexküll is surprising. In 1934 in his major work *The Organism* he paraphrased Uexküll's account of the milieu or environment (*Umwelt*) as follows: "Its existence and its 'normal' performances are dependent on the condition that a state of adaptation can come about between its structure and the environmental events, allowing the formation of an 'adequate' milieu".[41]

I would argue that Goldstein was especially receptive to Uexküll because he had already independently recognized in 1913, in *Über Rassenhygiene*, a slim volume advocating eugenics, what Uexküll would explore in more detail: that modern technology and capitalism had placed demands on the nervous system for which humans had not been prepared.[42] He explained in a passage that could have been written by Uexküll that "[w]e have finally learned to recognize the cause of nervous deviations in the specificity of our environment (*Milieu*). We have arrived at the result that so-called nervous degeneration has its cause in the disconnect between our capacities and the demands that the progress of culture put on us".[43] The argument was that in the face of the new conditions produced by the modern world, in particular the stress of modern industrialization, the only way to restore harmony in our relations to the environment would be for culture to change or for humans to change themselves. The conclusion was that it would be so difficult to change the culture that as a result human beings would have to change themselves through racial hygiene.[44]

To be sure, the debt Canguilhem owed to Goldstein and Uexküll's notion of the environment does not in any way entail a commitment on his part to the racial politics they developed. The question is whether Canguilhem, with his knowledge of that history, should have been more insistent when it came to posing the question of the relation of the narrative to racism, as Fanon proposed. On the few occasions when Canguilhem addressed the use of vitalist biology within National Socialism, he rehearsed the standard response: "The self-interested conversion of certain biologists to Nazism proves nothing against the quality of either the experimental facts themselves or the suppositions accepted to account for these facts—suppositions to which these biologists, prior to their conversion, had believed they owed their scientific support. One is not obliged to locate within biology, as a logically inevitable consequence, the attitude that certain biologists adopted out of a lack of character and philosophical resoluteness".[45] No doubt they can be disassociated in this way, and Canguilhem made things easier for himself than he might have done by using the too easy example of Hans Driesch. As Anne Harrington has pointed out, Driesch's work was certainly used by some Nazis, but he himself was opposed to the Nazi regime.[46] Indeed, Driesch actually wrote in opposition to von Uexhüll's *Staatsbiologie* on grounds similar to those raised by Holle: the state is not an organism.[47]

Canguilhem used similar arguments to drive a wedge between biological theories and the accounts of society that they are sometimes said to legitimate. One sees it in his account of Auguste Comte.[48] But it is even more blatant in another rare mention of racism in *Du développement à l'évolution au XIXe siècle*, a contribution to *Thales* that was subsequently reissued as a separate text. Canguilhem there dismissed the tendency of political theories to annex biology illegitimately: "Hitlerian racism is only one such case, albeit singularly detestable".[49] Canguilhem's example here—Charles Maurras, founder of Action Française—was again ill chosen for the purpose insofar as the latter rejected scientific racism. On the decisive question of whether it is the human being or the culture that should change, Canguilhem rejected the psychosocial definition of the normal in terms of adaptedness on the grounds that it assimilates society to the environment understood as "a system of determinisms".

By contrast, a society is "a system of constraints which, already and before all relations between it and the environment, contains collective norms for evaluating the quality of these relations".[50] But it is a shock to find that Canguilhem, a historian of science, took so little interest in the relation between racism and science. It is the responsibility of intellectual historians to reflect on how ideas are connected and how they spread even when they are not tied by relations of strict logical entailment. We understand very little how one idea calls for another to shape a worldview, but this is precisely what is needed in an examination of how terms like "the normal" or the "the abnormal" function in the background practices of extermination, sterilization or other forms of discrimination.

So what would Fanon have made of Canguilhem's *The Normal and the Pathological* had he lived long enough to read the 1966 version, which was no longer limited to the biological but extended to include the social? No doubt he would first of all have been worried about the dangers of any such expansion, but, as we have seen, that was, of course, a concern Canguilhem shared, as the essays collected in *The Knowledge of Life* show. He would also have found indications in *The Normal and the Pathological* that Canguilhem was aware that ideas of the normal could be used oppressively, as when he observed that in a certain period the bourgeoisie as "a normative class" had won the power to identify "the function of social norms, whose content is determined, with the use that that class made of them".[51] And, most significantly, Fanon would surely have welcomed Canguilhem's recognition of the concern Fanon had identified in 1952: "To define abnormality in terms of social maladaptation is more or less to accept the idea that the individual must subscribe to the fact of such a society, hence must accommodate himself to it as to a reality which is at the same time a good".[52] This, as we have seen, was Uexküll's error and also Goldstein's in 1913. But Canguilhem's response that "the organism is not thrown into an environment to which he must submit, but he structures his environment at the same time that he develops his capacities as an organism"[53] leaves unaddressed who gets to do that and how.

I have not, of course, been exploring these issues in order to pass judgement on Canguilhem whose credentials as an active member of the French Resistance are beyond question. My interest here lies in the way ideas of the normal and abnormal function within racism. Canguilhem's deepest insight in this context lies in his account of the way norms keep reinventing themselves: "The concept of normalization excludes that of immutability, includes the anticipation of a possible flexibility".[54]

The problem we have in addressing racism is the way in which across all the societal changes that inevitably takes place, the same structures are being reconstituted. So in the United States slavery is abolished, but it is replaced by sharecropping and the chain gang, and following the success of the civil rights movement, being black is criminalized in what has been called "the new Jim Crow".[55] We see it in the way society abandons the laws which legally enforce segregation and yet in housing and in education many places remain as segregated as before. And lynching is replaced by legal execution distributed unequally between the races. It is for this reason that I insist that we do not allow ourselves to lose sight of Foucault's question of whether the failure to be on one's guard about the way mechanisms and concepts function does not sometimes commit one unwittingly to a certain racism.

BIBLIOGRAPHY

Alexander, Michelle. *The New Jim Crow: Mass Incarceration in the Age of Colorblindness.* New York: New Press, 2010.

Bernasconi, Robert. *Nature, Culture, Race.* Stockholm: Södertörn University, 2010.

———. "Heidegger, Rickert, Nietzsche, and the Critique of Biologism". In *Heidegger and Nietzsche*. Edited by Babette Babich, Alfred Denker and Holger Zaborowski, 159–80. Amsterdam: Rodopi, 2012.

———. "Where Is Xenophobia in the Fight against Racism?" *Critical Philosophy of Race* 2 (2014): 5–19.

Buchanan, Brett. *Onto-Ethologies. The Animal Environments of Uexküll, Heidegger, Merleau-Ponty, and Deleuze.* Albany: State University New York Press, 2008.

Canguilhem, Georges. "Milieu et normes de l'homme au travail". *Cahiers internationaux de Sociologie* 3 (1947): 120–36.

———. "Note sur la situation faite en France à la philosophie biologique". *Revue de Métaphysique et de Morale* 52 (1947): 322–32.

———. *Essai sur quelques problèmes concernant le normal et la pathologique*. Paris: Les Belles Lettres, 1950.

———. *La formation du concept de réflexe*. Paris: Presses universitaires de France, 1955.

———. *Ideology and Rationality in the History of the Life Sciences*. Translated by Arthur Goldhammer. Cambridge, MA: MIT Press, 1988.

———. *The Normal and the Pathological*. Translated by Carolyn, R. Fawcett. New York: Zone Books, 1989.

———. *A Vital Rationalist. Selected Writings from Georges Canguilhem*. Translated by Arthur Goldhammer. New York: Zone Books, 1994.

———. *The Knowledge of Life*. Translated by Stefanos Geroulanos and Danielle Ginsburg. New York: Fordham University Press, 2008.

———. *Œuvres complètes*. Vol. 1. Paris: Vrin, 2012.

———. *Writings on Medicine*. Translated by Stefanos Geroulanos and Todd Meyers. New York: Fordham University Press, 2011.

Canguilhem, Georges, et al. *Du développement à l'évolution au XIXe siècle* (1060). Paris: Presses universitaires de France, 1962.

Chamberlain, Houston Stewart. *Natur und Leben*. Edited by Jacob von Uexküll. Munich: F. Bruckmann, 1928.

Driesch, Hans *Philosophie des Organischen*. Leipzig: Engelmann, 1921.

Duroux, Françoise. "L'imaginaire biologique du politique". In *Georges Canguilhem. Philosophe, historien des sciences*, 49–57. Paris: Albin Michel, 1993.

Esposito, Roberto. *Bios: Biopolitics and Philosophy*. Minneapolis: University of Minnesota Press, 2008.

———. *Immunitas*. Translated by Zakiya Hanafi. Cambridge: Polity, 2011.

Fanon, Frantz. *Toward the African Revolution*. New York: Grove Press, 1967.

———. *Black Skin, White Masks*. Translated by Richard Philcox. New York: Grove Press, 2008.

Foucault, Michel. "Bio-histoire et bio-politique". In *Dits et Écrits*, Vol. 3, 95–97. Paris: Gallimard, 1994.

———. *"Society Must Be Defended"*. Translated by David Macey. New York: Picador, 2003.

Gayon, Jean. "Le concept d'individualité dans la philosophie biologique de Georges Canguilhem". In *L'épistémologie française, 1830–1870.* Edited by Michel Bitbol and Jean Gayon, 431–63. Paris: Presses universitaires de France, 2006.

Goldstein, Kurt. *Über Rassenhygiene*. Berlin: Julius Springer, 1913.

———. *The Organism*. New York: Zone Books, 1995.

Harrington, Anne. *Reenchanted Science: Holism in German Culture from Wilhelm II to Hitler*. Princeton, NJ: Princeton University Press, 1996.

Heidegger, Martin. *Mindfulness*. Translated by Parvis Emad and Thomas Kalary. London: Continuum, 2006.

Holle, H. G. "Review of J. v. Uexküll, *Staatsbioogie*". *Deutschlands Erneuerung* 4, no. 7 (1920): 468.

Jones, Donna V. *The Racial Discourse of Life Philosophy: Négritude, Vitalism, and Modernity*. New York: Columbia University Press, 2010.

Krieck, Ernst *Leben als Prinzip der Weltanschauung und Problem der Wissenschaft*. Leipzig: Armanen, 1938.
Le Blanc, Guillaume. "From Matter to Materiality According to Canguilhem". Translated by Nicholas de Warren. *Graduate Faculty Philosophy Journal* 22 (2000): 255–70.
Macey, David. *Frantz Fanon*. New York: Picador, 2010.
Mader, Mary Beth. *Sleights of Reason: Norm, Bisexuality, Development*. Albany: State University New York Press, 2011.
Métraux, Alexandre. "Georges Canguilhem als Architekt einer Philosophie des Lebens". In *Mass und Eigensinn. Studien im Anschluss an Georges Canguilhem*, 317–46. Paderborn: Wilhelm Fink, 2005.
Ribot, Theodule. *English Psychology*. New York: D. Appleton, 1874.
Roberts, Dorothy. "Race and the New Biocitizen". In *What's the Use of Race?* Edited by Ian Whitmarch and David S. Jones, 259–76. Cambridge, MA: MIT Press, 2010.
Rose, Nikolas. *The Politics of Life Itself*. Princeton, NJ: Princeton University Press, 2007.
Roudinesco, Elisabeth. *Philosophy in Turbulent Times*. Translated by William McCuaig. New York: Columbia University Press, 2008.
Ruffié, Jacques. *De la biologie à la culture*. Paris: Flammarion, 1976.
Spengler, Oswald. *Der Mensch und die Technik. Beitrag zu einer Philosophie des Lebens*. Munich: Beck, 1932.
Uexküll, Jacob von. *Bausteine zu einer biologischen Weltanschauung*. Munich: F. Bruckmann, 1913.
———. *Staatsbiologie. Anatomie-Physiologie-Pathologie des Staates*. Berlin: Gebrüder Paetel, 1920.
———. *Theoretische Biologie*. Berlin: Gebrüder Paetel, 1920.
———. *Theoretische Biologie*. Berlin: Julius Springer, 1928.
———. "Houston Stewart Chamberlain 1855–1927. Die Persönlichkeit". In *Bücher des Verlages F. Bruckmann AG*. Munich: Bruckmann, 1928.
———. *Staatsbiologie. Anatomie-Physiologie-Pathologie des Staates*. Hamburg: Hanseatische Verlagsanstalt, 1933.

NOTES

1. Michel Foucault, *"Society Must Be Defended"*, trans. David Macey (New York: Picador, 2003), 263.
2. Foucault, *"Society Must Be Defended"*, 263.
3. Nikolas Rose, *The Politics of Life Itself* (Princeton, NJ: Princeton University Press, 2007), 155–86.
4. Michel Foucault, *"Bio-histoire et Bio-politique"*, in *Dits et Écrits*, Vol. 3 (Paris: Gallimard, 1994), 95–97.
5. Jacques Ruffié, *De la biologie à la culture* (Paris: Flammarion, 1976), 467.
6. Robert Bernasconi, *Nature, Culture, Race* (Stockholm: Södertörn University, 2010), and "Where Is Xenophobia in the Fight against Racism?", *Critical Philosophy of Race* 2 (2014): 5–19.
7. Roberto Esposito, *Bios: Biopolitics and Philosophy* (Minneapolis: University of Minnesota Press, 2008), 179–84.
8. See Martin Heidegger, *Mindfulness*, trans. Parvis Emad and Thomas Kalary (London: Continuum, 2006), 250. See also Robert Bernasconi, "Heidegger, Rickert, Nietzsche, and the Critique of Biologism", in *Heidegger and Nietzsche*, ed. Babette Babich et al. (Amsterdam: Rodopi, 2012), 159–80.
9. Frantz Fanon, *Black Skin, White Masks*, trans. Richard Philcox (New York: Grove Press, 2008), 150.
10. Fanon, *Black Skin*, 121.
11. Casual readers of the expanded edition of *The Normal and the Pathological* that first appeared in 1966 might be surprised by this last phrase, but it was true of the original edition of the book that Fanon knew—that is, Georges Canguilhem, *Essai sur*

quelques problèmes concernant le normal et la pathologique (Paris: Les Belles Lettres, 1950). It might seem surprising that Fanon was already reading Canguilhem's thesis in 1952, just as it has surprised some that he already knew Lacan's 1932 thesis at that time. It is likely that he learned of them from either François Tosquelles or Paul Balvet, who between them were the inspiration behind the Saint-Alban Psychiatric Hospital where Fanon worked and where Canguilhem himself worked in 1944. See David Macey, *Frantz Fanon* (New York: Picador, 2010), and Elisabeth Roudinesco, *Philosophy in Turbulent Times,* trans. William McCuaig (New York: Columbia University Press, 2008), 19–20.

12. Fanon, *Black Skin,* 122.
13. Fanon, *Black Skin,* 124.
14. Fanon, *Black Skin,* 130.
15. Fanon, *Toward the African Revolution* (New York: Grove Press, 1967), 40.
16. Canguilhem, *A Vital Rationalist: Selected Writings from Georges Canguilhem,* trans. Arthur Goldhammer (New York: Zone Books, 1994), 363.
17. Theodule Ribot, *English Psychology* (New York: D. Appleton, 1874), 17.
18. Canguilhem, *Ideology and Rationality in the History of the Life Sciences,* trans. Arthur Goldhammer (Cambridge, MA: MIT Press. 1988), 136–38, and Georges Canguilhem, *Knowledge of Life,* trans. Stefanos Geroulanos and Danielle Ginsburg (New York: Fordham University Press, 2008), 50.
19. Canguilhem, *The Normal and the Pathological,* trans. Carolyn R. Fawcett (New York: Zone Books, 1989), 175.
20. An early dismissive mention of race can be found at Georges Canguilhem, *Œuvres complètes,* vol. 1 (Paris: Vrin, 2012), 309. I am grateful to Alex Feldman for this reference.
21. Canguilhem, *A Vital Rationalist,* 364.
22. Canguilhem, *Knowledge of Life,* 115.
23. It must be understood that I am not assimilating Foucault and Canguilhem's notions of norm. On their differences, see Mary Beth Mader, *Sleights of Reason: Norm, Bisexuality, Development* (Albany: State University New York Press, 2011), 57–65, and Roberto Esposito, *Immunitas: The Protection and Negation of Life,* trans. Zakiya Hanafi (Cambridge: Polity, 2011), 141–44. I am merely indicating that the question of how to free certain mechanisms and concepts from racism that Foucault momentarily raises can profitably be posed in relation to Canguilhem's concept of the normal.
24. Georges Canguilhem, *Writings on Medicine,* trans. Stefanos Geroulanos and Todd Meyers (New York: Fordham University Press, 2011), 76.
25. Georges Canguilhem, "Milieu et normes de l'homme au travail", in *Cahiers internationaux de Sociologie* 3 (1947): 120–36. See also Georges Canguilhem, "Note sur la situation faite en France à la philosophie biologique", *Revue de Métaphysique et de Morale* 52 (1947): 322–32; and Guillaume Le Blanc, "From Matter to Materiality According to Canguilhem", trans. Nicholas de Warren, *Graduate Faculty Philosophy Journal* 22, no. 1 (2000): 255–70.
26. Kurt Goldstein, *The Organism* (New York: Zone Books, 1995). See also Alexandre Métraux, "Georges Canguilhem als Architekt einer Philosophie des Lebens", in *Mass und Eigensinn. Studien im Anschluss an Georges Canguilhem* (Paderborn: Wilhelm Fink, 2005), 317–46; and Jean Gayon, "Le concept d'individualité dans la philosophie biologique de Georges Canguilhem", in *L'épistémologie française, 1830–1870,* ed. Michel Bitbol and Jean Gayon (Paris: Presses universitaires de France, 2006), 435.
27. Brett Buchanan, *Onto-Ethologies: The Animal Environments of Uexküll, Heidegger, Merleau-Ponty, and Deleuze* (Albany: State University New York Press, 2008).
28. Canguilhem, *Knowledge of Life,* trans. S. Geroulanos and D. Ginsburg (New York: Fordham University Press, 2008), 103, 110, 112.
29. Canguilhem, *Knowledge of Life,* 111.
30. Canguilhem, *Knowledge of Life,* 42–43.
31. H. G. Holle, "Review of J. v. Uexküll, *Staatsbioogie*", *Deutschlands Erneuerung* 4, no. 7 (1920): 468.

32. Jacob von Uexküll, *Staatsbiologie. Anatomie-Physiologie-Pathologie des Staates* (Berlin: Gebrüder Paetel, 1920), 54.
33. Uexküll, *Staatsbiologie. Anatomie-Physiologie-Pathologie des Staates* (Hamburg: Hanseatische Verlagsanstalt, 1933), 79.
34. Uexküll, *Staatsbiologie*, 71.
35. Uexküll, *Staatsbiologie*, 78.
36. Uexküll, *Theoretische Biologie* (Berlin: Gebrüder Paetel, 1920), 243–48, and *Theoretische Biologie* (Berlin: Julius Springer, 1928), 225–28.
37. Uexküll, *Bausteine zu einer biologischen Weltanschauung* (Munich: F. Bruckmann, 1913).
38. Houston Stewart Chamberlain, *Natur und Leben*, ed. J. von Uexküll (Munich: F. Bruckmann, 1928), 93–187. See also Jacob von Uexküll, "Houston Stewart Chamberlain 1855–1927. Die Persönlichkeit", in *Bücher des Verlages F. Bruckmann AG* (Munich: Bruckmann, 1928), 9–13.
39. Ernst Krieck, *Leben als Prinzip der Weltanschauung und Problem der Wissenschaft* (Leipzig: Armanen, 1938), 88, 92.
40. Oswald Spengler, *Der Mensch und die Technik. Beitrag zu einer Philosophie des Lebens* (Munich: Beck, 1932), 12–22.
41. Goldstein, *The Organism*, 106.
42. Kurt Goldstein, *Über Rassenhygiene* (Berlin: Julius Springer, 1913), 38–50.
43. Goldstein, *Über Rassenhygiene*, 63.
44. Goldstein, *Über Rassenhygiene*, 66.
45. Georges Canguilhem, *Knowledge of Life*, 72.
46. Anne Harrington, *Reenchanted Science: Holism in German Culture from Wilhelm II to Hitler* (Princeton, NJ: Princeton University Press, 1996), 190–94.
47. Hans Driesch, *Philosophie des Organischen* (Leipzig: Engelmann, 1921), 573.
48. Canguilhem, *Knowledge of Life*, 44–46. See also Françoise Duroux, "L'imaginaire biologique du politique", in *Georges Canguilhem. Philosophe, historien des sciences* (Paris: Albin Michel, 1993), 52–53.
49. Canguilhem et al., "Du développement à l'évolution au xixe sièle", in *Thalès*, Année 1960, Vol. 2 (Paris: Presses Universitaires de France, 1962), 62.
50. Canguilhem, *The Normal and the Pathological*, trans. Carolyn R. Fawcett (New York: Zone Books, 1989), 282.
51. Canguilhem, *The Normal and the Pathological*, 246.
52. Canguilhem, *The Normal and the Pathological*, 282–83.
53. Canguilhem, *The Normal and the Pathological*, 284.
54. Canguilhem, *The Normal and the Pathological*, 247.
55. Michelle Alexander, *The New Jim Crow: Mass Incarceration in the Age of Colorblindness* (New York: New Press, 2010).

NINE

History and the Politics of "Life"[1]

Claudia Stein and Roger Cooter

Does history have anything to contribute to the current discussion on "life"? It might not seem so. Overall, historians do not speak directly to the vital issues of the day. With regard to biopolitics and ethics, for example, they tend to trail some way behind philosophers, sociologists and anthropologists who have far more to say. And yet, especially since the linguistic-cum-somatic turn of the 1980s, historians have written a great deal about life—manifestly through attention to the historicization of "the body". Moreover, they have contributed significantly not only to the literature on bioethics but also to that centring on what is to be human.[2] Today, more than ever, history-writing is awash with reference to life, not least with regard to the now huge appeal of the history of emotions, and the burgeoning literature on neurohistory.[3] But what exactly is the contribution of historians to the contemporary discussions of life? What are their methodologies, and how do these fit interdisciplinary agenda? Does history-writing in this area have potential for scholars in other fields? This chapter reviews some the recent literature with these questions in mind.

The history of the body became fashionable among historians in the 1980s. Suddenly an object that had hitherto been no one's particular interest (even in the history of medicine) became everyone's preoccupation. The somatic turn was broadly a means to explicate and illustrate how biological entities, concepts and categories like "the body" had served in the past to rationalize and naturalize an understanding of the world that was increasingly coming to be felt by many late twentieth-century intellectuals as fragmented. The concepts and categories of "modernity" were seen as historical constructs or representations of reality "made up" at particular moments. The history of the natural sciences became a favored place to explore how "life" was a social construct (much to the alarm of those who were to launch the "science wars" against such a view). The hero of the day was of course Michel Foucault, the philosopher who insisted that nothing was ever outside of history and, in particular, that "life" was itself a meaningless empty category that only gained meaning through the historical discourses in which it was nested (and hence was not *sui generus* but a correlate of particular knowledge/power nexuses).

More recently, "life" in history-writing has taken on new forms and generated new intellectual orientations and understandings. This is most evident in the current fashion for the history of emotions and affect. But it is also apparent through new empiricist turnings to neurohistory, on the one hand, and applications of the methodology of Actor Network Theory (ANT), on the other (the latter proposing a new social theory to accompany the new interest in life). In general, in line with a new generation of scholarship in the social sciences coalescing around the assertion that the very grounds of life itself are changing, especially through neurobiology,[4] historians are awakening to the suggestion that our times are being radically transformed in the wake of biomedicine, biotechnology and neurobiology. Historians trim their sails accordingly, reassigning their agenda. Some, at the would-be cutting edge, seek to equip themselves with the latest "breakthroughs" in the life sciences, celebrating along the way the "final overcoming" of perceived worn out disciplinary boundaries between the human and natural sciences.

This latter is well exemplified in *Echo Objects* by the historian of art, medicine and the body, Barbara Maria Stafford.[5] This is a book, she informs us, that she wrote after attending weekly seminars on computational neuroscience. Although Stafford was never a card-carrying anti-essentialist postmodernist, she was well aware of how the deconstructivist history of the historicized body had unfolded in postmodernity (as is evident, for example, in her *Body Criticism: Imaging the Unseen in Enlightenment Art and Medicine*[6]). In *Echo Objects*, however, she relates how neuroscience convinced her that "those of us in the humanities and social sciences" have acquired "wonderful new intellectual tools to re-imagine everything from autopoiesis [self-organization] to mental imagery".[7] Converted, she was, she says, compelled "to rethink the major themes of [my] life's work" and then, evangelically, to press this "rethink" on art, cultural and literary historians so that they would "consider seriously *the biological underpinnings* of artificial marks and built surfaces".[8] Neuroscience, she came to believe in this teleological configuration of her own self-understanding, "enables us . . . to comprehend . . . reflex tendencies from the inside out". Old problems could look new after being "sieved through the cognitive turn", and "traditional cultural assumptions by which many of us have long lived" could be turned "upside down".[9]

> As scholars of the myriad aspects of self-fashioning we can usefully enlarge, and even alter, our humanistic understanding of culture, *inflecting it with urgent discoveries in medicine, evolutionary and developmental biology, and the brain sciences.* In other words, the role of culture is not just to stand outside, critiquing science, nor is science's position external, and acting on culture. Rather, we are discovering at the most profound levels *that our separate investigations belong to a joint project, at last.*[10]

Other scholars, prominent in anthropology, architecture and urban planning, human and spatial geography, literature, media and cultural studies, art history and general history, have made similar claims for the neuro-turn in their fields, usually with the same appeal to interdisciplinarity—or "transdisciplinarity in the age of the brain", as one recent title has it.[11] One such enthusiast for the application of neurobiology to history is the Harvard medievalist Daniel Lord Smail, the person to whom the term "neurohistory" is attributed. In his *On Deep History and the Brain* (2008) and in subsequent overtly "neurohistorical"forays, Smail argues that through the latest research on the brain we can understand anew the nature of power in human societies. By these means, he thinks, we can add "a new interpretive dimension to our understanding of cultural transformations; it offers a way to incorporate the neurosciences into history without having recourse to historically sterile versions of evolutionary psychology (notably approaches that rely on strong theories of massive modularity and evolved dispositions)".[12]

In essence, Stafford and Smail seek to add a nonintentionalist dimension to the study of the past. Behind this is the belief, shared by many intellectuals today, that the role of reason and rationality in the reconstruction of human action in the past has been overrated. As a result of the reasoning of Kantian, neo-Kantian and Habermassian philosophers (the argument goes), human action is "too flat or disembodied an account of the ways in which people actually form their political opinions and judgments".[13] The would-be biological correction or "deepening" of the study of the past is believed to result in a "truer" (because more "complete") history. By transporting contemporary biological knowledge back into the past, these historians hope to "thicken" and broaden the bases of historical evidence, especially with regard to those periods where written sources are wanting or scarce, such as for prehistory or the Middle Ages. (Interestingly, outside of their consideration lies the Foucauldian-inspired idea of the 1980s and 1990s that the "discoveries" of the life sciences are themselves historical constructs and cannot therefore be used as neutral and universal tools to access and interpret the past.)

None of this, it should be said, entails a return to biological determinism, or at least is not *perceived* as such. While there is a curious return to the idea of scientific knowledge as essentialist and transhistorical (as in the work of Stafford and Smail), biology is no longer understood as crudely determining individual and collective human action. In common with most of today's bio-enthusiasts in the human sciences the dreadful idea of "biology-as-destiny" is refuted in favour of the idea of "life's" emancipatory qualities.[14] Life is thought of "as a creative space, a field of potentiality", something "unpredictable", "energistic", and shot through with the "dynamics of birth and creativity".[15] As Ruth Leys has pointed out in a penetrating critique of affect theory, scholars embracing biology today "hope to avoid the charge of falling into a crude reductionism by positioning themselves at a distance from the geneticism and determinism that was a target of the previous phase of cultural history".[16] Today, "life" is no longer regarded as a construct of culture, but as its very foundation. Combining neurobiological evidence with the ideas of dissenting philosophers of nature (from Spinoza to Deleuze and Guattari), scholars such as the human geographer Nigel Thrift argue for the centrality of biology in representations of culture, past and present. Meaning delivered through the study of texts is not enough, according to Thrift in his articulation of post- or "non-representional theory":[17] "It has become increasingly evident", he argues, "that the biological constitution of being . . . has to be taken into account if performative force is ever to be understood, and in particular, the dynamics of birth (and creativity) rather than death".[18] But as Leys further points out, the *appearance* of nonbiological reductivity is just that. In fact, although the negotiations between the new anti-intentionalism and biology are complex, philosophically opaque and quantitatively uneven, they all share a reliance on experimental science. Such writings purport not to be biologically driven, but if one investigates their evidence closely they can be seen as relying entirely on the natural sciences and their methods.

The enthusiasm for the importance of unintended "pre-conscious" human action is thought to necessitate a new kind of social theory. At least since the social theory was thought to be guided by human intentions—by the conscious decision making of humans. But now, in order for "life itself" to unfold and expand—to be "creative"—a new theory of the social is required. The new biological actors need a new social theory, one that transcends the old (intentionalist) "structures", "interests", "habitus" and so on. The new theory needs actions and behaviours to be "determined by affective dispositions that are independent of consciousness and the mind's control".[19] It is here that ANT steps into the breach, facilitating the reimagining of "the social" in terms of the interaction of "things", nonhuman as much as human. Human agency is not wholly removed through ANT, but it is deprived of its pride of place and purpose.

In truth, ANT is less a "theory" than an observational practice intended to challenge the assumptions and expectations of traditional (post–c.1800) sociology. With respect especially to laboratory "objects" and "things" (such as bacteria), ANT refuses to seek out meaningful structures or patterns, to prioritize any particular agency or to lend causal power to one thing over another in accounting for the stabilization of anything. In the face of a sociology of human interactions which presupposes something inherent to the world and which believes it capable of being revealed or "discovered", ANT posits a sociology of symmetric interaction between "things, or objects, or beasts"[20] deemphasizing the role of human actors and human agency. As one of its interlocutors explains:

> Human lives are so bound up with artifacts that even so apparently simple an action as putting on our clothes produces a complex collective, a hybrid actor. The repertoire of the clothed man or woman is extended and enacted through the collective that includes his or her clothes, mobile phone, laptop computer, car, and so on, as well as his or her body itself. Responsibility for action, in this depiction, is shared among the various actants as competence and responsibility are the properties of sociotechnical composites.[21]

Agency is thus irreducible to people or to technical artifacts; it "is hybrid actants composed of humans and non-humans that act, and the act and the actant cannot be separated".[22] Never still, actants are forever entering into new relations at the same time as transforming existing ones. It is this galaxy of parts constantly in action that constitutes Bruno Latour's "sociology of associations" or "sociology of networks" or, more commonly now (although usually without explicit debt to Latour), the investigation of "assemblages".

The "social", so far as it can even be spoken of within the Latourian scheme of "things", is neither a fixed entity nor something that has innate substance, power or force able to shape the lives of individuals, as assumed in conventional studies of gender, race, class, nations and states. Rather, "the social" is but the "affect" *of* "the social" as it emerges hand-in-hand with its empirical description or its "unfolding" by analysts. Inherently fragile, it is ever fugitive, and therefore inexplicable. Humans can explain it, but only one-sidedly through recourse to such reductionist analytical categories as "economics" and "ideology". What they ought to do instead, says Latour, is investigate the specific constellations and conglomerates consisting of humans and things (the actants in their networks of conversation)—and this not only in the human sciences but also in the arts, ecology and everywhere. Society, in the old sense, does not, therefore, exist for Latour, because in his view the social is ever in the making through the associations of actants; there are no would-be universal laws and human interests to be revealed, as the old sociology believed. Moreover, sociotechnical collectives need continually to be reconstituted in order to be kept together, their inherent fragility being threatened at every moment with destruction and change—processes that Latour fondly regards as "creative".

These ideas have been reiterated and refined in Latour's *Reassembling the Social: An Introduction to Actor-Network Theory* (2005). Behind them lies the conviction that the world today is surprisingly unpredictable, ever changing and highly contingent—"an enormous ocean of insecurity broken up by some little islands of stabilized forms".[23] With its rapid information flows (as on Facebook or through email) and material exchanges (as on eBay), its celebrated "breakthroughs" in biology and neurology, its ethnic pluralism, multitasking, financial flexibility and so on, it has the *appearance* of being infinitely more complicated and problematic than it ever was in the past. Given, too, the end of communism and the rise of global neoliberal capitalism, it is a world, according to Latour, that conventional human-action centred sociology in its relentless search for patterns cannot account for—unlike the "sociology of associations", which doesn't try. The only way left for scholars to get a hold on what is going on is to describe the ever-changing constellations between human and nonhuman actors—that is, to become anthropologists of the present.

There are as yet few examples in history-writing of how this new making-up of the present and "life" within it is to be written into history and, in particular, how the history of "the social" can be rewritten to accommodate the new orientation towards "life". However, one historian who has made the attempt is Patrick Joyce (a historian also known for having previously piloted poststructuralist approaches to history-writing). According to him, in his recent *State of Freedom: The Social History of the British State since 1800* (2013),[24] the older tool kits of the social sciences are now not enough to explain the working of power and social life. In order to do so, he maintains, we need to borrow from the new social sciences.

> New approaches are particularly helpful because a number of them allow us to think in the more creative vein of processes rather than of structures, and so of the state in "network" terms, as something like an "assemblage" which is held together (sometimes very uncertainly) at particular key sites or nodes and through the actions of key actors and processes human and non-human. There is an emphasis on the state as heterogeneous and multiplex, so that where once the state and its characteristics were seen in *a priori* terms, they are now seen as outcomes or achievements, often insecure ones, and as the product of specific historical, social and material processes.[25]

The crucial intellectual move, according to Joyce, is the turn away from older notions of society as a coherent totality. Involved now is

> looking beyond the familiar division of the world into elements that are permanently and essentially on one or the other side of a line dividing the natural and the social, the human and the non-human, the material and cultural—divisions that are all predicated on the misleading divide between the immaterial and the material. . . . The general idea therefore

> is that the social does not lie outside the actors and networks in which it is located (say in "society" or "nature"), but is the outcome of these.[26]

Joyce thus buys into the new theoretics that enable the new discussions of "life" to thrive; it's a rewrite of history from the reconceptualized present that shores up that reconceptualization. Now believing that we should understand people's individual and collective actions as coming from "below the level of their conscious awareness",[27] he calls for a rewriting of the past that demonstrates the continuities with this newly conceptualized social present. At the level of historical practice this is a curious return to a pre-postmodern position. It is what historians have practiced ever since the invention of their craft as an academic discipline in the early nineteenth century, and which was deplored by Nietzsche for its unwitting complicity with contemporary regimes of truth.[28] The most cursory glance at the history of history-writing reveals that historical narratives have indeed long played a key therapeutic role in boosting individual and national confidence and self-esteem, providing purpose and meaning and confirming identities. As one recent commentator has put it:

> History in modern times has fulfilled the role of meaning-giver—a secular replacement for those who have accepted the death of a god previously perceived in terms of Providence, or of a "destiny" made conveniently manifest to some. Histories, that is to say, have been written to underpin a "now" that needs justification—that needs to be seen as a meaningful, purposeful outcome of what has gone before; histories, both personal and public, have contrived to make some sense of human life, importantly endowing it with a sense of direction of its own.[29]

It is the belief that history gives direction that has permitted it to continue to be seen as a vital and central part of education—even, or perhaps especially, in our current supposedly multicultural, diversity-embracing times. Carefully constructed histories by professional historians are believed to provide stable foundations for societies. As Nietzsche once pointed out, such history-writing offers "the happiness of knowing oneself not to be wholly arbitrary and accidental, but rather growing out of the past, its heir, flower and fruit".[30] This is not only what historians themselves consider their job, but also that which the wider public believes, as one sees daily in the media. Are not historians persons who identify "true stories" of the past and report them in neutral, non-judgemental ways?

The sense of urgency whereby Joyce conjures the need for new theoretical approaches to the study of the past suggests to us that he believes past and present are now out of synch. He is hopeful that once we get the theory and methodology of the present "right"—that is, when we see it through the lens of ANT and allied theories, and apply it to the investigation of the past, the past and the present will come in line. They will be linked by a single storyline that will explain why the present is the way it is. Joyce's rewriting of the past thus helps to confirm—indeed "invent"—a present that is understood as the unintended outcome of endless associations and networks of human and nonhuman actors. To be sure, these means to describe the present are innovative (within history-writing), but the idea of a continuous storyline isn't; it is scarcely different from the endless histories of the emergence of nation-states that celebrate human agency, which Joyce so deplores.

But is this all that historians can do? Are historians simply condemned to rewrite the past in the light of the present? Must they only hold up mirrors to today's discussions of "life", which so far have been conducted mainly by philosophers, ethicists and social scientists? Can they themselves participate *in* these debates beyond providing interesting historical details?

We claim that they can, by exercising (rather than abandoning) the strengths they have gained over the past half century or so—namely, the ability to read "backwards from what seems natural, obvious, self-evident, or universal in order to show that these things have their history, their reasons for being the way they are, their effects on what follows from them and that the starting point is not a (natural) given but a (cultural) construct, usually blind to itself".[31] Historians need not therefore simply offer up a past to elaborate a new present and, least of all, provide tales of continuity that serve once more, as Nietzsche warned, to help "the dead bury the living".[32] The object, rather, must be to submit a new scrutiny to what is taken for granted in the present—to wit, the current assumption in the social sciences that human action is largely driven by unconscious decisions and the unintended outcome of interactions with nonhuman actors. With Foucault's concept of history-writing as critique in mind (his "history of the present"), historians ought to inquire into the conditions of possibility for the emergence and unfolding of . . . views. Why, today, do Latour's ideas on human and nonhuman agency and the role of preconsciousness seem the "right" solution to the "right" problem? Why this specific regime of truth, this particular knowledge/power nexus?

For Foucault, it appeared that different epochs posed problems and found solutions to them, the problems and the solutions seeming inevitable and necessary, while others were overlooked or rejected. His "history of the present" called into question the self-proclaimed inevitability of any moral and social system, past or present. Impelled by the desire to comprehend the evaluative frameworks of any historical present, he asked about the multiple sources of power that make things seem the way they seem. History-writing only becomes effective, he thought, to the degree that "it introduces discontinuity into our very being . . . depriv[ing] the self of the reassuring stability of life and nature . . . [and]

uproot[ing] . . . traditional foundations [that] . . . relentlessly disrupt [history's] pretended continuity".[33] Discontinuities were to be savoured precisely in order to provide or expose *alternatives* to the always seemingly overwhelming present. Important for the current debate on "life" is that Foucault considered our current understanding of ourselves as the effects of historical processes of problem solving, processes which articulated relations of power as they identified objects of knowledge.

The question for the historian is not so much one of knowing what life *is*, but rather how it is that, today, the idea of "life" has come to occupy such prominence in contemporary intellectual culture, and how the current discourses on "life" come to fill it with meaning. Historians can also strive to make visible the premises upon which the organizing categories of the current debate on "life" are based, which would include, above all, illuminating the domains of knowledge/power more or less silently eclipsed, marginalized or erased in the process.

To do this fully would require connecting today's concern with "life" to the wider domains in which discussions over "life" are situated, not least the politics, economics and culture of neoliberalism. We have elsewhere attempted this and have no wish here to repeat ourselves.[34] Perhaps more germane is to observe how the current discourses on "life" among philosophers and social scientists threaten to compromise the very space required in order to pursue Foucault's history of the present. As evidence of this, we might cite the tendency among certain intellectuals to play down the importance of Foucault in favour of attending more to the ontological concerns of "life" of Foucault's one-time mentor, Georges Canguilhem.

Nikolas Rose, for example, in his *The Politics of Life Itself* (2007), seizes upon Canguilhem (with an aside on the over-familiarity of Foucault's history of the present) with the aim of reinforcing his faith in the creative force of "life itself" today.[35] Canguilhem, in his anti-mechanistic and anti-positivist historical investigation into vitalism, claimed that the normativity of biological life was essentially external to social life. Biological life cannot be grasped by logic, he argued in his various immediate post–World War II writings, but instead, follows a logic of its own—a logic that is ever-creative and transformative, as well as ever-inventing and destructive of its own norms. Although Canguilhem was aware of "life" in a social context, his works of the late 1940s and 1950s never sought to explain this in relation to biological life.[36] His emphasis, rather, was on the idea of life's meaning being created out of life's own vitality not, *out of* social and cultural norms, as Foucault would argue. "Life" is not an empty category but is filled by its ever-changing vitality, he thought. Today, in an intellectual world enamored of discussions of "life", where the idea of preconsciousness gains ground over human consciousness in decision making, and where "the social" in its traditional sense is held to be obsolete, it is not hard to see why Canguilhem's star shines while Foucault's fades. Nor is it hard to see how comfortably the idea of "life emerging out of life itself" sits with neoliberalism's celebration of growth, flexibility, ever-changing openness and the idea of individuals as ever making and remaking their own norms.

We have been concerned in this chapter with the role of history *in* and *for* contemporary discussions of "life". Historians, we have seen, have been much involved with these discussions. But while involved, they have failed to locate themselves within them, and hence have only legitimized, rather than opened the way for an informed critique of the discourses in question. They have either plied the conventional assumption that history looks only to the past, and not the present, or (through neurohistory, on the one hand, and histories of continuities between the past and the biological present, on the other) helped to eclipse the space for critical understanding of what is now one of the main enchantments of the present ("biological life"), exploring its blind spots and its politics.

The contribution of historians to contemporary debates on life and society, we submit, ought to be less to the question of what *is* life or how it can be assessed and defined, than to *how* the idea of "life" today has come to occupy such an overwhelming place in intellectual culture. Historical investigation can be an inquiry into both the grounds of our current knowledge about "life" and the analytical categories that establish its "inevitability"—that is, to understand that the current discussion is itself an *interpretation* of reality, not reality itself. Historians can contribute by identifying the sources of current values—how they came into being, the relationships they have constituted, the power they have secured and, most importantly, the actual knowledge/power they have eclipsed or are eclipsing.

BIBLIOGRAPHY

Baker, Robert M., and Laurence M. McCullough. *The Cambridge World History of Medical Ethics*. Cambridge: Cambridge University Press, 2009.

Clarke, Adele E., Laura Mamo, and Jennifer Ruth Fosket. "Biomedicalization: A Theoretical and Substantive Introduction". In *Biomedicalization: Technoscience, Health, and Illness in the U.S.*, edited by Adele E. Clarke, Laura Mamo and Jennifer Ruth Fosket, 1–44. Durham, NC: Duke University Press, 2010.

Callard, Felicity, and Constantina Papoulias. "Biology's Gift: Interrogating the Turn to Affect". *Body and Society* 16 (2010): 29–56.

Cooter, Roger, and Claudia Stein. *Writing History in the Age of Biomedicine*. New Haven, CT: Yale University Press, 2013.

Foucault, Michel. "Nietzsche, Genealogy, History". In *Language, Counter-Memory, Practice*, edited by Donald Bouchard. Oxford: Blackwell, 1977.

Johnson, Barbara. Translator's introduction to Jacques Derrida, *Dissemination*. Chicago: University of Chicago Press, 1981.

Joyce, Patrick. *State of Freedom: The Social History of the British State since 1800*. Cambridge: Cambridge University Press, 2013.

Latour, Bruno. *We Have Never Been Modern*. Translated by Catherine Porter. New York: Harvester Wheatsheaf, 1993.

———. *Eine neue Soziologie für eine neue Gesellschaft*. Frankfurt am Main: Suhrkamp, 2008.

———. *Pandora's Hope*. Cambridge, MA: Harvard University Press, 1999.

Leys, Ruth. "The Turn to Affect: A Critique". *Critical Inquiry* 37 (2011): 434–72.

Macherey, Pierre. "Für eine Naturgeschichte der Normen". In *Spiele der Wahrheit. Michel Foucaults Denken*, edited by François Ewald and Bernhard Waldenfels, 171–92. Frankfurt am Main: Suhrkamp, 1991.

McMaster, Tom, and David Wastell. "The Agency of Hybrids: Overcoming the Symmetrophobic Block". *Scandinavian Journal of Information Systems* 17 (2005): 175–82.

Muhle, Maria. *Eine Genealogie der Biopolitik: zum Begriff des Lebens bei Foucault und Canguilhem*. Bielefeld: Transcript, 2008.

Nietzsche, Friedrich. *On the Advantage and Disadvantage of History for Life*. Translated by Peter Preuss. Indianapolis: Hackett, 1980.

Papoulias, Constantina, and Felicity Callard. "Biology's Gift: Interrogating the Turn to Affect". *Body and Society* 16 (2010): 29–56.

Parson, Allan. "Human and Non-Human Agency". https://sites.google.com/site/praxisandtechne/Home/architeture/performativity/human-and-non-human-agency. Accessed 20 March 2014.

Plamper, Jan. *Geschichte und Gefühl: Grundlagen der Emotionsgeschichte*. München: Siedler, 2012.

Reddey, William M. "Neuroscience and the Fallacies of Functionalism". *History and Theory* 49 (2010): 412–25.

Ruffing, Reiner. *Bruno Latour*. Stuttgart: Fink, 2009.

Smail, Daniel Lord. "Daniel Lord Smail's faculty webpage". http://history.fas.harvard.edu/people/faculty/smail.php. Accessed 20 March 2014.

———. *On Deep History and the Brain*. Berkeley: University of California Press, 2008.

Smith, Roger. *Being Human: Historical Knowledge and the Creation of Human Nature*. Manchester: Manchester University Press, 2007.

Southgate, Beverley. "'Humani nil alienum': The Quest for 'Human Nature'". In *Manifestos for History*, edited by Keith Jenkins, Sue Morgan and Alun Munslow. London: Routledge, 2007.

Stafford, Barbara Maria. *Body Criticism: Imaging the Unseen in Enlightenment Art and Medicine*. Cambridge, MA: MIT Press, 1991.

———. *Echo Objects: The Cognitive Work of Images*. Chicago: University of Chicago Press, 2007.

Thrift, Nigel. "Pass It On: Towards a Political Economy of Propensity". *Emotions, Space and Society* 1 (2008): 83–91.

———. *Non-Representational Theory: Space, Politics, Affect*. London: Routledge, 2008.

NOTES

1. This chapter builds on previous work, particularly Roger Cooter and Claudia Stein, *Writing History in the Age of Biomedicine* (New Haven, CT: Yale University Press, 2013).

2. For the historical literature on bioethics, see *The Cambridge World History of Medical Ethics*, ed. Robert M. Baker and Laurence M. McCullough (Cambridge: Cambridge University Press, 2009); on being human, see, for example, Roger Smith, *Being Human: Historical Knowledge and the Creation of Human Nature* (Manchester: Manchester University Press, 2007).

3. For an excellent overview of both, see Jan Plamper, *Geschichte und Gefühl: Grundlagen der Emotionsgeschichte* (München: Siedler, 2012).

4. Adele Clarke et al., "Biomedicalization: A Theoretical and Substantive Introduction", in *Biomedicalization: Technoscience, Health, and Illness in the U.S.*, ed. Adele Clarke et al. (Durham, NC: Duke University Press, 2010), 1–44.

5. Barbara Maria Stafford, *Echo Objects: The Cognitive Work of Images* (Chicago: University of Chicago Press, 2007).

6. Barbara Maria Stafford, *Body Criticism: Imaging the Unseen in Enlightenment Art and Medicine* (Cambridge, MA: MIT Press, 1991).

7. Stafford, *Echo Objects: The Cognitive Work of Images*, 1.

8. Stafford, *Echo Objects*, 1, italics added.

9. Stafford, *Echo Objects*, 175–76.

10. Stafford, *Echo Objects*, 1–2, italics added.

11. See Melissa Littlefield and J. Johnson, eds., *The Neuroscientific Turn: Transdisciplinarity in the Age of the Brain* (Ann Arbor: University of Michigan Press, 2012).

12. Daniel Lord Smail, "Daniel Lord Smail's faculty webpage", http://history.fas.harvard.edu/people/faculty/smail.php (accessed 20 March 2014). See also Daniel Lord Smail, *On Deep History and the Brain* (Berkeley: University of California Press, 2008). For a critique, see William M. Reddey, "Neuroscience and the Fallacies of Functionalism", *History and Theory* 49 (2010): 412–25.

13. As summarized in Ruth Leys, "The Turn to Affect: A Critique", *Critical Inquiry* 37 (2011): 436.

14. Leys, "Turn to Affect", 441.

15. Constantina Papoulias and Felicity Callard, "Biology's Gift: Interrogating the Turn to Affect", *Body and Society* 16 (2010): 31.

16. Leys, "Turn to Affect", 441.

17. Nigel Thrift, *Non-Representational Theory: Space, Politics, Affect* (London: Routledge, 2008). See also Nigel Thrift, "Pass It On: Towards a Political Economy of Propensity", *Emotions, Space and Society* 1 (2008): 83–91.

18. Thrift, quoted in Papoulias and Callard, "Biology's Gift", 31.

19. Leys, "Turn to Affect", 443

20. Bruno Latour, *We Have Never Been Modern*, trans. Catherine Porter (New York: Harvester Wheatsheaf, 1993), 13.

21. Allan Parson, "Human and Non-Human Agency", https://sites.google.com/site/praxisandtechne/Home/architeture/performativity/human-and-non-human-agency (accessed 20 March 2014). Parson is here summarizing the work of Tom McMaster and David Wastell, "The Agency of Hybrids: Overcoming the Symmetrophobic Block", *Scandinavian Journal of Information Systems*, 17 (2005): 175–82. For these scholars, the main text is Bruno Latour's *Pandora's Hope* (Cambridge, MA: Harvard University Press, 1999).

22. McMaster and Wastell, "The Agency of Hybrids", 179.

23. Latour, *Eine neue Soziologie für eine neue Gesellschaft* (Frankfurt am Main: Suhrkamp, 2008), quoted in Reiner Ruffing, *Bruno Latour* (Stuttgart: Fink, 2009), 86.

24. Patrick Joyce, *State of Freedom: The Social History of the British State since 1800* (Cambridge, MA: Cambridge University Press, 2013).

25. Joyce, *State of Freedom*, 19.

26. Joyce, *State of Freedom*, 19–20.

27. Joyce, *State of Freedom*, 203.

28. Friedrich Nietzsche, *On the Advantage and Disadvantage of History for Life*, trans. Peter Preuss (Indianapolis: Hackett, 1980).

29. Beverley Southgate, "'Humani nil alienum': The Quest for 'Human Nature'", in *Manifestos for History*, ed. Keith Jenkins et al. (London: Routledge, 2007), 75, italics added.

30. Quoted in Joan W. Scott, "History-Writing as Critique", in *Manifestos for History*, 23.

31. Barbara Johnson, translator's introduction to Jacques Derrida, *Dissemination* (Chicago: University of Chicago Press, 1981), xv, quoted in Scott, "History-Writing as Critique", 24.

32. Nietzsche, *On the Advantage*, 8.

33. Michel Foucault, "Nietzsche, Genealogy, History", in *Language, Counter-Memory, Practice*, ed. Donald Bouchard (Oxford: Blackwell, 1977), 140.

34. "The New Poverty of Theory: Material Turns in a Latourian World", in *Writing History in the Age of Biomedicine*, ed. Cooter and Stein, chap. 10.

35. Quoted in Roger Cooter and Claudia Stein, "Cracking Biopower" in *Writing History in the Age of Biomedicine*, ed. Cooter and Stein, 193.

36. On Canguilhem see Maria Muhle, *Eine Genealogie der Biopolitik: zum Begriff des Lebens bei Foucault und Canguilhem* (Bielefeld: Transcript, 2008), 105–60. On the similarities between Foucault and Canguilhelm in regard to their understanding of normativity, see also Pierre Macherey, "Für eine Naturgeschichte der Normen", in *Spiele der Wahrheit. Michel Foucaults Denken*, ed. François Ewald and Bernhard Waldenfels (Frankfurt/Main: Suhrkamp, 1991), 171–92.

Part III

Pathology and Ageing

TEN

Life of Pain: Remarks about Negativity and Effort in Georges Canguilhem

Giuseppe Bianco

Life's miseries pain is deep.[1]

PAIN AND PATHOLOGY

In *The Normal and the Pathological*, Canguilhem provides a definition of health based on an insightful formulation he borrows from the surgeon René Leriche. Health would be—he writes—"life lived in the silence of the organs".[2] What Canguilhem means by this is that, when the human being isn't touched by disease, his body "functions" in harmony with the environment, so that he is incapable of perceiving its tacit presence. In contrast, commotion is the product of man's disease, or of his risking disease. The primary language of the body is thus the language of pain. It is within and through the experience of pain that one grasps the fact of possessing a body endowed with organs, whose behavior is susceptible to change. Canguilhem adds, "The state of health is a state of unawareness where the subject and his body are one. Conversely, the awareness of the body consists in a feeling of limits, threats, obstacles to health".[3] This allows Canguilhem to assert that a living being's self-awareness depends on pain and illness. Without undergoing painful ordeals, not only does man, like every other living being, fail to acknowledge his illness, but he also doesn't have any reason to assert a "normal" identity, which thus remains silent, implicit or rather nonexistent. Outside of the world's manifest *opposition* and *resistance*, as experienced through pain, life has no means of knowing itself. Pain and illness are thus the modalities through which the living has awareness of itself or, at least, through which it has awareness of what, after the fact, becomes health and normality as ideal states for it to restore.

The idea of health as silence of the organs, and of pain as the language of disease transforms both the idea one has of normality and the idea one has of medical knowledge and technique. From this negative definition of health, Canguilhem develops a theory of the organism as the power to produce new norms and a theory of medicine as technique involving the priority of the sick person. Without infirmity, the living would be incapable of deciding what health means for it. Health is thus posterior to disease and does not precede it ontologically; without the disease and its language, medicine cannot exist. Consequently, on the one hand, physiology is subsequent to pathology, since it is the morbid condition that decides on and evaluates what health is, and which requires therapeutic intervention; on the other hand, science is subsequent to technique, because knowledge is generated as a result of attempts at technical intervention on the living by the very living itself. In opposing Descartes, Canguilhem upholds that technical instruments aren't "solidified theorems".[4] It is not so much the case that technology is posterior to science, as mere application, but it is rather that science is the product of the technical "creative impulse" which is, in itself, a prolongation of life. The failures of life's creative activity generate scientific knowledge. Just as in the face of an obstacle, a living being feels pain and recognizes through reflection that it finds itself in a morbid condition, since it thinks that its own power is insignificant; in the same way, faced with an obstacle, a technical creation can fail to function and call for an understanding of the reasons for its failure, thus generating knowledge. Canguilhem's doctrine articulates two inseparable tenets. The first tenet consists in a theory of the living as axiological polarity: the living is this being which incessantly evaluates that which is good or bad for itself. The second tenet consists in a theory of knowledge privileging technique over science: the technique is an extension of the living being's creative activity and science is nothing but the reflexive product of technical breakdown.

These ideas have met with considerable success. On the one hand, in therapeutics, they profoundly transformed the idea of the type of relation existing between caregiver and patient; on the other hand, in political philosophy, they have infiltrated the debate on "social suffering". The goal of these pages is not to decide on the topicality of the conception of pain proposed by Canguilhem, but rather to search for its origin. I will follow the conceptual trace according to which the living lacks awareness of itself except in the event of pain—namely, when it experiences a contradiction and an exertion upon something that resists it. This aspect led several commentators to argue that there is, in Canguilhem's work, a prioritization of negative values—namely, of displeasure, of pain, of disease. It is as if it were pain, the perceived obstacle, the danger of illness—that is, *negativity*—which determined the identity of the living subject. From this point of view, health is a positive value engendered by the living after the fact.

I have already emphasized how Canguilhem borrows the definition of health as "silence of the organs" from René Leriche (1879–1955), a professor at the Collège de France who occupied from 1937 onwards a position in clinical medicine, previously held by Charles Nicolle. Leriche, a surgeon, is one of the representatives of humanist medicine who aspired to reinstate the relationship between patient and sufferer in medicine. Yet everything leads one to think Leriche wasn't the first author to have focused Canguilhem's attention on negative values. In the second edition of *The Normal and the Pathological*, Canguilhem emphasizes how the conceptions of Leriche and Kurt Goldstein, far from being incompatible with one another, are actually complementary. However, as with Goldstein, who informs all thinking on life comprised in *The Normal and the Pathological* (but whom Canguilhem didn't actually read until late),[5] it is probable that his previous researches had also already prepared his encounter with the theory of Leriche.

CANGUILHEM AND ALAIN

Leriche's theory occupies an ambiguous place in the construction of *The Normal and the Pathological.* On the one hand, Canguilhem praises Leriche's conception of pain and of pathology, conceived as a primary fact, just as much as his theory of medicine, conceived as a technological act. On the other hand, he criticizes Leriche's implicit conception of a merely quantitative difference between the pathological and the normal condition. The chapter on Leriche is actually located at the end of a brief history of the conceptualizations of the relation between the normal and the pathological, after two chapters devoted to the conceptions of Auguste Comte and Claude Bernard. Canguilhem tracks down, as much in the work of these last two thinkers as in the work of René Leriche, a common idea concerning the living, what he calls, following Comte, "Broussais' principle". Doctor François Joseph Victor Broussais (1772–1838) had advanced, in his *De l'irritation et de la folie* (1836), the argument for the identity between the normal and the pathological, in accordance with quantitative variations. In Broussais's vision, the sick man is not different from the normal man; he is only a "less" normal man. Diseases are the result of irritations caused by aggressions coming from outside the body. These irritations mess up the functioning of the body, conceived on the model of a machine subjected to the law of reflex action. When an irritation is especially serious, the organism falls prey to sickness. This theory, which implies the priority of the "normal" for the determination of the pathological, and the priority of science—physiology—over its technical applications—pathology—had been rendered famous by Comte, who had highlighted its importance for medicine as well as sociology. Comte had started from an analogy between the physical and the social body in order to arrive at the notion of there being only graduated differences between peace, a normal condition of society, and wars and revolutions, its pathological states. Consequently, just as in the case of a disease resulting from an irritation, which disturbs the normal state, man cannot change the normal social order by means of revolutions, which would alter its essential configuration. All man can do is treat these irritations. After Comte, this theory had been picked up by Claude Bernard, to whom Canguilhem devotes a chapter in *The Normal and the Pathological*, but also by the discipline of psychology, in particular by Théodule Ribot (1839–1916), and by the social sciences, especially by Emile Durkheim (1858–1917) and Lucien Lévy-Bruhl (1857–1939). The two sociologists deployed organicist metaphors—thus speaking of normal and of "sick" societies—and conceived politics as a therapeutic intervention able to profit from sociological knowledge.

Ribot, Durkheim and Lévy-Bruhl were not the only ones who appropriated Broussais's and Comte's conceptions. As I recently argued,[6] despite his abhorrence of psychologists and sociologists, the Kantian philosopher Emile Chartier, better known as Alain (1888–1951), had also thought diseases and passions to be the result of irritations. These irritations cannot modify the structure of the organism, but they can alter its manners of performance, by lowering them below "normal" parameters. According to Alain, in order to heal, or in order to free oneself from passions, it is enough to reason, to think wisely, to distract the irritated part of the body from its irritation. The power of reason, writes Alain, concentrates "in a firm judgment, against death, against disease, against a dream, against a deception".[7] On the basis of the very same philosophy of judgement, Alain became a firm defender of pacifism. According to Alain, the philosopher, just like the politician, the society's doctor, can abruptly change neither the individual nor society as a whole, but has to treat irritations by distracting the patient's attention from the irritated part. Canguilhem's DES dissertation, written by a young Canguilhem in 1926, dealt with *The Theory of Order and Progress in the Work of Auguste Comte* [*La théorie de l'ordre et du progress dans l'œuvre d'Auguste Comte*].[8] The topic, chosen under the influence of Alain, was supervised by his friend, the philosopher and sociologist Célestin Bouglé (1870–1940). The study addresses the theory of order and progress, but underlines the importance of applying Broussais's principle to the development of society.

In the middle of the 1930s, and against the backdrop of rising fascism, Alain's thought appears, to some of his students, incapable of accounting for a rapidly changing world. According to Alain, fascism is an authoritarian regime like any other, consisting of an abuse of power, and in order to fight it one has to invalidate all recourse to violence, since one must never seek to "overthrow political powers", but only try to "deflate" them. Once he recognized the fact that it was impossible to subdue Hitler, once he realized that one could no longer accept the doctrine of full pacifism elaborated by Alain during World War I, Canguilhem devoted himself to an attempt at updating the philosophy of his teacher, which will wind up in a reform and an implicit critique. In 1936 Canguilhem joined the Faculty of Medicine, looking forward to finding in this discipline "an introduction to concrete human problems".[9] In 1935, Canguilhem's "problem" was, without any trace of doubt, the problem of achieving effective political action, which should have thus followed the Comtean motto: "to know is to predict and to predict is to control". If "in order to act, it is necessary at least to localize",[10] then medicine is a "technique", an operative knowledge that targets action. Actually, in *The Normal and the Pathological*, the monograph on medicine that Canguilhem published in 1943, the stated goal is to "to integrate some of the methods and attainments of medicine into philosophical speculation".[11] On the one hand, in *The Normal and the Pathological*, Canguilhem problematizes the foundations of Alain's anthropology, which included a mechanist conception of the biological body, a quantitative conceptualization of the rapport between pathological and normal states, and the parallelism between the organic and the social body. On the other hand, Canguilhem elaborates a theory of the organism as normative power, and a theory of technique as a creation.

NORMS AND TECHNIQUES

Canguilhem had precisely written, in the first pages of the "Introduction" to *The Normal and the Pathological*, that he expected medicine to provide "an introduction to concrete human problems". These problems were those of the relations between "science and technology" and that of "norms and the normal".[12] Before Canguilhem, Alain had solved these problems by grounding his project in the interpretation of Kantianism offered by Jules Lachelier (1832–1918) in *Le fondement de l'induction* (1871), later perfected by Jules Lagneau (1851–1894), Alain's teacher at the Vanves high school. According to Alain, the mind is the center of emanation as much for judgements of fact as for judgements of value. Technology is thus posterior to science, it is the mere application of judgements of fact emitted by human beings. If science fails in its technological applications, it means that man has erred in his judgement. The absolute priority of the transcendental subject is the main reason why both Alain and Canguilhem distrust the attempts to reduce the human being to a matter of fact.

Mind, human being's characteristic trait, is, on the contrary, the one which constitutes facticity through its power of judgement. But in order to reason well, Alain claims, it is necessary to will reasoning. Without the will, no reasoning nor any action is possible.

As much as Canguilhem, Alain only felt contempt for the solutions provided by sociology to the problem of the rapport between science and technology, and to the problem of values. Emile Durkheim, in his famous 1911 essay, "Judgments de valeur et jugement de fait" (included in the volume that Célestin Bouglé published in 1925, *Philosophie et sociologie*[13]), had tried to reduce judgements of value to judgements of fact. According to Durkheim, value consists in a relation between an object and an ideal, so it is in fact reducible to factual judgement. The value produced by a society must be preserved, since it guarantees the social bond protecting society from anarchism, and so, from all pathology. Thus, in 1903, in *Philosophie morale et science des mœurs*, a work written in the wake of Durkheim's own research, Lévy-Bruhl defines sociology as an *a posteriori* science of facts of civilization, and detaches the sociology of morals from moral philosophy. In the preface to the book's third edition, from 1927, Lévy-Bruhl explains, under the inspiration of Durkheim, that within judgements of value it is objects which are taken into account, but in relation to our instincts, desires and "natural" sympathies.

Obviously, this conception of value was inadmissible for a student of Alain, who grounded values in the free activity of the subject, independently of matters of fact. Speaking of his philosophical formation in the 1930s, Canguilhem emphasizes this incompatibility between the Durkheimian and the Kantian conception of value in an article published in the 1980s:

> The philosophers of truth [the sociologists] lent them [to judgments of value] the appearance of necessity. It is by means of Durkheimian sociology that the judgment of value has become, in France, a banal philosophical question. By defining value as the one ideal that reason aims at fulfilling, and by introducing society as the origin of that ideal, Durkheim unsettled all the rationalists attached to the transcendental independence of the rules of reasoning.[14]

Canguilhem's positions were only compatible with those of Célestin Bouglé, "ambivalent Durkheimian",[15] familiar with Alain and with the *Revue de métaphysique et de morale*. Ever since his work on *Les idées égalitaires* (1899), Bouglé, while defending the importance of sociology, had highlighted the intentional character of human phenomena, the importance of judgements of value and their irreducibility to factual judgements. In 1905, in the volume *La démocratie devant la science*, he had opposed all attempts to naturalize social phenomena, and had criticized the application of the concepts of health and pathology to society. In 1922, Bouglé grouped a series of lectures in a volume entitled *Leçons de sociolo-*

gie sur l'évolution des valeurs.[16] Shortly thereafter, Canguilhem presented a short lecture on Marx's conceptualization of value in Bouglé's course.

LE SENNE: PAIN AND VALUES

In 1934, Canguilhem wrote a brief review of the *Essai sur la constitution de la pensée grecque*, which was Pierre-Maxime Schuhl's doctoral thesis, inspired by Lévy-Bruhl and Durkheim.[17] In his work, Schuhl poses exactly the question of the nexus tying together society, science and technology. Schuhl's approach, which aims at rooting technology in the evolution of the economical and social structure of a specific human group, risks—if pushed too far—reducing culture and values to mere sociocultural determinants. Canguilhem then points to a different possibility of rearranging the relations between science, technique and society: according to the renewed conception, the techniques need to be considered as veritable *value-choices* in themselves. To that effect, Canguilhem formulates the problem not on the basis of a sociology of techniques, but on a "systematic study of values". This suggestion seems to refer to the work of René Le Senne (1882–1954), who was the author of *Le devoir* (1930)[18] and *Obstacle et valeur* (1934).[19]

Canguilhem had reviewed the first of the two books in 1933.[20] *Le devoir* was Le Senne's main dissertation, supervised by Frédéric Rauh (1861–1909), but largely influenced by the idealism of Octave Hamelin (1856–1907). In his thesis, the philosopher takes to task the reductionist character of sociology. For instance, after a long and detailed discussion of the event of the discovery of argon by Rayleigh and Ramsay, Le Senne critizes Durkheim, emphasizing that sociological conditions are indeed necessary but not sufficient to explain scientific discoveries. Even behind the production of scientific knowledge, there lurks a series of judgements of value which, as such, depend on the free choice of the subject. Thus, Le Senne concludes, "to reduce morality to a science would be to suppose that the means are imposed upon us as conditions, which would dispense with having to actively will them".[21] Canguilhem was particularly struck by the significance that Le Senne grants to human will in all dimensions of human life, including in simple factual judgements, the grounds of scientific truth. In his discussion, Canguilhem writes that Le Senne's treatment of the discovery of argon constitutes "a masterpiece of philosophical analysis, as much through the rigor of the interpretation, as through the art of vividly rendering the lived drama of the philosopher's consciousness", and thus "these pages are worthy of becoming classics of the literature: students of philosophy can find there other ways of reasoning than those contained in the *Introduction to the study of experimental medicine*"[22] by Claude Bernard.

While Canguilhem condemns Durkheim for having attempted to reduce judgements of value to factual judgements, he praises Le Senne for having attempted to reduce judgements of facts to judgements of value. The philosopher had made an original attempt—namely, transposing in the language of values the concepts that the critique of knowledge had always considered as logical necessities. In *Le devoir*, Canguilhem writes, Le Senne considers knowledge to be "a duty like many others", and categories to be "less . . . forms of intellectual constraint than guidelines"; consequently, the table of categories is considered by Le Senne to be a kind of "axiology", a kind of "Decalogue".[23]

But where do the values embodied in judgments come from? Le Senne starts from the idea of a consciousness ontologically marked by negativity: without "an uneasiness", without "an unsatisfied need", "the affirmation" of consciousness is impossible. More precisely, Le Senne sees as the principle of all life functions the consciousness of *pain*. At the origin of "all the geneses" of the functions of consciousness "lies a suffering".[24] Le Senne condenses the fact of the priority of pain in a unique variation on the Cartesian "I think", according to which "I suffer, therefore I am": "to suffer—he explains—is first of all to know that one is suffering, and simultaneously, in the Cartesian manner, to discover one's own existence in that knowledge".[25] The pain Le Senne discusses isn't a purely mechanical phenomenon; it is, even if Le Senne doesn't use the exact word, *existential*—it engages the whole of consciousness and cannot be avoided. Pain is a "fact of consciousness", and it is directly related to the volitional aspects of consciousness.

> Exactly as it happens with perception in the case of conditioned reflex, affective pain tends toward automatization in the pain which one can name organic, since it is associated with the essential functions of the organism, such as hunger and thirst. But by remaining a form of pain, it remains psychological. . . . The impoverishment of the organism in water or food supplies is one fact, the pain of hunger and of thirst another; and just as the essential aspect of knowledge is the irreducibility of any translation to its object, hunger and thirst do not reveal a condition of the organism otherwise than by altering it quantitatively and qualitatively.[26]

When, faced with the "novelty of the environment", neither instinct nor habituation are sufficient any longer, the human being becomes aware, through pain, of her own "inadaptation". This awareness, writes Le Senne, isn't only "reparative" but also "synthetic and creative", and it allows for "the anticipation of failure".[27] However, without failure, without pain, without crisis, without contradiction, the subject wouldn't reason—she wouldn't produce any value. Le Senne retrieves the contradiction within the painful nature of life, which unveils itself even in the perception by means of which "we simultaneously and correlatively recognize both our impotence to perceive and our willingness to perceive".[28] Thus Le Senne concludes with the following, decisive remark: "the sense of the real is the experience of conflict".[29]

Such statements could have only impressed the young radical philosopher, Canguilhem. Despite the discrepancies between the philosophies of value of Alain, an atheist, and of Le Senne, whose philosophy resulted in the conceptualization of a transcendental absolute, Canguilhem looked favourably on a voluntarist philosophy according to which "contradiction [is] the originary and definitive fact of consciousness".[30] Canguilhem ultimately concluded his report of 1934 with a sentence that could have found its place in *The Normal and the Pathological* or in *Knowledge of Life*:

> Just as the laws of resistance become an object of interest only because I clash with my environment, so the laws of life interest me for the sole reason that I must die, and so the laws of society interest me only because disorder and injustice disgust me.[31]

EFFORT: FROM MAINE DE BIRAN TO CANGUILHEM

The reader of Canguilhem the epistemologist and historian of science will inevitably be surprised by Canguilhem's admiration for *Le devoir* by Le Senne, who was a Catholic. Le Senne cofounded with Louis Lavelle the series "Philosophie de l'Esprit", and authored in 1942 a *Traité de morale générale* and, in 1946, a *Traité de characteriologie*. In a posthumously published biographical note, Le Senne writes that his philosophical project implied "a radical opposition to Sartre and to Marxism".[32] Even if, from the 1940s onward, his name vanishes from Canguilhem's texts, René Le Senne was one of the contemporary philosophers for whom the young Canguilhem showed the greatest respect in the 1930s.[33] A professor in Chambéry (1910–1914) and Marseille (1914–1923), he later taught at Louis Le Grand high school between 1926 and 1929, and then at the University of Nancy, before becoming professor in moral philosophy at the Sorbonne in 1942. In 1927 Le Senne wrote a positive report on Canguilhem's training as a teacher, which allowed him to obtain his "*agrégation*".[34]

Le Senne, who during the 1930s wasn't yet known as a Catholic writer, as a moral philosopher or as the main French representative of characterology, was probably very close to the Alainists. During the 1930s, Alain wrote to Elie Halévy that he had been in touch with Le Senne, "one of our good mentors",[35] for some time. The author of *Le devoir* influenced the thought of another Alainist, Simone Weil, who was his pupil at the Victor Duruy lycée before meeting Alain at Henri IV.[36] Canguilhem's eulogistic review was published in *Méthode*, a journal in which students close to Alain, such as Raymond Aron and Jean Hyppolite, had published articles. *Méthode*, which was aimed at professors of philosophy within the secondary education system, had been created by the Alainist Georges

Bénézé (1888–1978), Hyppolite's khâgne professor at Poitiers, and author in 1936 of two theses extremely influenced by Lachelier and Hamelin: *Allure du transcendantal*[37] and *Valeur*.[38] It is in the aftermath of the publication of his two works—the content of which was perfectly aligned with Le Senne's[39] treatment of values—that Bénézé began a short epistolary correspondence with the latter.[40] At exactly the same time, Canguilhem wrote to Le Senne[41] in order to help his colleague and friend Camille Planet. The main dissertation Planet was on the verge of completing, *Valeur et réalité*, hadn't been well received by his supervisor André Lalande (1867–1963), author in 1928 of a *Psychologie des jugements de valeur*.[42] Lalande had objected to Planet's thesis, which was essentially influenced by German works in the philosophy of values (*Wertphilosophie*), by Lachelier and by Le Senne, on the grounds that it was a rather abstract construction lacking documentation. On the contrary, Le Senne thought highly of Planet's thesis, which he had read. In fact, Planet had attempted a project analogous to his senior's earlier enterprise—namely, that of providing an axiological interpretation of experience, of grounding factual judgements in value judgements.

Le Senne's thought, which rendered the theory of knowledge dependent on moral philosophy, belonged to the same family of doctrines as the philosophy of Alain (who believed that in order to think, it is necessary to will to think, and who conceived morality as nothing other than philosophical consciousness) or as the fact of "knowing oneself to be mind and, as such, absolutely obliged".[43] Canguilhem had read and appreciated Octave Hamelin, from whom, like Le Senne, he had borrowed the expression "functions of the mind [*fonctions de l'esprit*]" as a replacement for the Kantian "categories of thought". Le Senne mentions Alain in *Obstacle et valeur*, and in *Le devoir* he cites and praises Jules Lagneau, who had been Alain's teacher and whom Canguilhem adored. He mentions Lagneau when discussing the problem of the relation between effort and representation. According to Le Senne, Lagneau had realized two very important aspects: "on the one hand, contradiction is at the heart of effort; on the other hand, effort encompasses a rational demand to exist".[44] Now these aspects evince from Lagneau's work the decisive influence of the philosophy of effort of Maine de Biran. In fact, Le Senne writes that it is impossible not to "assimilate him to Biran". Lagneau rehearses Maine de Biran's idea, according to which the subject grasps "his activity in correlation with the resistance of a nonself [*non-moi*]".[45] It is this very resistance, "the pressure of social and organic contradictions",[46] which will allow him to discover not just the "hyperorganic" force of the Ego but also the supreme value—God. Thus the "glory of Maine de Biran" consisted in "having recognized that the Ego could only be located in effort; and to be localized in effort meant to be localized in contradiction".[47]

As recent studies have shown,[48] the Biranian concept of effort is one of the centerpieces of the interpretation of Kant as formulated by Lachelier in *Du fondement de l'induction* and bequeathed to Lagneau, Hamelin, Alain and finally to Canguilhem. This reading identifies the *cogito*, interpreted in light of the concept of effort, with the Kantian definition of mind as "originary synthetic unity of apperception", resulting in a notion of subjectivity as an intentional activity of synthesis of a resisting experiential manifold. Of course, Canguilhem doesn't miss the opportunity to criticize Maine de Biran in the course of his famous lecture "What Is Psychology?"[49] He considers Maine de Biran to be one of the founders of modern psychology. But for Biran psychology is not a science, but a mere "adaptive technique" meant to normalize man. Nonetheless, there is a long tradition, inspired by Maine de Biran, which runs from Lachelier to the young Canguilhem. According to this tradition, "consciousness requires the conflict between a force and a resistance".[50] The human being has no self-consciousness outside the confrontation with an obstacle. For all these thinkers, at the origin of the consciousness of the living lies effort, negativity, pain.

In the *Traité sur la decomposition de la pensée* (1804), Biran forges the concept of hyper-organic force: this force, which reveals itself through effort, is the sign of volition. This concept was inspired by the doctor Marie François Xavier Bichat (1771–1802), who distinguishes between two kinds of nervous system, two forms of sensibility, two types of life. For Bichat, an *organic* sensation is a sensation whose designated organ is simultaneously its place of origin and its terminus, whereas in *animal* sensation the origin and the terminal point are distinct. Biran shifts this division in order to produce his own definition of the hyper-organic force as separated from what he calls "the organic unconscious".[51] This conception will serve as the ground for the reception of Kant, among whose inheritors one will count the young Canguilhem. At the end of the 1930s, at the time when Canguilhem writes his medical thesis, a curious reversal takes place. Through a gesture seemingly implying a return to Bichat—who, in his *Physiological Researches on Life and Death* [*Recherches physiologiques sur la vie et la mort*] (1801), had defined life as resistance, as "the ensemble of functions resisting death"—the philosopher repositions at the heart of the living everything that Alain and Le Senne had placed in the transcendental subject. It's now the living, which is the source of the will, norms, values and judgements. But this turn only became possible after the encounter with Kurt Goldstein, with *Wertphilosophie* and, more broadly, with the "German discourses"[52] on life.

BIBLIOGRAPHY

Alain, Élie Halévy, and Florence Halévy. *Correspondance avec Élie et Florence Halévy*. Paris: Gallimard, 1958.

Alain. *Quatre-vingt un chapitres sur l'Esprit et les passions*. Paris, Alcan, 1921.

Alain. *Lettres à Sergio Solmi sur la philosophie de Kant*. Paris: Paul Hartmann, 1946.

Azouvi, François. *Maine de Biran: La science de l'homme*. Paris: Vrin, 1997.

Bénézé, Georges. *Allures du transcendantal*. Paris: Vrin, 1936.

———. *Valeur. Essai d'une théorie générale*. Paris: Vrin, 1936.

Bianco, Giuseppe. "The Origins of Georges Canguilhem's 'Vitalism': Against the Anthropology of Irritation". In *Vitalism and the Scientific Image in Post-Enlightenment Life Science, 1800–2010*, edited by Charles T. Wolfe and Sébastien Normandin, 243–67. London: Springer, 2013.

Black Flag. *Damaged*. Los Angeles: SST Records, 1982.

Bouglé, Céléstin. *Leçons de sociologie sur l'évolution des valeurs*. Paris: Alcan, 1922.

Canguilhem, Georges. *The Normal and the Pathological*. Translated by Carolyn R. Fawcett and Robert S. Cohen. New York: Zone Books, 1998.

———. "La problématique de la philosophie de l'histoire au début des années 30". In *Raymond Aron, la philosophie de l'histoire et les sciences*, edited by Alain Boyer, Jean-Claude Chamboredon and Philippe Raynaud. Paris: Presses de la rue d'Ulm, 2005.

———. "What Is Psychology?" Translated by Howard Davies. *Ideology and Consciousness* 7 (1980): 37–50.

———. *Œuvres complètes*. Paris: Vrin, 2012.

———. *The Knowledge of Life*. Translated by Stefanos Geroulanos and Daniela Ginsburg. New York: Fordham University Press 1992.

Durkheim, Emile. *Philosophie et sociologie*. Paris: Presses universitaires de France, 2014.

Kuhn, Rolf. "Dimensions et logique interne de la pensée de Simone Weil". In *Simone Weil: philosophe, historienne et mystique*, edited by Gilbert Kahn, 331–71. Paris: Aubier-Montaigne, 1978.

Lalande, André. *Psychologie des jugements de valeur*. Paris: Imprimerie nationale, 1928.

Le Senne, René. "Note bio-biliographique de René Le Senne (1882–1954)". *Les études philosophiques* 10 (1955): 361–63.

———. *Le Devoir*. Paris: Alcan, 1930.

———. *Obstacle et valeur*. Paris: Alcan, 1934.

Lefève, Céline. "La lecture épistémologique de la psychologie de Maine de Biran par Georges Canguilhem". In *L'envers de la raison. Broussais, Canguilhem, Foucault*, edited by Pierre F. Daled, 35–52. Paris: Vrin, 2008.

Roth, Xavier. *Georges Canguilhem et l'unité de l'expérience. Juger et agir, 1926–1939*. Paris: Vrin, 2013.

Schmaus, Warren. "Kant's Reception in France: Theories of the Categories in Academic Philosophy, Psychology, and Social Science". *Perspectives on Science* 11 (2003): 3–34.

Schmidgen, Henning. "Georges Canguilhem et 'les discours allemands'". In *Philosophie et médecine en hommage à Georges Canguilhem*, edited by Anne Fagot-Largeault. Paris: Vrin, 2008.

Vogt, Pierre. "Un durkheimien ambivalent: Célestin Bouglé 1870–1940". *Revue française de sociologie* 20 (1979): 123–39.

NOTES

1. "Life of Pain", in Black Flag, *Damaged* (Los Angeles: SST Records, 1982).
2. Georges Canguilhem, *The Normal and the Pathological*, trans. Carolyn R. Fawcett and Robert S. Cohen (New York: Zone Books, 1998), 91.
3. Canguilhem, *The Normal and the Pathological*, 91.

4. Georges Canguilhem, "Machine and Organism", in *The Knowledge of Life*, trans. Stefanos Geroulanos and Daniela Ginsburg (New York: Fordham University Press, 1992), 76.

5. "We have tackled not only the exposition but also the reading of Goldstein last" (Canguilhem, *The Normal and the Pathological*, 191).

6. G. Bianco, "The Origins of Georges Canguilhem's 'Vitalism': Against the Anthropology of Irritation", in *Vitalism and the Scientific Image in Post-Enlightenment Life Science, 1800–2010*, ed. Charles T. Wolfe and Sébastien Normandin (London: Springer, 2013), 243–67.

7. Alain, *Quatre-vingt un chapitres sur l'Esprit et les passions* (Paris: Alcan, 1921), 9.

8. The DES, acronym for "diplome d'études supérieures" is the dissertation that French students were meant to write at the end of their third year at university.

9. Canguilhem, *The Normal and the Pathological*, 33–34.

10. Canguilhem, *The Normal and the Pathological*, 39.

11. Canguilhem, *The Normal and the Pathological*, 34.

12. Canguilhem, *The Normal and the Pathological*, 34.

13. Emile Durkheim, *Philosophie et sociologie* (Paris: Presses universitaires de France, 2014).

14. Georges Canguilhem, "La problématique de la philosophie de l'histoire au début des années 30", in *Raymond Aron, la philosophie de l'histoire et les sciences*, ed. Alain Boyer, Jean-Claude Chamboredon and Philippe Raynaud (Paris: Presses de la rue d'Ulm, 2005), 25.

15. P. Vogt, "Un durkheimien ambivalent: Célestin Bouglé 1870–1940", *Revue française de sociologie* 20 (1979): 123–39.

16. Céléstin Bouglé, *Leçons de sociologie sur l'évolution des valeurs* (Paris: Alcan, 1922).

17. Georges Canguilhem, "P.-M. Schuhl, *Essai sur la formation de la pensée grecque*" (1936), in *Œuvres complètes* (Paris: Vrin, 2012), 485–87.

18. René Le Senne, *Le Devoir* (Paris: Alcan, 1930).

19. René Le Senne, *Obstacle et valeur* (Paris: Alcan, 1934).

20. Georges Canguilhem, "R. Le Senne, *Le devoir*" (1933), in *Œuvres complètes*, 441–44.

21. Le Senne, *Le Devoir*, 379.

22. Canguilhem, "R. Le Senne, *Le devoir*", 444.

23. Canguilhem, "R. Le Senne, *Le devoir*", 442.

24. Le Senne, *Le devoir*, 379.

25. Le Senne, *Le devoir*, 379.

26. Le Senne, *Le devoir*, 59.

27. Le Senne, *Le devoir*, 25.

28. Le Senne, *Le devoir*, 26.

29. Le Senne, *Le devoir*, 27.

30. Canguilhem, "R. Le Senne, *Le devoir*", 443.

31. Canguilhem, "R. Le Senne, *Le devoir*", 443.

32. R. Le Senne, "Note bio-biliographique de René Le Senne (1882–1954)", *Les études philosophiques* 10 (1955): 361–63.

33. Le Senne is quoted by Canguilhem in the 1938 lecture "Technical activity and creation" and in the *Traité de logique et de morale* written with Camille Planet and published the following year (republished in Canguilhem *Œuvres complètes*).

34. Cf. Xavier Roth, *Georges Canguilhem et l'unité de l'expérience. Juger et agir, 1926–1939* (Paris: Vrin, 2013).

35. Alain, Élie Halévy, Florence Halévy, *Correspondance avec Élie et Florence Halévy* (Paris: Gallimard, 1958), 410.

36. On the influence of Le Senne on Weil, especially with regard to the her formulation of her philosophy of values, cf. Rolf Kuhn, "Dimensions et logique interne de la pensée de Simone Weil", in *Simone Weil: philosophe, historienne et mystique*, ed. Gilbert Kahn (Paris: Aubier-Montaigne, 1978), 331–71.

37. Georges Bénézé, *Allures du transcendantal* (Paris: Vrin, 1936). The curious title of the book could be at the origin of the mysterious expression "*allures de vie*" that Canguilhem uses in *The Normal and the Pathological* when designating the forms taken by the behaviour of a living being. As Alain will explain later, in one of his letters to Sergio Solmi where he cites Bénézé, the "allure du transcendental" is meant to signify the different modalities through which the Ego unifies a plethora of experiences, by subsuming them under the scheme. Cf. Alain, *Lettres à Sergio Solmi sur la philosophie de Kant* (Paris: Paul Hartmann, 1946), 37. The "*allures de vie*" would thus refer to the different ways in which life is unified by the living.

38. Georges Bénézé, *Valeur. Essai d'une théorie générale* (Paris: Vrin, 1936).

39. In the first sentence from *La valeur*, Bénézé explicitly writes that consciousness "is indispensable for sustaining as much the judgment of value as the judgment of fact". Cf. Georges Bénézé *La valeur* (Paris: Vrin, 1936), 1.

40. Cf. Fonds René Le Senne, Manuscrits de la bibliothèque Victor-Cousin, MSVC, 377–84.

41. Cf. Roth, *L'unité de l'expérience*.

42. André Lalande, *Psychologie des jugements de valeur* (Paris: Imprimerie nationale, 1928).

43. Cf. Alain, *Lettres à Sergio Solmi*, 34.

44. Le Senne, *Le devoir*, 256.

45. Le Senne, *Le devoir*, 256.

46. Le Senne, *Le devoir*, 257.

47. Le Senne, *Le devoir*, 254.

48. W. Schmaus, "Kant's Reception in France: Theories of the Categories in Academic Philosophy, Psychology, and Social Science", *Perspectives on Science* 11 (2003): 3–34; Roth, *Georges Canguilhem et l'unité de l'expérience*.

49. Georges Canguilhem, "What Is Psychology?", trans. Howard Davies, in *Ideology and Consciousness* 7 (1980): 37–50. For an analysis of the place Biran occupies in Canguilhem's appraisal of psychology in 1956, see Céline Lefève, "La lecture épistémologique de la psychologie de Maine de Biran par Georges Canguilhem", in *L'envers de la raison. Broussais, Canguilhem, Foucault*, ed. Pierre F. Daled (Paris: Vrin, 2008), 35–52.

50. Canguilhem, "What Is Psychology?", 46.

51. For the relations between Biran and Bichat, see François Azouvi, *Main de Biran: La science de l'homme* (Paris: Vrin, 1997).

52. Cf. Henning Schmidgen, "Georges Canguilhem et 'les discours allemands'", in *Philosophie et médecine en hommage à Georges Canguilhem*, ed. Anne Fagot-Largeault (Paris: Vrin, 2008), 49–62.

ELEVEN

Human Life and Subjectivity: Learning from Foucault

Piergiorgio Donatelli

In this chapter, I will deal with problems tied to the notion of human life, especially to procreation, sexuality and dying; they are also related to the concerns of bioethics, but it will also be clear that the way I treat them is a long way from what is commonly conceived as the agenda of bioethics. I will not discuss any such problems in detail: instead, I will suggest a general scheme of how to read such problems and how to place them within a larger scheme which concerns the problems we have more generally with human life.

I want to suggest the idea that we largely lack the space of ethical experience and thought related to human life. This is a crucial fact of our moral culture and is itself the result of the kind of clash which has characterized modernity, the clash between modernity and tradition.[1] I will give a very sketchy description of this clash. As I see it, we have on the one side the traditional conception of human life, which held the idea that there is only one way of having a space for ethical life, a space made up of three main components: (1) a teleological anthropology; (2) a social order structured around intrinsic distinctions, ranks and hierarchies; and (3) a divine law conception of ethics.

The defenders of modernity have fought a battle in order to liberate themselves from this framework, from this articulated background of human action. The way the battle was fought, and the way it was represented in their eyes, was along the following lines: What was at stake was to sweep away entirely the traditional background, and with it to reject the very idea that there is a dense background which makes sense of human action, consisting of things like a particular notion of human life, of social meanings and attitudes attached to human experiences such as birth and death, of perceptions of the body and of a whole fabric of attitudes and things conceived as appropriate. The latter may include the shared solemnity and mysteriousness of death, the reverence for pregnancy, the perception of women's body as a special and "other" (to the male as the "standard") kind of being and, along with these, virtues like chastity, the absolute prohibition of suicide and many other things.

So the idea was to sweep all that away and to bring these areas of life under new liberal and democratic rules, such as respect for individual freedom and the promotion of individual and social well-being. This latter picture involves (1) the falling away of the old background, tied to these three components, and especially the fall of a divine law conception of ethics with the Reformation; (2) the rise of science and technology, which give a description of human life that removes from nature, and from human nature in particular, both teleology and the human significances attached to natural phenomena; and (3) technological progress that in the last decades has created the conditions for choice in matters of human life where these hadn't existed in previous centuries.

In this widely held picture that I am describing, moral rules, aimed at protecting freedom and promoting interests, operate at a different level over a material which is perceived as neutral—that is, they protect freedom and promote interests which come from areas of human life regarded as lacking in moral significance, areas which are neutrally offered to us by scientific description. I cannot really go very much into this now, but this picture is dominant among many contemporary defenders of liberalism. If we go through the history—which has its beginnings with the early moderns (such as Hobbes and Locke) and reaches down to our contemporary thinkers—we see how influential this picture is.[2]

Here I will just mention two of the authors I'm thinking of, without further comment. If you take a utilitarian like Peter Singer[3] or a defender of rights such as Judith Jarvis Thomson,[4] I believe one can see how they detach entirely the work done by moral reflection from experience in these areas of human life: from what giving birth, pregnancy and dying are as a matter of facts, attitudes and human concepts which make them something relevant and salient for us. They place moral reflection in one place, governed by a certain normative scheme (utilitarian or deontological), and human life in another, and ask the sciences for details when they are in need of some precision.

To go back to our story, this picture was actually shared by both the defenders of democratic modernity and the defenders of the old world, the world of the *ancien régime*. The defenders of the old world saw and have seen in the new liberties, and in the technology which made them possible in new areas of life, a way of losing what is for them the *only* background capable of expressing human and ethical life; whereas defenders of modernity saw and have seen in such a loss of the old background the *only* possibility of achieving a social scene liberated from the old impositions, a scene now opened for the first time to choice and freedom.

As I have said already, I should go into more detail than is possible here. I can only allude to a picture which, for example, John Stuart Mill was already offering as a diagnosis of the battle between conservatives and reformers at the beginning of the nineteenth century, a picture that I find still faithful to what is happening today. When conservatives, such as philosophical Catholics, maintain that by allowing freedom of choice at the beginning or at the end of life we risk losing our humanity entirely, they argue that there is only one framework, one whole modality of existence, which makes room for human relationship and ethical life.[5] They don't allow for transformations or changes in the whole *arrière-plan* which accounts for different ways of being human and living ethically, ways which can be counted as significant or even as radical improvements.

At the same time, when a liberal reformer like Peter Singer holds that the way human life is, and the very fact of having a perspective that is shaped by the fact that we are human beings, are entirely irrelevant to the moral point of view—or actually that our human perspective stands in the way of our achieving a properly moral impartial point of view—he is continuing the clash I have described. He argues that evolutionary science and technology have liberated us from our local historical and human perspectives and that they have done this for the good, allowing the moral point to view to be placed somewhere else, freed from the particularities of the human background.

It is my contention that this reconstruction of modern moral culture leaves us with problems. I think it does create problems for conservatives, but I am interested now in the problems it creates for liberals, for people who care for freedom and progress. Put very roughly, on this picture—but it isn't just a picture of our way of living, it is our very way of living reflective aware of itself along such lines—ethical thought is separated from the language and experience which come from the fact that we are human beings who are shaped by a human experience with sexuality, dying and so on. In this view, progress was achieved as a liberation from any preestablished form of experience with human life, and the interpretation of human life was given over to specialized enterprises which deal with it cognitively and practically (the sciences, the medical technologies and so on), so we are left only with the elaboration of our interests and goals as if in a vacuum.

According to the liberal picture, this is good. Liberals, like Singer, think that the only way to liberate human life from the oppressiveness of the traditional background, which left no room for freedom and individual interests at all, is to hand human life over to science and technology, which offer us a morally neutral notion of it over which we can deliberate freely. Singer is keenly aware of the limits that human nature so described poses to human deliberation, but they are limits posed on it from the outside: human nature does not shape in any sort of way what moral and human deliberation is for us.[6]

I want to argue that this picture and this entire way of living as unsatisfactory because it leaves us with problems, and the problems are the following. We have achieved liberty at the expense of having lost the sense of what we should do with it, of how freedom is something precious and deeply human, of how individual interests are significant because they express who we are. We have acquired freedom and the importance of promoting individual interests (say, the two key concepts of autonomy and of quality of life) at the price of losing their depth, at the price of detaching them from a sense of why we should really care to be free and to be individuals with preferences and interests of our own. And this is because we don't have the language and the entire framework, the background, to express ethical choices as choices which come from ourselves, shaped by our subjectivities. So I come to my first conclusion: there is work that we need to do which concerns the constitution of our subjectivity in these areas of life, in sexuality, in procreation and in death. I also want to remark that this work concerns the fact that we might not see that we have problems here, we might not really see that we have problems which concern our subjectivity, conceived as the field where choices are really possible for us, where it makes sense to choose.

What kind of philosophical instruments do we need in order to be able to pursue this project, which comprises both the critical reconstruction of the moral culture which is in our past and the devising of the work that needs to be done now? I will once again sketch out a line of reasoning. As was made clear in my brief summary, the idea of the clash that I presented requires the notion of background or framework, the idea that moral considerations are the expression of frameworks of human life. Briefly put, I hold that the traditional background (organized around the three axes I mentioned) is shaped in a way, which doesn't leave room for freedom. But liberals do not recognize that freedom and individuality require their own framework in order to grow and flourish. There is a limited number of authors that I find congenial in order to develop this project. One of them is Michel Foucault.

I find Foucault crucial in order to describe this contrast and to lead this description to the goal I am pointing to—that is, to the work of opening new fields of experience congenial to the growth of subjectivity, to the goal of getting hold of human life in the first person—but, as will be clear from the very scattered considerations I will put forward, he is not the only one. Another important author we need to learn from is John Stuart Mill, a modern author who in his own way worked on certain problems that Foucault discovered subsequently (around the notions of truth, courage, scandal, practices of the self) and developed in his last lecture courses.

In my diagnosis, we need to elaborate ways through which our subjectivity can be expressed. I will start by making a few comments on the experience of dying—or, as we should say, on life with the prospect of our death. The free exercise of one's subjectivity in this area has a number of enemies. The traditional conception doesn't leave room for subjectivity: it inscribes an area of life such as dying in modalities of living in which we conform to some paradigm, such that, for example, the idea of dying with dignity is seen as involving a specific and given way of dying which excludes a space of experimentation and expression of one's personal vision and response to life. But the liberal conception is also unsatisfactory, as it protects freedom and rights, defends our right to the means which enable us to be free, like the negotiation of the uses of medical technology in the service of our deepest desires, and yet allows science and medicine to get hold of these areas of life entirely without seeing the need for the elaboration of a subjective perspective.

In neither case do we see the emergence of an area of life, like dying, where freedom can grow—an area where human life is something which we can get hold of from a subjective point of view, where our subjectivity can flow naturally, where we make use of, and activate, our *primary human potentialities*. In other words, an area in which we are engaged with the *primary moral work*, as I would like to call it—the work through which we give form to our life from our own point of view, using a large array of resources which have not been previously selected and shaped by a certain understanding of what living is and should be in such conditions. Such primary potentialities are employed in the traditional framework in order to subjugate, to create individuals who conform. But in the new situation created by modernity such human primary potentialities are not seen at all. This can be registered at various points. The liberal conception doesn't see that science and science-oriented technologies like medicine are not morally neutral ways to get hold of the notions around human life. Science and technicians shape life (procreation, sexuality, illness, ageing) from their points of view: from the point of view of interests and goals, which are only *partially* the interests and the goals that we have as human beings who are concerned with our subjectivity. In such a perspective, life is shaped by the point of view of the theoretician, the technical point of view of healing an organ, the social point of view of health policies and risk calculation. These are all significant points of view. I am not interested at the moment in following Foucault in unmasking such points of view; I only wish to argue that they are points of view in the third person, and they are not shaped by an interest to give form to one's life as subjects who live and encounter experience not as theoreticians, as physicians, as social scientists, politicians and so on, but as human beings primarily engaged in giving form to their lives. So the liberal view doesn't see at all that there is a problem with what is involved in living life as subjects in areas such as dying and ageing. From my point of view, this means that there is work to be done here, a task for those who wish to defend freedom and individuality.

Foucault is someone who may help us to articulate this problem philosophically. He is someone who envisages the idea of this primary moral work. This is my way of reading him: he invites us to consider the practices which allow for the constitution of subjectivity. In the first instance, he sees this dimension of the self and ethics; he gives a description of it which undergoes a transformation in the last years of his life and becomes more complex. He turns to classical Greek philosophical culture, where he finds, in various ways, models which enable us to think of forms and practices of the self as ways of constituting subjectivity. His work helps us to recognize how our relationship to reality, discovering its truth and responding to it depend on ways of construction and formation of the self, on practices of the self. This approach, which we find in *The Hermeneutics of the Subject*,[7] is rendered more complex and articulate in the last courses. Here the idea of *parrhesia* or truth-saying achieves an independent place with respect to the notion of the care of the self, of which it remains a component. Foucault shows how it isn't just that truth-saying presupposes practices where we take care of ourselves in determinate modalities, but that truth-saying as such can be different things under the same model which makes the relationship between truth-saying and care of the self explicit. So we have, for example, the various lines of development starting from Plato. There is the line which starts from the *Alcibiades* where taking care of oneself is directed to the definition of the soul, and the life of truth is a kind of life where we are in contact with the world of truth displayed by metaphysics. And then there is the line starting from the *Laches* where the life of truth is directed to *living* the true life, where truth faces proofs and is exemplified in one's whole bodily life. Within this second model he develops his fascinating considerations on Cynicism and on the way in which this tradition brings the issue of the true life to its limit and overturns it.[8]

The issue of Cynicism—as a path taken within the perspective of the true life—is deeply interesting, but I wish now to go back to the *Hermeneutics of the Subject*, and to how the issues treated there are also taken up in *The Courage of Truth*—that is, to the story of how the relationship between the practices of the self and truth-saying are gradually obscured over the centuries, the story of how at a certain point Christian metaphysics, and later on modern science, impose themselves as forms of truths which hide the constitutive dimension of the practices of the self. In the language of *The Courage of Truth*, there is a confiscation (*confiscation*) that religion makes of the issue of the true life and an invalidation (*annulation*) which science makes of it.[9] At the same time, these crucial dimensions which belong to the history of European and Western moral culture employ and activate certain practices of the self. But they are not practices which are elaborated and employed as a field of freedom which belongs to the individual, a field where forms of subject modulation can be activated: they are presented and employed as the only appropriate ways to be in contact with truth, and such ways are—to simplify—those of obedience, in the pastoral model, and of scientific objectivity. I am, of course, offering a terribly simplified picture roughly extracted from Foucault's story, a story which allows for many more elements to be connected in various ways. For example, he shows how the sedimentation of a certain metaphysics of the self can be variously connected with different stylizations of the self and vice versa: I will not here elaborate on what this means for my reading of Foucault.[10]

The point that I am interested in highlighting in Foucault's reconstruction is the following: he unmasks, and makes available to us, a field of constitution of the subject, a field where an analysis is offered, as he writes, of the "the different forms by which the individual is led to constitute him or herself as subject",[11] a field of the activity of self-formation which can account for the sedimentation of dimensions such as Christianity and modern science. In this light—as I read him—he unmasks Christianity (or rather a significant line in it) in two ways. On the one hand, it is viewed as a way of constituting an individual through the development of a very worrisome concern for his or her own soul, in order to make it open to the dictates of the pastor who transmits the will of God in the form of a realm of metaphysical truths.[12] On the other hand, it becomes as a way of hiding all of this—that is, of hiding behind metaphysics and divine law that work which is being done to render human beings individuals who will be dedicated to these sorts of activities, who will be shaped by them, shaped by this entire background.

These are two sides of the same conceptual phenomenon: individuals are rendered obedient, and obsessed by the concern to decipher themselves, to the extent that the fact that they are making a definite and very particular use of larger potentialities is hidden from them. What is hidden is the possibility of having access to other families of practices, or other modalities, leading in other directions. Other ways of giving form to their lives are obscured—that is, the very fact that *they are giving form to their lives* is hidden. In the traditional background the resources for individuals to describe themselves in this way are not really available: they would instead describe themselves as being obedient to their pastor, doing what is required by God, confessing their sins and so on.

In an analogous way, Foucault suggests that we can read the scientific outlook as a peculiar form of disciplining of the self aimed at producing individuals qua theorizers and technicians, a discipline which hides in its own way the work being done in order to make of individuals the type of theoreticians who will be shaped by the scientific worldview. There is an analogous removal (*effacement*), to use the language of *The Courage of the*

Truth,[13] that is operated by both religion and institutionalized modern science (in their own diverse ways) regarding the question of the true life. In the language of the *Hermeneutics of the Subject* (in the first lecture), this appears as the question of the conditions of spirituality, the conditions of self-activation and self-modulation through which the field and the subject of metaphysics and knowledge emerge.

Thus I read Foucault, first of all, as someone who does the work of unmasking in order to make available again for us what I have called the primary moral work, centring on the field of practices which are employed in order to constitute us as certain kinds of individuals, those practices which are at the roots of the religious-metaphysical and the scientific traditions. Foucault's work on the field of the practices of the self—which I call in my language the primary moral work—can be conceived as work on the practices which constitute a certain kind of subject and which at the same time open certain dimensions of knowledge, and both of these (and we should add the connection they bear to how one is governed) are ways of shaping and presenting areas of experience, breeding grounds (*foyers d'expérience*) of new fields of experience.[14] Going back to the Greeks, Foucault discovers this primary moral work, articulating in particular two significant ways of carrying it out, two models according to which it can be pursued, the Stoic and the Cynic.

This is central to my own interests and project in contemporary ethics. Foucault points to something that I find crucial. In my view—going back to the scheme that I offered at the beginning—it is of fundamental importance that the battle against the traditional world has been fought, and thus the battle against the way in which Christianity kept together a vision of society, a teleological metaphysics and a divine law conception of ethics; the same battle that has given democracy and science a prominent place in contemporary societies. These were fundamental goals to achieve within the battle for freedom. But, as I have suggested, this is not enough or, more radically, the way the battle has been fought and represented could actually favour the enemies of freedom. In order to further engage in the battle for freedom we need to see the issue as Foucault sees it: as an issue which concerns the practices of the constitution of the individual as a free subject. In this light, we may appreciate how the traditional background was organized in such a way as to produce subjugated individuals. Yet defenders of freedom and individuality today do not see that science and technology can work as forms of disciplining of the self for particular goals—that are only partially *our* goals (tied to the primary moral work)—which leave in the shade our larger capacities for self-transformation and self-constitution.

Secondly, contemporary liberals relocate ethical work at the level of moral rules, codes and normative discussion in a way, which also fails to activate the larger bases of self-constitution; on the contrary, it contributes to hiding them from view. A certain liberal and democratic tendency which insists on the importance of rules and institutions seems to believe that the crucial work is that of depositing democratic achievements, putting them aside as a safe conquest. Yet this leaves in shadow the active work which is required at each moment in order to do something with such achievements which will relate them to our own self; it doesn't make visible the way in which having acquired freedom, and having liberated ourselves from superstitious and cruel conceptions of life, is something that requires to be revitalized at each moment by a sense of the live possibility of making something of oneself. It is this that such rules and laws and conceptions make possible (and which made such rules and views crucial in the first instance). In the perspective that I am suggesting, a democratic achievement is never simply acquired, it is never simply added to the deposit of democratic achievements, it doesn't simply belong to the stock of achieved progress made in the various fields.

Being able to visualize the issue of self-constitution, of the primary moral work, not only helps us to unmask religious metaphysics and scientific technology as peculiar ways of self-constitution which hide from us our larger primary human potentialities but also helps us to focus on a further and connected issue. The perspective of the practices of the self is also a perspective on the *mobility of the self* (I will use this expression). Highlighting the bases of activation and constitution of the self as open possibilities for the individual connects the concept of freedom to the kind of continuous work through which we have access to such practices. Foucault develops this idea in various ways: for example, by showing the different ways in which, according to the Stoic and the Cynic, taking care of oneself, or the issue of the true life, requires a battle and a kind of continuous *exercise*. I wish to underline this point and connect it to the issue of what we consider to be the democratic acquisitions of our societies. Foucault says at one point that liberation is crucial but that it is not enough, we also need the practices of freedom.[15]

I should like to radicalize that point. By focusing on the dimension of the mobility of the self, as I have called it, we can appreciate that democratic acquisitions conceived as mere acquisitions may themselves become enemies of freedom and democracy, because they do not help us to engage in the battle for subjectivity, as they leave in the shade the aspect of the struggle, the exercise and the work which needs to be undertaken in order to constitute oneself as a subject of freedom: as the sort of subject to whom such democratic acquisitions are speaking. Achievements in terms of freedom, justice and so forth, are obviously crucial, but if they are received as mere acquisitions they can themselves become enemies of freedom (and justice, etc.), because they become an obstacle to our need to come into contact with the areas of experience that such democratic acquisitions are supposed to protect, promote and foster in their various ways.

In this perspective, Foucault's insistence, late in his work, on the Cynic tradition, as opposed to the Stoic, is important and totally crucial to us. We have the need to reactivate the primary moral work as work on oneself which is at the same time political work, a form of work in which we are engaged in a struggle against the conformist character of our societies, against the tendency they have to wither our interest in finding new modes of life, against the inclination of passively absorbing modes of life that are devised for us by others—and all of this in the shadow of those rights which should defend and promote them. We have the need to engage in a battle which is at the same time a struggle against our inclination to passivity, as well as against the conformist tendency of society, an ongoing struggle to find new areas at the margins, at the outskirts of the centre, where we can experience our life in ways which activate our faculties. We need to engage in life experiments which are carried out at the same time by individuals and groups and which deliver experience which can be circulated in order to reactivate democratic acquisitions, to bring life to them again, to transform them, to give them a life which allows for the opening of new fields of experience instead of letting them work as obstacles to new openings. In this light it is important to follow the Cynic lesson as elaborated by Foucault, a lesson which teaches the need to engage in a struggle against habits and conventions, and which puts at its centre the necessity of scandal, of experimenting life as a form of exposure of oneself, the exposing of one's total physical, biological, bodily and thus emotional and visional life to the risks of scandal[16] (I would suggest that in a different style and in a very different epoch, this is very close to the diagnosis and the proposal offered by Mill, for example, in *On Liberty*: the necessity of a battle against the tendency to passivity and conformity and the need to be eccentric, strange, each one more individual than the other, as he writes[17]).

I should now like to go back to the beginning of this chapter. If we wish to defend dying, sexuality, giving birth to children as fields where we experience life in the first person, where we have the possibility to give form to our life as something personal, unique, interesting and beautiful (the beauty which comes with the interest and wonder individual characters in novels may have, for example), we need to adopt the kind of analysis that I outlined previously in commenting on Foucault. Keeping these three areas open as fields of life experimentation and subject constitution is a personal and social issue, and it has enemies. Clearly there are the old enemies who hold to the traditional background. But this task has new enemies as well, that have to do with the way in which a certain democratic conception leaves to science and technology the task of getting hold of, and managing human life, confiscating or removing from view, the articulation of our experience in the first person. Such experience thus finds itself deprived of a space of its own—a space that is different from that of the mere choice which is opened by new technological possibilities and protected by democratic rules, a space where choice could be actually seen to operate in the void.

In advanced democratic societies we are free to dispose of ourselves, in procreation, at the end of life, in sexuality; yet at the same time we might not have the sense that we can express our subjectivity in these areas. As we have never engaged in the task of giving form to our life in such areas, we do not have the words, the experience, the techniques of self-constitution, we are deprived of much of what we need. In order to open these fields of experience, we must see them as areas where struggle is needed, struggle against the conceptions of life that tend to prevail, against the sense—also ingrained in the dominant notion of rights and democratic protections—that there is really no work to be done in order to become subjects. I don't have space here for this kind of analysis but it seems to me that it is really crucial in order to examine things like sexuality, procreation and dying. I will add just one further consideration.

There is in fact a significant asymmetry between the beginning and the end of life. I believe we haven't yet really opened a field of subjective experience in the case of dying, or we are just barely starting now. This is a field of experience where medical technology has at the same time transformed and, to a certain extent, opened dying as an area of experience (the making of dying into a very long process, pervaded through and through by technology and thus, in advanced democratic societies, by an innumerable chain of choices protected by rights and good medical practice); yet it is also a field where technology and medical competence presents as an obstacle, as it tends to shape this area of life exhaustively from the point of view of medical intervention. We haven't really begun, or have just started, to open the field of life vis à vis the prospect of our death as an area where we can give form to our life, an area to be shaped through and through by our subjectivity, a place where we can impress our vision, unique and personal, to what is happening to us.

Conversely, some work has been done (especially in the 1970s) to open the field of sexuality, and many democratic achievements have followed. The field of sexuality is in itself incredibly open to continuous changes. As Foucault himself has taught us, it is quite astonishing how what was considered one of the crucial moments from the perspective of the traditional background, capable of giving voice to our deepest and most universal nature, has been and continues to be the place of so many dramatic changes and inventions. However, here I only wish to suggest that work was done a few decades ago in order to open a field of experience around sexuality in which it makes sense to be subjects who give form to our lives. But the new perspectives opened and the new democratic acquisitions deposited in laws and in the habits of sexual freedom and intimacy were left alone, as it were, isolated from the life in the first person, from the kind of struggles which led to such acquisitions in the first place, from the need to rethink for oneself what such acquisitions mean. The result has been twofold (at least), it seems to me. On the one hand, such acquisitions have been weakened and are now open to the attacks of conservatives. This is clear for example with women's freedom over their bodies: laws regulating abortion are still in place but a new language is growing which seems to have an easy job of eroding such liberties. I have in mind now especially the appeal to conscience clauses which is being employed worldwide in order to erode the freedom of access to abortion.[18] On the other hand, the very fact of having rights which defend freedom in such areas can become an obstacle to experimentation, to the sense that there is a need to fight for one's subjectivity—whereas we are simply encouraged to receive and absorb a certain modality of existence.

I think this is pretty clearly happening with sexual freedom for women, for lesbians and gays, where such freedom is now a territory which can be easily invaded and exhausted by the most conformist views which ask nothing from the individual but to conform—and one key concept here is that of "family". The role of democratic acquisitions is particularly clear in the case of lesbian, gay, bisexual, transgender (LGBT) communities, which have done so much in order to experiment and to provide content to the very idea that sexuality is an area of subjectivity as opposed to an area of conformity. Protected by democratic acquisitions, which are clearly fundamental, like gay marriage, this area of life risks being reabsorbed by traditional views, which makes it hardly hospitable to subjectivity.

The moral problems I have briefly discussed here are not in the first instance problems about what we ought to do, about which rules to adopt, they are rather problems about *which problems we should have*, they concern the very field of experience and moral life. We might discover that we do not have a sense that there is a field of experience in which to live in the first person, where we can give form to our life and shape ourselves. This shift in focus involves a radical transformation of what we take to be the tasks and methods of philosophical ethics. I think what is involved may be characterized in terms of the following theoretical tasks: (1) the task of rethinking ethics as a question which concerns the frameworks that make a certain moral interrogation possible (a problem which is in a sense also Wittgensteinian);[19] (2) the task of rethinking ethics as something which concerns the modes of self-transformation, modes through which we shape our life as a field of experience: something tied to the tradition of philosophy as a practice of life;[20] (3) the task of rethinking ethical experience as something which requires to be kept alive, to be taken care of, and which lives constantly on the verge of being deformed or lost: an idea expressed within the tradition which Stanley Cavell has gathered under the title of moral perfectionism.[21] So I see this modulation of philosophical work as one of rethinking ethics as a field of experience, as a field which is activated by the potentialities of the self, by work on oneself, an unspecialized kind of work—what I have called primary moral work—and also as thinking of such work as always on the verge of being defeated, as struggle and as constant exercise.

We need to think of ethics as a question of the problems we should have and thus as a question about fields being opened to view (as work on *problematizations*, in Foucault's language).[22] We need to consider such fields as the result of the primary moral work, a work of self-constitution which mobilizes resources that are open to view only from the nonspecialized human perspective: from the point of view of a nonspecialized encounter with experience. Finally, we need to consider the opening of these fields as a form of work where we swim against the current, work which opens itself to the paradox of wishing it to result in the circulation of ideas and examples, in its having political significance and endurance, yet at the same time in being work against what is deposited as mere acquisition, habits, traditions, given models on how to live.

BIBLIOGRAPHY

Anscombe, Gertrude Elisabeth Margaret. "The Dignity of the Human Being". In *Human Life, Action and Ethics*, edited by Mary Geach and Luke Gormally, 67–73. Exeter: Imprint Academic, 2005.

———. "Modern Moral Philosophy" (1958). In *Ethics, Religion and Politics. Collected Philosophical Papers*, Vol. 3, 26–42. Oxford: Basil Blackwell, 1981.

Cavell, Stanley. *Conditions Handsome and Unhandsome: The Constitution of Emersonian Perfectionism*. Chicago: University of Chicago Press, 1990.

———. *Cities of Words: Pedagogical Letters on a Register of the Moral Life*. Cambridge, MA: Harvard University Press, 2005.

Davidson, Arnold Ira. "Ethics as Ascetics: Foucault, the History of Ethics, and Ancient Thought". In *The Cambridge Companion to Foucault*, edited by Gary Gutting, 123–48. Cambridge: Cambridge University Press, 2005.

Donatelli, Piergiorgio. "Lo sfondo del liberalismo e gli inizi della vita umana". *Iride. Filosofia e discussione pubblica* 26 (2013): 225–35.

———. *La vita umana in prima persona*. Roma-Bari: Laterza, 2007.

———. "Mill's Perfectionism". *Prolegomena* 5 (2006): 149-64.

———. "Coscienza, libertà e professioni sanitarie". *Notizie di Politeia* 27 (2011): 95–105.

Foucault, Michel. *The Hermeneutics of the Subject*. Translated by Graham Burchell. New York: Palgrave Macmillan, 2005.

———. *The Courage of Truth. The Government of Self and Others II*. Translated by Graham Burchell. New York: Palgrave Macmillan, 2011.

———. *The Government of Self and Others*. Translated by Graham Burchell. New York: Palgrave Macmillan, 2010.

———. "The Ethics of the Concern for Self as a Practice of Freedom". Translated by Peter Aranov and Dameil McGrawth, in *Foucault Live (Interviews, 1961–1984)*, edited by Sylvière Lotringer, 432–49. New York: Semiotext(e).

———. *The Use of Pleasure. Volume 2 of The History of Sexuality*. Translated by Robert Hurley. New York: Pantheon, 1985.

Jarvis Thomson, Judith. "A Defense of Abortion". *Philosophy and Public Affairs* 1 (1971): 47–66.

Singer, Peter. *Practical Ethics*. Cambridge: Cambridge University Press, 1993.

———. *Rethinking Life and Death: The Collapse of Our Traditional Ethics*. New York: St. Martin's Griffin, 1996.

———. *The Expanding Circle. Ethics and Sociobiology*. Oxford: Oxford University Press, 1982.

———. *A Darwinian Left. Politics, Evolution and Cooperation*. London: Weidenfeld and Nicolson, 1999.

NOTES

1. I elaborated this view in some detail in my *La vita umana in prima persona* (Roma-Bari: Laterza, 2012).

2. See P. Donatelli, "Lo sfondo del liberalismo e gli inizi della vita umana", *Iride. Filosofia e discussione pubblica* 26 (2013): 225–35.

3. Peter Singer, *Practical Ethics: Second Edition* (Cambridge: Cambridge University Press, 1993); *Rethinking Life and Death: The Collapse of Our Traditional Ethics* (New York: St. Martin's Griffin, 1996).

4. Judith Jarvis Thomson, "A Defense of Abortion", *Philosophy and Public Affairs* 1 (1971): 47–66.

5. See, for example, Gertrude Elizabeth Margaret Anscombe, "The Dignity of the Human Being", in Mary Geach and Luke Gormally, eds., *Human Life, Action and Ethics* (Exeter: Imprint Academic, 2005), 67–73. The larger view is presented in her "Modern Moral Philosophy" (1958), in Gertrude Elizabeth Margaret Anscombe, *Ethics, Religion and Politics: Collected Philosophical Papers*, Vol. 3 (Oxford: Basil Blackwell, 1981), 26–42.

6. See, for example, Peter Singer, *The Expanding Circle. Ethics and Sociobiology* (Oxford: Oxford University Press, 1982); *A Darwinian Left. Politics, Evolution and Cooperation* (London: Weidenfeld and Nicolson, 1999).

7. Michel Foucault, *The Hermeneutics of the Subject*, trans. Graham Burchell (New York: Palgrave Macmillan, 2005).

8. Michel Foucault, *The Courage of Truth (The Government of Self and Others II)*, trans. Graham Burchell (New York: Palgrave Macmillan, 2011), 159–60.

9. Foucault, *The Courage of Truth*, 235.

10. Foucault, *The Courage of Truth*, 164–65.

11. Michel Foucault, *The Government of Self and Others*, trans. Graham Burchell (New York: Palgrave Macmillan, 2010), 5.

12. Foucault, *The Courage of Truth*, 333–34.

13. Foucault, *The Courage of Truth*, 235: "there has been confiscation of the problem of the true life in the religious institution, and invalidation of the problem of the true life in the scientific institution. . . . The question of the true life is always gradually removed from philosophical reflection and practice".

14. Foucault, *The Government of Self and Others*, 3.

15. Foucault, "The Ethics of the Concern for Self as a Practice of Freedom", trans. P. Aranov and D. McGrawth, in Sylvière Lotringer, *Foucault Live* (New York: Semiotext(e), 1989), 433.

16. Foucault, *The Courage of Truth*, 280, 234.

17. See Donatelli, *La vita umana in prima persona*, chap. 5; *Introduzione a Mill* (Roma-Bari: Laterza, 2007); "Mill's Perfectionism", *Prolegomena* 5 (2006): 149–64.

18. See Pierpaolo Donatelli, "Coscienza, libertà e professioni sanitarie", *Notizie di Politeia* 27 (2011): 95–105.

19. I explored this approach in my *La vita umana in prima persona*.

20. Arnold I. Davidson was crucial in order to make this perspective visible. See, for example, in connection to Foucault, "Ethics as Ascetics: Foucault, the History of Ethics, and Ancient Thought", in Gary Gutting, ed., *The Cambridge Companion to Foucault* (Cambridge: Cambridge University Press, 2005), 123–48.

21. See Stanley Cavell, *Conditions Handsome and Unhandsome: The Constitution of Emersonian Perfectionism* (Chicago: University of Chicago Press, 1990); *Cities of Words: Pedagogical Letters on a Register of the Moral Life* (Cambridge, MA: Harvard University Press, 2005).

22. See Michel Foucault, *The Use of Pleasure. Volume 2 of The History of Sexuality*, trans. Robert Hurley (New York: Pantheon, 1985), 10–11.

TWELVE

Ethics of Care and Face Transplants: After Levinas, Deleuze and Guattari[1]

Marjorie Gracieuse

I am no doubt not the only one who writes in order to have no face.
—Michel Foucault, *The Archaeology of Knowledge*

In this chapter, I shall contrast Emmanuel Levinas's moral, metaphysical and humanist conception of the face, placed under the sign of transcendence, with the semiotic and political approach of "faciality" (*visagéité*) elaborated by Felix Guattari and Gilles Deleuze. These two perspectives, although radically distant from one another, can nevertheless help us to dissipate a certain form of idealism that is sometimes deployed in the discursive field of bioethics, and more particularly in the context of face transplants. The face seems to appear as a hybrid reality. On the one hand, it is a material interface between the outside and the inside of corporeality. On the other hand, the face presents itself as an expressive surface, whose becoming somehow escapes the orders of representation and perception, since it conceals a force or power that is not reducible to its appearance and social coding. Thus, should we follow Levinas, and think of the face as bestowed with a metaphysical sense, which would both transcend and constitute our humanity? Or should we apprehend the face as the material expression of a purely immanent and amoral vitality, which Deleuze does not hesitate to name "desire" in one of his lecture courses?[2]

First, I shall investigate Levinas's conception of the face, by measuring the presuppositions and limits of this central idea, particularly with respect to "the call of the face", in which Levinas sees a call for care (*un appel au soin*) and compassion with the Other's suffering. I will then make use of some ethical aspects of Deleuze's philosophy in order to propose an alternative vision of care, which I will illustrate in the specific case of face transplants. This will lead me to explain the reasons that led Deleuze and Guattari to detach the face from the social transcendence of the signifier and to promote an a-signifying politics of the face, which is now understood as terrain of experimentation with the infinite potentialities of matter.

FACE, TRANSCENDENCE AND CARE IN LEVINAS

Emmanuel Levinas sought to renew the traditional meaning of the human face, while conserving a lexicon inherited from ontology as well as theology. In his writings, he invites us to think of the face beyond its physical and perceptible appearance, leading us to abandon the idea that a physical destitution of the face, on the occasion of an accident or a tumour for instance, would be equivalent to an actual privation of one's face (*visage*) or to the loss of one's identity. The relationship to the "Other" (*autrui*) as "Face" (*visage*) inaugurates, for Levinas, a new sense of ethics and humanity. It not only embodies a radical transcendence but also enacts an experience that bears a moral teaching to which one cannot remain indifferent.

The very source of the ethical relation originates in the Other; the face presents itself as "an ethical *fact* that owes nothing to values", since it is in fact values that "owe everything to it".[3] Here, the term *fact* is employed by Levinas in the quasi-Kantian sense of *Faktum.* It is relative to a transcendental law insofar as it expresses a "call" and denotes a commandment that Levinas devotes himself to describing in its infinite ambiguity and metaphysical tenor. The face, then, transcends my being; it imposes upon me its solemn authority and teaches me the sense of a radical alterity or difference that cannot be judged in relation to a preposited sameness or logical identity. By virtue of its ethical force, the face defies all representations or images I can form of it, since it incarnates the presentation of a metaphysical infinity that resists all cognitive projects of appropriation:

> the way in which the other presents herself, exceeding the idea of the other in me, we here name face. This mode does not consist in figuring as a theme under my gaze, in spreading itself forth as a set of qualities forming an image. The face of the Other at each moment destroys and overflows the plastic image it leaves me.[4]

Levinas thereby does not define the face as a thing, nor as an intentional object. Rather, the face is characterized by its evanescent presence: it gives itself as it withdraws. It is not reducible to its mere physical aspect, since it is constituted of a quasi-metaphysical flesh that is in a sense inalterable, even in cases when the Other might be considered as "disfigured". This essential inalterability entails that the face is not seen in the proper sense of the term. Thus, what makes the transcendental unity, or the identity of a person, for Levinas, does not reside in the composite of flesh and bones that we commonly call "the face". Indeed, the latter is not, strictly speaking, something that could be "transplanted" or "grafted". Rather, the face is a pure *manner* or *way* of appearing; its flesh bears an intangible dimension, a nonphenomenal presence that paradoxically gives a person her ethical integrity and personal identity.

This is why Levinas speaks of the face as "ethical resistance", because its mode of presentation resists any attempt to enclose the Other in a represented identity or idea. Its apparition or "epiphany" forces me to apprehend the Other not as *alter ego*—that is, from my self-centred position in the world—but rather from the other's inassimilable difference, which disrupts the ordinary quietude of my egological existence. This manner of appearing leads Levinas to present the face as that which breaks through material existence: it embodies an always-singular otherness that exceeds sensory perception and makes sense in and of itself:

> It is this presence for me of a being identical to itself that I call the presence of the face. The face is the very identity of a being; it manifests itself in it in terms of itself, without a concept. The sensible presence of this chaste bit of skin with brow, nose, eyes, and mouth, is neither a sign allowing us to approach a signified, not a mask hiding it. The sensible presence, here, de-sensibilizes itself in order to let the one who refers only to himself, the identical, breaks through directly.[5]

The impossibility of ignoring the Other's face also characterizes its radical proximity, since the face, by its sole presence, is a "call" that affects and invests me with what Levinas calls a "responsibility". Indeed, the Other is described in his writings as the one for whom and before whom I am always responsible. Each face carries a certain ambiguity such that, disfigured or not, it remains an enigma that generates in me a pure affect or lived sense of respect, under the mode of a pre-verbal commandment. The latter comes from an ever-withdrawn transcendence, which is not of the order of the law, but of a God beyond being, "transcendent to the point of absence", Levinas writes, and who awakens in each of us a "Desire for the Good".

The very origin of ethics is thus founded on a nonreflective, nonegoistic affect, which Levinas calls desire[6] and which he conceives as a nonerotic, involuntary movement towards the Other. This "pure passion" exceeds thought; its movement is not intentional, but precisely prevents any decisive and voluntary "reprise" from the part of the subject. This metaphysical desire does not operate out of any lack: it depends on excess or superabundance, strives towards indeterminacy, towards an otherness that is always further away than any particular empirical objects.

This is why desire does not have satisfaction as its end, and in fact has no end. It is infinite desire and desire of the infinite, unceasingly pushing subjectivity beyond repose and awakening in it a desire to care for the Other.

We see that this silent commandment expressed by the face is always already, according to Levinas, a call for solicitude and fraternity: it moves me to devote myself infinitely to the Other and to enter in proximity with her. If I *must* accept this radical responsibility, it is because this responsibility originates from the other's silent or explicit call, and not because it arises from a decision I have made. Levinas even compares this "call of the face" to an "election" whose implacable truth is experienced in the flesh prior to any empirical knowledge. Thus, the call generates in the subject who receives it an affect of "care" (*souci*) that must be understood both as solicitude and as concern for the Other. Levinas even speaks of the face as that which reveals "the medical vocation of each human being".

Care, understood as feeling and disposition, is indeed that through which the human opens up to sociality and overcomes the animal effort of life that characterizes all living beings. This capacity of being worried for the Other, for her mortality before being worried for oneself is not only Levinas's critical response to Heidegger's notion of *Sorge* as ontological feeling of anxiety and concernful solicitude for existence but also an objection to a certain philosophical vitalism, from the perspective of which the essence of the human could be defined as "bare life" or as "care of being" (*souci d'être*).

It comes as no surprise, then, that this vocation to care for the Other is not and cannot be an effect of my will. Rather, it comes from a metaphysical and infinite desire that the Other awakens in me, and which Levinas interprets as coming from an immemorial and infinite debt towards the Other, a debt which persecutes each subjectivity and "accuses me beneath the level of prime matter",[7] beneath the very materiality of my vital and living body. Responsibility, according to Levinas, is thus to be understood not as a choice, but as an assignation that takes subjectivity hostage. This feeling of obligation to care for the Other reveals the profound asymmetry of any genuinely ethical relation, since for Levinas the Other is always "higher" than me, her "call" pervades in and beyond all that can be said and done.

The Other's call must be understood as the *a priori* condition of an ethics that can never be fully actualized, since this "appeal to care" for Levinas *always already* exceeds my affective and sensitive capacities. Indeed, by means of its very vulnerability to suffering and bare exposure to murder, the face of the Other does not give me power, but on the contrary, defies my "power to be able to". This situation of powerlessness when faced with the Other's suffering generates a profound pathos in me but, at the same time, it also elicits an ethical generosity in my being, an unconditional attention to the Other that takes the form of an inescapable affect.

Hence the paradox at the core of the Levinassian ethics: the Other awakens in me a desire for hospitality and caring but, at the same time, she dispossesses me of my power to act and affects me with passivity. By means of her very vulnerability, the Other obsesses or even persecutes me continuously. Levinas goes as far as to say that it is for this reason that I'm never done with the Other: even if I respond to her calling, or choose not to, my attitude will never be enough to overcome this radical vulnerability and structural susceptibility to suffering that concerns us both. This is why for Levinas, since sociality is inscribed in my flesh and the Other's flesh, vulnerability concerns *all* subjectivity, and not simply those we would ordinarily call "patients".

Thus, for Levinas, care is not solely a responsibility that I have towards the Other, but it is also that in which my dignity and humanity are founded. That foundational role of care is the reason why my humanity is not an essence: it is both that which is always in question and that to which I must respond: an absolute responsibility towards the Other, based on an obligation that anachronously precedes any commitment. Humanity is therefore the very difficult task of substituting oneself for the Other, of being-for-the-Other. Levinas hears in the call of the face a call for compassion, which he defines, in his seminal article entitled *Useless Suffering,*[8] as "the nexus of human subjectivity" and "supreme ethical principle". Contrary to sympathy, in which he sees a culturally constructed moral sentiment, compassion is anterior to natural and cultural processes, according to Levinas: it is a nonreflexive affect, constitutive of the very structure of subjectivity and entirely directed towards the Other. Whereas Levinas recognizes that the Other's suffering is ultimately "unassumable" and "useless", it is also that which pushes me to speak with the Other and accompany her verbally in her distress. Accompanying the other in suffering, deploying a charitable love towards the Other, is for Levinas the very sense of human dignity and the finality of ethics as religion or "first philosophy":

> To suffer with the other, to suffer for the other. This is the very beginning of a difficult humanity; it is a relentless task. This finality of suffering is a finality of a superior order, which is not simply animal, that is, which is not related to the biological accomplishment of my destiny, as living being. This finality is that of suffering for the Other and thus that of loving her, or rather, it entails to love the Other and therefore to suffer for the Other. The human dignity that concerns us all comes from this.[9]

We see that the Levinasian conception of ethics goes far beyond the face and the body simply construed in their biological, organic materiality: it destroys the old ideal of personal autonomy, recalling the radical heteronomy of all subjectivity, making of interpersonal solidarity, through discourse and proximity in suffering, the imperative of ethics. Care is thus a duty that I owe to the Other, an Other who, for Levinas, does not have any duty towards me, but only rights. However, can we really speak of "responsibility", when the latter is neither measured by my physical capability nor the consequence of a voluntary engagement? Furthermore, must we pose the problem of ethics in such "dolorist" terms and theological overtones? Are there not other modalities for the ethical rapport, and for care in particular, that will not be doomed to self-sacrifice, and whose sense would be the product of the relation of care, and not grounded in an immemorial and nonpolitical transcendence?

THE SOCIAL VISIBILITY OF FACES AND THE LIMITS OF LEVINASIAN ETHICS

We have seen that in Levinas's philosophy, the Other is someone towards whom I have a responsibility that goes beyond what I may or may not have done to her, and also beyond what I can or will do for her. However, this infinite responsibility towards the Other has nonetheless a limit that Levinas himself recognizes, since it can only take place between two subjects, in the "face-to-face" situation that functions as a paradigm for all ethical relations. But as soon as a third party enters the scene, this initiates the domain of political relations, within which, Levinas argues, "I am no longer infinitely responsible for the other and consequently, no longer in an asymmetrical, unequal relation".[10] Such is the reason why, in the political context of citizenship, I can live myself as the Other's equal. However, this equality is precisely not the condition of ethics, but only the political and cultural field within which transcendence as asymmetry and source of love can emerge, as a response to the call of faces, which Levinas interprets as a "call for Justice".

The problem of justice for Levinas is thus inseparable from the promotion of a mode of political organization that would allow for the realization of transcendence within the material immanence of power relations. This entails that Justice is something we need to *care* about, it always remains to be done or, better, rendered. However, the problem with such a division between ethics and politics is precisely that any genuine relation of care seems to be possible only on the occasion of exceptional moments of *rupture* that disrupt, by their intrinsic heterogeneity, the order of power and citizenship.

Power is indeed considered in purely negative ways by Levinas, meaning that it is by essence murderous of the Other, as the "order of being" where self-interest reigns and continuously hinders the emergence of genuine ethical relationships of love:

> Here is extreme contemporaneousness or immanence. . . . Being's interest takes dramatic form in egoisms struggling with one another, each against all, in the multiplicity of allergic egoisms which are at war with one another, and are thus together. War is the deed or the drama of the essence's interest.[11]

Political immanence is thus a necessary evil, within which the anarchical force of a purely ethical Desire can never reign but only let itself be heard and experienced in evanescent moments. This is why for Levinas, if we all are fundamentally capable of caring for the Other, we are also all guilty of never being vigilant enough to preserve the political and economical conditions that allow for the proliferation of ethical relations.

It is not surprising, therefore, that in later works, Levinas privileges liberal democracy as the most propitious regime for the preservation of these ethical moments of care. By virtue of the degree of freedom of speech and action that this political regime insures, the liberal state is structurally haunted by the infinite debt towards the Other: "that is perhaps the very excellence of democracy, whose fundamental liberalism corresponds to the ceaseless deep remorse of justice: legislation always unfinished, always resumed, a legislation open to the better".[12] Contrary to what we might think, liberalism for Levinas is not the regime of individualism that would favour autonomy and the cultivation of an egoistic care of oneself at the expense of the Other. Rather, it is a political regime that allows the Face to reveal its transcendent power beyond the socioeconomical divisions, letting the fundamental Goodness of Desire express and accomplish itself infinitely, allowing for moments of Justice to pierce through the realm of political antagonisms and economical war.

However, it is legitimate to question Levinas's philosophical decision to trust this passion that he calls "Desire", since he conceives it as essentially deprived of self-interest and even above life itself. Even if we accept the Levinasian belief, according to which Desire constitutes the intrinsic goodness of human essence, as nonerotic love, this belief does not prevent desire from being culturally alienated, exploited and institutionalized in the political order, in the immanence of power relations. Besides, by stating that Desire is a luxurious need,[13] that it is an affect that can only be experienced in a being who is already happy and whose basic, vital needs are satisfied, Levinas seems to make of ethical Desire a kind of privilege or luxury that ultimately depends on socioeconomical conditions.

In the context of economical liberalism, which engenders individuals who measure the actions and passions—that is, all the singular events that compose a life according to their market-value and return on investment—charity tends to be denatured. It becomes a private righteousness or is reduced to economical philanthropy—that is, to an activity reserved for those who have the economical and temporal capital to make of charitable existence their moral and/or economical imperative. Moreover, how can we deny that the "Face", the way Levinas understands it, is in fact invisible, silenced and unheard in a world where faces are codified, represented or even turned into principles of distinction or discrimination, here referred to a political or religious group, there turned into a normative criterion for the political and economical benefit of some over others?[14]

Thus, the difficulty with Levinas's perspective lies in the fact that he tends to decontextualize the specific act of caring from its political and social conditions of effectuation, since he conceives of care as a universal, *a priori* principle, dependent on a metaphysical relation (nonerotic love, disinterested desire or charity) capable of regulating in and of itself all ethical relations and therapeutic practices. This depoliticized vision of care, if it has the advantage of making us imagine what would be an ideal vision of charity and love for one's neighbour, does not tell us much about the conditions under which a concrete and transformative relational activity with what is essentially "without relations"—namely, the Face—can effectively take place. Indeed, in Levinas's depoliticized conception of the face, its transcendent and universal significance appears more like an object of belief or faith in what it means to be human (for him: to be an essentially good *creature*) than a historically formed and thus contingent cultural image whose ethical sense remains nonetheless undecided as long as it has not been empirically encountered and explored.

Besides, by insisting on the unchosen nature of what he calls "responsibility", Levinas abstractly detaches the "face-to-face" relation from the material and relational becoming that concerns both the caring agent and the suffering Other. In so doing, Levinas cannot account for the political and plural reality of care, which is not a preexisting subjective property or innate capacity, but rather a culturally coded scheme of action and performance, constitutive of the very agency of the subjects that engage in it. It is therefore in the immanent realm of action and power relations that the very materiality of care and the social function of the face must be politically analysed, ethically conquered and perpetually reinvented.

"LOOKING AFTER SENSE": A DELEUZIAN APPROACH TO CARE

By valorizing the radical vulnerability of the Other and the radical asymmetry of the rapport that ties me to her, Levinasian ethics seem to overlook the fact that it is always with the active collaboration of the Other, with her vital power, that I can find the resources for helping her as well as for acting ethically. If the suffering of the Other disempowers me at first, it might also give me a strange power; not a power over the Other's vulnerability, but the potency of becoming-able to care for the Other and finding unsuspected resources to respond to the situation of suffering. Faced with the other's suffering, we can indeed feel powerless. However, this feeling of powerlessness is also an experience at the limit of the liveable whose physical, and not simply transcendental experience compels us to become-active. Care is therefore not reducible to an infinite debt that makes me passive and that I cannot pay back. On the contrary, it is a force of empowerment, which can be transmitted to the Other and contribute to helping her. This "becoming-able" to help the Other does not simply come from my intrinsic vitality, but is produced by the problematic relation itself—that is, it springs from this dissymmetrical rapport in which the Other can in turn find resources, at least giving a new sense and value to her suffering, if not alleviating it.

Instead of pre-supposing, as Levinas does, a transcendent sense and direction of the relation of care, it might be more accurate to affirm that care only bears a sense and a value by virtue of its vital components, which do not pre-exist their conditions of emergence and production. If we can indeed think of desire as an anarchical force that evolves outside the forms of the Subject and the State, this virtual and informal nature of desire should not prevent us from realizing that desire is not spontaneously or naturally good, but always already culturally channelled. From this perspective, if the domain of political relations is indeed that of perpetual agony and alienation, it is not because the majority of humans remain deaf to the Other's call. Rather, it forces us to question the reasons for which they are actually deprived of the economical and political means through which they could learn and understand the necessity of a caring relation to the Other.

In fact, the harassment or persecution does not come from the Other, but from a more and more invasive set of policies and social codes of conducts that we must actively overcome if we are to genuinely encounter the Other and care for her. In that context, care can no longer be conceived as a universal "medical vocation" that would be equally present in each of us, but rather as a cultural potentiality that can only become effective when it finds its material and sociopolitical conditions of accomplishment. Between the potential of care and its optimal conditions of effectuation, there can only be political resistance and struggle against the codifications of conducts that prejudge the needs and exigencies of the ethical relation, particularly in the institutional context of medical practice.

If a disease, a wound, a state of distress and suffering can indeed be lived differently and acquire a new sense and value, it might be because care is not a natural or moral obligation and does not require the revelation of a transcendent command or duty. Rather, it demands a series of critical operations and political acts, which can become powerful enough to destroy the discursive schemes of signification and nondiscursive protocols of action that habitually shape our ordinary and institutional relations to others. Care is a formative experience, which endows the carers with a new practical knowledge that exceeds the normative clinical protocol of the medical act as much as it does established moral beliefs.

In the context of face transplants, if we take the specific case of the disfiguration of the face, for instance, we see that the problem of the recipient of the graft is not to retrieve a "lost" identity, but rather to learn how to live one's face differently—that is to say, as altered not simply in a material sense but also in her social and collective identity. Indeed, for the patient, conquering the ability to express emotions does not depend simply on cellular regeneration, but on the development of an immanent capacity to integrate the new dynamisms of the graft in order to reinvent a new sense of self. This capacity is precisely the object of a collective creation. The ethical relation at stake in medical care does not happen between two well-defined subjects but, on the contrary, involves a composition of powers (*puissances*) which struggle against the power of suffering, which the patient expresses, but to which her vital power is not reducible. In fact, the relation of care is ethical only insofar as it generates new affects in the carers and the patient, affects that become perceptible only inasmuch as they are actualized by expressions, gestures and words.

Thus, it is through a process of individual and collective individuation that a suffering subjectivity can develop new capacities: responding to one's wound, she becomes able to give a new face to life itself. This responsiveness, this capacity to respond *to* one's own wound, is the object of a perpetual apprenticeship. Furthermore, this perpetual apprenticeship concerns not only the patient in her corporeal immanence but also the caring team itself as a social group capable of collectively empowering the patient. For the patient, this approach implies a reprise, a reaffirmation of all the new potentialities, forms and norms of life that the wound makes possible, as a new material configuration. Of course, the wound of the Other is not mine, but the individuation of the wound itself is in fact immediately collective, since it is always with the conventional apparatus of culture that one judges oneself. Such judgements can render one's wounds even more painful: hence Deleuze's thoughtful interrogations with respect to the social codifications of a wound: "Which war is not a private affair? Conversely, which wound is inflicted by war, and coming from society as a whole? Which private event does not have all its coordinates—that is, all its impersonal, social singularities?"[15]

This approach, which makes of care a vital problem of social and collective sense-production, might benefit medical ethics as it invites us to refrain from reducing the body to an individual organism and to reconsider the resources that Deleuze calls, following Gilbert Simondon, a "pre-individual field"—that is, a virtual field of potentialities, contemporary with any actual form of subjectivity, but whose becoming extends to the social milieu to which an individual belongs. Moreover, it enables us to understand in what sense what is needed in medical relations of care is not so much a so-called humanization of medicine but a perpetual attention to the urgent and vital dimension of medical care, to which one must respond by an act of power, a deliberate performance and positive action over the Other, which implies as much the *care of oneself* (as attention to, and development of, one's capacity to help and serve the Other) as the *care for the Other* (as perpetual attention to the very becoming and signs of suffering).

From this perspective, we see that the philosophy of bioethics can be something other that the moralization of medical practice—that is to say, a clinical critique of social norms that places ethics to the test of life itself, since this enterprise is not rooted in a legal and fixed codification of medical practice, but attentive to the powerful and transformative vital signs of what Georges Canguilhem called "the exigencies of the living".

THE FACE AS POLITICS: BECOMING-CLANDESTINE

With Deleuze and Guattari, it is no longer a matter of saying what the face is, but of interrogating the conditions under which the face can escape the linguistic regime of the "subject function" that is dominant in capitalism. Deleuze and Guattari allow us to apprehend the polyvocality of the face by emphasizing its self-differential materiality, without reducing its sense to a personal identity or transcendent signifier. According to them, the face is never given but produced by a cultural mode of domination that in fact reached its paroxysm with the advent of capitalism. It was Felix Guattari who created the concept of "faciality" (*visagéité*), in his book *The Machinic Unconscious*, published in 1979, in order to give a name to the abstract image of faces that the empire of significance requires to regulate and control the expressive singularities of material corporeality. Indeed, what is commonly called the "face" functions as an immaterial, perceptive scheme that we now use unconsciously in order to classify faces and code their expressive features. This is why the face plays a role in society that is similar to a "pure signifier", since we don't know what it signifies exactly, but we continue to believe that it signifies something.[16]

The face is above all a surface onto which schemes of power come to inscribe themselves, as various strata of knowledge that ossify its intrinsic becoming by granting it an alleged identity. Indeed, the face is described by Deleuze and Guattari as a "white wall" onto which a variety of conventional significations rebound, channelling the vitality of the face and submitting it to the game of political and social significance: "Expression should be understood not simply as the face and language, or individual languages, but as a semiotic collective machine that pre-exists them and constitutes regimes of sign".[17]

This predefined system of signs and meaning that constitutes the dominant perceptive scheme of our culture acts like a "plane of transcendence" with respect to the material immanence of the body, insofar as it is a regime of knowledge that functions like a sieve or diagram through which we perceive others and ourselves according to the binary opposition subject/object. For that reason, this world in general, and the face "as we know it" constitutes for Deleuze a veritable system of judgement that prevents any new mode of understanding ourselves from emerging. This particular regime of power that Deleuze and Guattari call "capitalism" functions thanks to an economical codification of living materiality that turns bodies into docile productive forces and criteria of socioeconomical distinction. It also shapes and produces "the face" as the site of an alleged personal identity, abolishing the body's becoming in favour of a control and overcoding of bodies:

> The head, even the human head, is not necessarily a face. The face is produced only when the head ceases to be a part of the body, when it ceases to be coded by the body, when it ceases to have a multidimensional, polyvocal corporeal code—when the body, head included, has been decoded and has to be overcoded by something we shall call the Face.[18]

This privilege of the face over the body derives from an historical "concerted abolition of the body" and is even, for Deleuze and Guattari, the material origin of the organic representation of political power, from the perspective of which a good organization of power would be hierarchically organized, with a "head" that would command to organs. We see that, far from being an anthropomorphic invariant, the face is the product of power relations and the very condition of their exercise: "the face is a politics", Deleuze and Guattari argue. As abstract cultural archetype, the face is "not a universal. It is not even that of the white man; it is White Man himself, with his broad white cheeks and the black hole of his eyes. The face is Christ . . . Jesus Christ superstar: he invented the facialization of the entire body and spread it everywhere".[19]

Such is the reason why, contrary to Levinas, Deleuze and Guattari seek to break with any ideal or transcendent codification of the face, insisting on its irreducible ambivalence and material multiplicity. The sense of the face is thus never given, but eminently problematic since it is precisely that which needs to be continuously produced and re-invented, as a way to combat the ready-made significations that are imposed upon it. Thus, the ethical and political problem of sociality is not that of revealing the transcendent sense of the human face, but of constructing a new sense of the face, which would not serve the normative regime thanks to which liberal democracy functions and hardens its socioeconomical hierarchies.

Therefore, instead of opting for the recurrent argument according to which social and cultural politics would always corrupt the human's alleged "good nature or will", Deleuze and Guattari choose to demonstrate in what sense it is our very perception and use of power that can alienate us or, on the contrary, empower us to invent new modes of existence. It is our very way of willing to "facialize" bodies that we must reflect upon, in order to invent another regime of desire and perception: "This is an affair not of ideology but of economy and the organization of power (*pouvoir*). We are certainly not saying that the face, the potency of the face (*la puissance du visage*), engenders and explains social power (*pouvoir*). Certain assemblages of power (*pouvoir*) require the production of a face, others do not".[20]

Thus, for Deleuze and Guattari, it is a question of wresting matter from the territorialities that are traced upon it by dominant discourses, in order to show that other divisions, other regimes of signs and valences of desire are possible. To the supra-historical transcendence invoked by Levinas, one can therefore oppose the infra-historical material immanence of a life that cannot be judged in the name of superior values and whose sense is purely virtual (that is, neutral in and of itself, since the virtual only acquires its political and ethical tenor once it is lived, problematized and actualized by bodies). From this perspective, we need to learn to perceive faces out of the "abstract machine of faciality" in order to invent a new use of the face: "If human beings have a destiny, it is rather to escape the face, to dismantle the face and facialisations, to become imperceptible . . . to make faciality traits themselves finally elude the organization of the face".[21]

CARING BY LOSING ONE'S FACE AND CHOOSING ONE'S LOVE: THE OTHER AS EVENT

To undo the face does not mean returning to a-signifying semiotics. Rather, it requires retrieving the intensive field of virtualities that presides over the facialization of the body, in order to experiment with what we are capable of feeling, thinking and doing to others and ourselves. In this sense, whereas Levinas promoted a dematerialized and metaphysical sense of the Face, Deleuze and Guattari choose the reverse path—namely, that of rematerializing the face by wresting its a-signifying, yet concrete, vitality and expressiveness from the abstract, transcendent signifiers that are imposed upon it and that bind it to a politics of identity. It comes as no surprise that, when it is deprived of its anthropological and judgemental veil, the face appears as made of a-signifying material singularities, which cannot be reduced to an identifiable totality, to this icon proper to the signifying regime.

Dismantling the face is thus a politics but also an ethics since it involves an encounter with the Other not as a face, but as event or block of intensities, with whom I can compose my forces and invent new types of aesthetical, political and ethical connections. From this perspective, we must acknowledge that all ethical relations are not reducible to the relation of care for the Other's suffering. In fact, the care for the Other is just a particular case of ethical life, a case that cannot be elevated to the status of paradigm because it overlooks the fact that all ethical relation to the Other immediately implicates an immanent relation to one's self—that is, to use Michel Foucault's expression, a care of one's self.

Deleuze's and Guattari's ethics of immanence provides thereby a convincing response to Levinas's conception, since it shows that the mode of the ethical relation I can adopt towards the Other is ontologically correlated to the ethical relation I have to myself. What makes the value of the relation of care is therefore not care in itself, but the various ways by which I can actualize it. The more I remain sensible to the Other's signs and affects, the more I become able to act adequately and compose my own forces with the concrete situation that the Other creates or into which she is "captured". Thus, it is not sufficient to have an ethical relation to the Other in order to care for her, since a careless hatred can still be considered as an ethical relation to the Other, but one of the weakest ones, one that does not transform or teach us anything, the lowest degree of ethics. Conversely, it is not sufficient to care for the Other in order to have an ethical relationship to her, since I can affectively or actively care for the Other without being able to generate in her new affects; my good intentions can indeed provoke psychological suffering or even physical pain. This is why we must think, against Levinas, the dissymmetry yet the *reciprocal becoming* at stake in the ethical relation—that is, as trans-individual composition. Care is a relational event within which the Other and I can become capable of creating transformative affects in each other, affects that are the real agents of care, its "individuating factors", since they can empower both parties (i.e., the joy of caring, on the one hand, and the joy of being looked after, on the other).

This is why, contra Levinas's conception of love as "beyond Eros", Deleuze and Guattari remind us, after Nietzsche, that the ethical relation conceived as a command to love one's neighbour is a cultural phenomenon which is in fact properly unethical, since by positing the sense of ethics in an *a priori* and transcendent mode, it negates the performative immanence and transformative event of the ethical encounter. If ethics has indeed everything to do with desire and love, it is because the latter remain to be invented and produced: there is no spontaneity of desire towards goodness, but only a constant learning and experimentation on the occasion of variable encounters with others.

For Deleuze and Guattari, therefore, the origin of ethics is neither the passion of love for the Other nor compassion in suffering, but rather what they call, like Levinas, "desire". However, for them, desire is not a movement from immanence towards transcendence, but a connective and material force, which continuously invents new types of relation to others in the not simply coercive but also inventive immanence of power relations. Desire can be attracted or repulsed, composes or flies away from others and has therefore no *a priori* direction or signification, if not that of exceeding all fixed determinations and ossified modes of relationships.

Of course, this desire can be economically disinterested, but it is always interested in life's potentialities: it is a vital force radically opened to its "outside" and striving for ever new connections with one's complex individuality and with others. To replace one's desire to be loved by a "force to love, a virtue that gives and produces, that engineers"[22] means to cultivate one's power to see the remarkable in the ordinary, the interesting in the apparently insignificant. It involves experimenting with the joy of creating and exploring new worlds in oneself and others. Although of involuntary origin, since desire is above all an appetite for the intensities of life, this vital force is also for Deleuze a *decisive* power that we can train into "a spiritual will" and that can empower us to better care for and understand others. But without this culture of desire that creates a truly ethical subjectivity, without this care of oneself as a constant attention to one's capacity to connect with others and form new multiplicities, one is doomed to remain at the level of conventional morality, which for Deleuze is the lowest degree of ethical life.

To become able to choose one's way of loving and caring does not mean to ultimately find oneself again, unchanged as it is the case in all narcissistic forms of love. On the contrary, it requires the abandonment of one's ego to the transformative power of chance and encounters, in order to remain passionately open to change and actively capable of introducing a difference in one's thought and one's life. Thus, becoming able to love without one's ego[23] means to love without the concept of love that is imposed upon us, in which we usually suffocate and yet within which we lock ourselves and others:

> I no longer have any secrets, *having lost my face*, form, and matter. I am now no more than a line. *I have become capable of loving*, not with an abstract, universal love, but *a love I shall choose*, and that shall choose me, blindly, my double, just as egoless as I am. One has been *saved* by and for love, by abandoning love and the ego. Now one is no more than an abstract line, like an arrow crossing the void.[24]

BIBLIOGRAPHY

Butler, Judith. *Parting Ways: Jewishness and the Critique of Zionism*. New York: Columbia University Press, 2012.

Deleuze, Gilles. Lecture course of the 2 February 1982, www.webdeleuze.com.

———. *The Logic of Sense*. Translated by Mark Lester and Charles Stivale. New York: Columbia University Press, 1990.

Deleuze, Gilles, and Felix Guattari. *Anti-Oedipus: Capitalism and Schizophrenia*. Translated by Mark Seem. New York: Viking Press, 1977.

———. *A Thousand Plateaus: Capitalism and Schizophrenia II*. Translated by Brian Massumi. Minneapolis: University of Minnesota Press, 1987.

Levinas, Emmanuel. *Entre-Nous: On Thinking-of-the-Other*. Translated by Michael B. Smith and Barbara Harshav. New York: Columbia University Press, 1998.

———. "Entretien avec Levinas". *Autrement* 102 (1994).

———. *Of God Who Comes to Mind*. Translated by Bettina G. Bergo. Stanford, CA: Stanford University Press, 2003.

———. *Totality and Infinity: An Essay in Exteriority*. Translated by Alphonso Lingis. Dordrecht/Boston/London: Kluwer Academic Publishers, 1991.

———. *Otherwise than Being or Beyond Essence*. Translated by Alphonso Lingis. Dordrecht: Kluwer Academic Publishers, 1993.

Protevi, John. "Love". In *Between Derrida and Deleuze*, edited by Paul Patton and John Protevi. London: Athlone Press, 2002.

NOTES

1. Translated by Nathifa Greene and Marjorie Gracieuse.

2. Gilles Deleuze, lecture course of the 2 February 1982: "there is an identity of the face and the affect . . . the affects traverses the entire body . . . when the face loses its function of individualisation and abandons its social role, there only starts the adventure of the face . . . *the face is desire*, that is, passional affect" (www.univ-paris8.fr/deleuze).

3. Emmanuel Levinas, *Of God Who Comes to Mind*, trans. Bettina G. Bergo (Stanford, CA: Stanford University Press, 2003), 147.

4. Emmanuel Levinas, *Totality and Infinity: An Essay in Exteriority*, trans. Alphonso Lingis (Dordrecht/Boston/London: Kluwer Academic Publishers, 1991), 50–51.

5. Emmanuel Levinas, *Entre-Nous: On Thinking-of-the-Other*, trans. Michael B. Smith and Barbara Harshav (New York: Columbia University Press, 1998), 33.

6. "The relation with the Infinite is not a knowledge but a proximity, preserving the excessiveness of the uncontainable which grazes the surface; it is Desire, that is, precisely, a thought that thinks infinitely more than it thinks. To solicit a thought that thinks infinitely more than it thinks, the Infinite cannot incarnate itself in a Desirable, cannot, being infinite, enclose itself in an end. It solicits through a face". Levinas, *Entre-nous*, 58.

7. Emmanuel Levinas, *Otherwise Than Being or Beyond Essence*, trans. Alphonso Lingis (Dordrecht: Kluwer Academic Publishers, 1993), 110.

8. Levinas, *Entre-Nous: On Thinking-of-the-Other*, 104.

9. Levinas, "Entretien avec Levinas", *Autrement* 102 (1994), 142. My translation.

10. Levinas, *Otherwise than Being or Beyond Essence*, 129.

11. Levinas, *Otherwise than Being or Beyond Essence*, 4.

12. Levinas, *Entre-Nous: On Thinking-of-the-Other*, 229.

13. Levinas, *Totality and Infinity*, 62.

14. On the problem of elucidating the conditions under which, for Levinas, a political "enemy" has a Face, Judith Butler astutely notes that "for Levinas, the prohibition against violence is restricted to those whose faces make a demand upon me, and yet these 'faces' are differentiated by virtue of their religious and cultural background. This then opens up the question of whether there is any obligation to preserve the life of those who appear 'faceless' within his view or, perhaps, to extend his logic, by virtue of not having a face, do not appear at all. We have not yet seen a study of the 'faceless' in Levinas". In Judith Butler, *Parting Ways: Jewishness and the Critique of Zionism* (New York: Columbia University Press, 2012), 204.

15. Gilles Deleuze, *The Logic of Sense*, trans. Mark Lester with Charles Stivale (New York: Columbia University Press, 1990), 152.

16. With religion, "the world begins to signify before anyone knows what it signifies; the signified is given without being known". Gilles Deleuze and Felix Guattari,

A Thousand Plateaus: Capitalism and Schizophrenia II, trans. Brian Massumi (Minneapolis: University of Minessota Press, 1987), 112.

17. Deleuze and Guattari, *A Thousand Plateaus*, 63.
18. Deleuze and Guattari, *A Thousand Plateaus*, 170.
19. Deleuze and Guattari, *A Thousand Plateaus*, 176.
20. Deleuze and Guattari, *A Thousand Plateaus*, 175.
21. Deleuze and Guattari, *A Thousand Plateaus*, 171.
22. Deleuze and Guattari, *Anti-Oedipus: Capitalism and Schizophrenia*, trans. Mark Seem, Robert Hurley and Helen Lane (New York: Viking Press, 1977), 333.
23. Deleuze and Guattari, *A Thousand Plateaus*, 35. "What does it mean to love somebody ? It is always to seize that person in a mass, extract him or her from a group. . . . Every love is an exercise in depersonalization on a body without organs yet to be formed, and it is at the highest point of this depersonalization that one can be named, receive his or her family name, acquire the most intense discernability in the instantaneous apprehension of the multiplicities belonging to him or her, and to which he or she belongs". See also John Protevi's "Love", in *Between Derrida and Deleuze*, edited by Paul Patton and John Protevi (London: Athlone Press, 2002).
24. Deleuze and Guattari, *A Thousand Plateaus*, 134. My emphases.

THIRTEEN

Ageing and Longevity

Paul-Antoine Miquel

My aim, in this chapter, is twofold. First, I want to characterize ageing as a *normative* biological process, rather than a purely physical one. This, in effect, means that "ageing" is not a pure wear-and-tear mechanism. In fact, it isn't even a mechanism, and I'll explain why I'm critical of the use of that word in biology. It is, therefore, not accurate, in my view, to define ageing either as a mere imbalance between exergonic and endergonic reactions in the metabolism, as "the free radical theory of ageing" does, or as an imbalance between the excessive formation of reactive oxygen species and the limited amount of antioxidant defences. As we'll see, ageing is a *process*, but it is also a *constraint*, or a set of biological constraints, values, functions and norms. More specifically, it is a *repulsive* constraint.

This means that ageing is not so much a norm, or function, as a normative property. (I will focus on the example of the mammal, and will give an illustration of this point through the example of vicious molecular circles in connective tissues and in mitochondria.) With the expression "normative property", I mean that ageing is not simply a regulative property. It is a property through which the structure and constraints of a living being are actually transformed. Ageing is not destruction or degradation, but *self*-destruction. Ageing indicates that biological systems have a tendency to lose their normative power: that is, their ability to change their norms and generate new ones, and to do so in a regulated fashion. I suggest we refer to ageing thus understood as a process of *alteration*. Second, I want to analyse the relations between ageing and longevity. Not all organisms age, as everybody knows. In worms, yeast and other organisms, ageing is opposed to longevity, which counteracts the process of self-destruction. If the dynamic of ageing is also self-regulated, how are we to understand the claim, often found in the literature, that longevity "counteracts" ageing?[1]

THE FREE RADICAL THEORY OF AGEING

Initially proposed by Denham Harman,[2] the "free radical theory" is based on the idea of an imbalance between exergonic and endergonic reactions in the metabolic process of cellular respiration. It considers this imbalance to be the main factor in the degradation of organisms, through cellular dysfunction, genetic mutations and diseases. In another paper,[3] Harman also insists on the role of mitochondrial machinery in ageing, since the mitochondrial DNA is located near the respiration complex and isn't easily repaired. Cellular respiration is a metabolic reaction that converts the nutrients of an organism ($C_6H_{12}0_6$) in ATP with the help of an oxidizing agent: oxygen (0^2). Aerobic reactions thus correspond to the following general and schematic equation:

$$C_6H_{12}0_6 + 6O_2 => 6CO_2 + 6H_2O + \text{Energy}$$

Mitochondria convert the energy released in adenosine triphosphate (ATP). The process through which ATP is synthesized by a flow of $H_{+,}$ and by phosphates (P_i) is called "oxidative phosphorylation". During this process, the endergonic reaction (synthesis of ATP by a flow of protons H_+) counterbalances the exergonic one (which releases energy through the creation of a chain of electrons). Finally, oxidants and electrons are reduced in CO2, ATP and water.

However, this equilibrium has an entropic cost: the creation of reactive oxygen species as byproducts. Oxygen is incompletely reduced, giving way to radicals such as the superoxide radical ($\cdot O_2^-$) or the hydroxyl radical ($\bullet OH^-$). At the same time, superoxide anion ($\cdot O_2^-$) is reduced to (H_20_2) through the action of superoxide dismutase. And (H_20_2) is reduced to water (H_20) by catalase. Superoxide dismutase and Catalase are enzymatic antioxidant defences. However, this reduction is also incomplete, since a part of ($\cdot O_2^-$) is converted to $ONOO^-$ and a part of (H_20_2) is also converted to $\bullet OH^-$. One could also argue that the first imbalance between the endergonic and the exergonic reaction in the process of oxidative phosphorylation is in a way prolonged or completed by the second imbalance between the creation of free radicals and the reaction of antioxidant defences. Nothing, in such biochemical reactions, seems to contradict the stochastic and molecular characterization of the second principle of thermodynamics, which reduces ageing entirely to a process of physical deterioration.

I won't go any further into the molecular analysis of such a process of degradation. My point lies elsewhere. Can we follow Leonard Hayflick[4] and accept that the relations between longevity and ageing are simply relations between a biological mechanism, controlled by genes and a purely stochastic and entropic process? The answer is no. To begin with, there are examples of organisms with few oxidant defences that live a long time, like parrots. But my point isn't purely empirical. It's also theoretical. Hayflick writes the following: "wear-and-tear must affect all machines in a world subject to the second law of Thermodynamics, and the machinery of living things cannot be an exception".[5] But we need to understand, first, that an organism is precisely not a machine, since a machine is a technological device characterized by artificial constraints. Second, we need to recall that, even at the thermodynamic level, organisms are *open* thermodynamic systems—that is, systems *far from equilibrium.* As all physicists today know (Schrödinger,[6] Bertalanffy,[7] Prigogine),[8] dissipative systems produce a permanent flow of negative entropy, even though the total entropic cost for the environment is averaged out. Consider a crystal: it is constantly growing, and not subject to wear and tear. Yet, unlike organisms, it cannot age. Ageing is a biological constraint that can be explained neither by entropy nor by the properties that characterize dissipative structures, such as flames, or organized structures, such as crystals.

AGEING IS NOT SIMPLY A REGULATORY CONSTRAINT

Let me mention two classical examples of regulatory constraints in biology. The first one is the so-called lactose operon, for the discovery of which Monod, Jacob and Lwoff were awarded the Nobel Prize. The second one is the metabolism, as understood by Rosen's model of organizational closure. Monod,[9] Jacob and Lwoff proposed a model of the operon that requires that we distinguish between regulatory and structural genes. The regulator gene codes the transcription and translation of a specific, allosteric protein. The function of that repressor protein is catalytic as well as regulatory. It blocks the transcription of three structural genes that code the synthesis of betagalactosidase and permease proteins. Yet betagalactosides, when present in the bacterium membrane, play also a regulatory role. They are substrates for a chemical reaction and also induce repressor proteins inactivation on the operon locus. Genes and proteins play obviously the role of biological constraints or functions that cannot be derived directly from a physical property. And the genome is already represented as a complex structure, since genes have also horizontal and regulatory activities. They are not defined exclusively by the vertical principle of "genic determination", which is part and parcel of the so-called central dogma of molecular biology.[10]

Following Rosen,[11] we need to distinguish between "material" and "efficient" causes in biology. For instance, a chemical reaction is the relation between an effect and a *material* cause. But the activation of a chemical reaction, or its inactivation by regulatory properties, is an *efficient* cause. Rosen tries to understand the metabolism as a closed organization that involves material and regulatory (or efficient) causes or—to use another expression, which Neovarelians seem to favour—"constraints".[12] At this very general level, the organism is also defined by the fact that it regulates all its activities and replaces all its material and regulatory elements. Let's call C* this very general level, at which the organism regulates the replacement of all material (P_1, P_2, P_3 . . .) as well as regulatory elements (C_1, C_2, C_3 . . .). "C*" symbolizes the biological invariant, or the principle of organizational closure. If we look more closely, we realize that C* is involved in a recursive procedure that's applied to itself indefinitely, and "organizational closure" is more like the fixed point that emerges through the procedure in question:

$$(1)\ C^{*} = \Phi\, C^{*}$$

This means that the principle of organizational closure, which symbolizes all material elements and constraints that govern the biological autonomy of an organism, is also self-generated. It is the result of a self-productive activity. As such, it's not simply a regulatory principle, on which material elements and biological regulative constraints depend.

Nevertheless, in the first example, biological constraints can be activated or inactivated in a way that's *reversible*. It's the case, for instance, of the repressor protein. One could argue that those regulative activities are controlled by genes and ultimately by something like a "computer programme" implemented in each cell genome of the organism.[13] This assumption was in fact widely shared and used by biologists between the early 1960s and the late 1980s. In the second, more sophisticated example, a biological constraint can be destroyed or replaced in a way that's also reversible. Through a recursive procedure, the organizational structure C* is replaced by itself, yet not altered by the fact that a constraint C_1 is destroyed, or replaced.

In ageing, by contrast, the structure of the organism is *irreversibly* altered: it can no longer work the way it used to; there is a *before* and an *after* the event. Such is the reason why I suggest we define ageing as a *repulsive* constraint, as well as a normative property. Ageing is a constraint through which the biological norms of an organism are irreversibly changed. Let's call this sort of constraint "chronological". It attests that, during the ageing process, time acts on the constraints themselves, and not only on material elements. Let me be very clear here: by "time" I mean the becoming of the system, rather than the human measuring of that becoming. In my view, this "becoming" is as real and material as the physical elements involved.

In *Investigations*, Stuart Kauffman[14] offers a renewed definition of the concept of *work*. Work, he claims, is not simply "the release of energy" in a certain determinative structure, as in classical physics. For instance, when a planet turns around the sun, some quantity of work is generated without altering the conservative structure of the system in question. Kauffman calls "work" something that, in biochemistry, happens in relation to the structure of thermodynamic cycles. Cycles can be virtuous or vicious. The autonomy of a living organism is directly expressed by the existence of virtuous cycles, through which work propagates.[15] Under certain "constraints", work produces new constraints, which in turn produce more work:

> in addition, the organism carries out one or more real work cycles, linking spontaneous and non spontaneous processes. It does in fact measure, detect, and record sources of energy, and *does work to construct constraints* on the release of energy, which when released in the constrained way, *propagates* to do more work, often constructing further constraints on the release of energy or doing work by driving further non spontaneous processes.

We can relate Kauffman's claim to a specific intuition that Canguilhem formulates in *The Normal and the Pathological*.[16] A living organism, Canguilhem suggests, does not simply obey certain norms. Equally, it doesn't simply accomplish certain functions, under certain "physiological constants". Physiological constants or constraints actually change over time in biology. They change within single living organisms as well as between living organisms. Canguilhem calls "propulsive" those constraints that are able to produce new constraints and through which the organism doesn't simply follow existing norms.[17] In other words, the organism has the ability to change its norms and invent new ones. Thus, a propagative constraint is nothing but a normative one. Canguilhem takes the example of the immune system[18] that confers the property in an organism to recognize and resist the presence of intruders.

According to Canguilhem, repulsive constraints are also normative, because what is *not* normativity is integrated in normativity and also belongs to its definition. This idea is not clearly explicit in Kauffman. But I propose to connect what normativity is not with what Kauffman calls the "vicious circles of work". Vicious circles of work are constraints through which biological constraints are destroyed. In this instance, work produces new, destructive constraints, through which all biological constraints are changed. Typically, this happens when development is replaced by ageing and when life is eventually replaced by death. Like a Janus mask, life has two faces. Ageing, disease and death are also life. Thus, we can define ageing as a normative self-destructive process. Ageing is not simply destruction, or entropy. It is not simply a physical change through which mechanical activity is lost. Ageing is a *normative* property, through which propulsive constraints are changed into repulsive ones.

VICIOUS CIRCLES

What would a catabolic characterization of ageing as an emergent self-destruction function, or norm, or constraint, look like? Following Maffini, Soto and Sonnenschein,[19] who approach cancer from the point of view of the relation between stroma and epithelium, and refuse the somatic mutation theory, I want to analyse ageing at the level of connective tissues, and not simply at the level of genes. Specifically, I want to focus on the relation between cells and the extracellular matrix in connective tissues and draw on the work of Ladislas Robert and Jacqueline Labat-Robert,[20] who, in the 1980s, provided a nontrivial example of the vicious circle of ageing.

I'll return to this example, but let me mention first that biologists have analysed other examples of vicious circles. The most famous one is the way in which the production of oxygen radicals during cellular respiration triggers somatic mutations of mitochondrial DNA which, in turn, induce respiratory chain dysfunction, that process itself leading to further mutations. Finally, such a biochemical vicious circle leads *exponentially* to more ROS expression, and to more mitochondrial damage. Unfortunately, such an exponential degradation is mostly blocked by cell apoptosis, or by other mechanisms.[21] Another interesting example concerns the relation between degraded proteins and the proteasome in eukaryotic cells, since proteasome dysfunction can also be induced by oxidative stress through interactions of protein aggregates.[22] However, this point needs to be clarified. For instance, what is the exact role of protein aggregates in neurodegenerative diseases? How is the balance between the action of chaperone molecules and the proteasome perturbed during ageing? Many questions have been posed, but few have been answered.

As Robert often mentions, up until the end of the 1970s, cells were considered to produce the extracellular matrix, and were thought to be independent of it. In connective tissues, the extracellular matrix (ECM) is composed of two classes of macromolecules proteoglycans and fibrous proteins, like collagen, elastin, fibronectin or laminin. Such proteins are transcribed, translated and expressed in cells. But the matrix is not inert; it interacts with cells and cytoskeletons, through integrin and other receptors inducing a "loop"[23] between (ECM) and the elements from which (ECM) is synthesized. *Yet this loop is not closed on itself.*

In 1989 Robert et al.[24] identified the elastin/laminin receptor. More connective tissues are generated during ageing. At the same time, they are altered, structurally and functionally. Also known is the influence of connective tissues in certain human pathologies, like cancer.[25] In connective tissues, degradation is controlled. It propagates, as a dynamic post-translational property that emerges during ageing. Thus, ageing is not simply a degradation of constraints; rather, a new global constraint emerges in this process of degradation.

Developing from the bottom up, elastin molecules induce in mammals an increasing fixation of calcium and lipid deposition that saturates elastin fibre, degraded by elastases produced in an age-dependent manner by smooth-muscles cells and fibroblasts. It is a typical posttranslational process, in which chemical agents interacting with elastin are not synthesized by genes. Robert et al. have shown[26] that elastin degradation products interact with the elastin receptor, inducing loss of calcium regulation and elastases productions, which in turn degrade more elastin proteins in a self-amplifying process. Gilles Faury[27] has also shown that, in addition to this receptor, elastin peptides produce no-dependent vasorelaxation in rat aorta rings. This effect has efficiency in young adults, but declines with age. In this case, another process is involved, through which the chronic overload of the elastin receptor triggers a release of elastases and of free radicals. The same occurs with fibronectin degraded by proteases in fragments that also increase the proteolytic activity. To this day, those complex self-amplified processes aren't fully understood. For instance, why are more connective tissues generated during mammalian ageing? Why does the rate of biosynthesis fibronectin[28] and tropoelastin RNA increase with age? But the point I wish to make lies elsewhere. The interactive loop between cells and (ECM) is altered by such self-amplified processes, in a dynamical and normative way, and in connection with the formation of atheroma, with the heart's loss of contractile efficiency, the decline of the lungs' respiratory capacity, the progressive hardening of vascular walls, I use the expression "dynamical way" to emphasize the fact that the system in question has a becoming, and time is a form of agency. By "normative", I mean that the system is acting on its own constraints and adaptive power, and not only on its material elements. As such, apoptosis is just a limited and molecular manifestation of a much more important and meaningful process of ageing. Apoptosis must be understood through ageing, and not the other way around.

LONGEVITY

Molecular biology has changed in the last few years, notably through the so-called epigenetic turn.[29] What does such an evolution mean for the problem I'm concerned with here? Let me begin by assessing the claim that ageing is nothing but an evolutionary property. This vision was notably present in Weismann's work at the end of the nineteenth century.[30] Weismann understood ageing as the result of the coupling action of natural selection and heredity. But why? An easy answer could be found in the "disposable soma theory".

The origin of species reveals that Darwin was already aware of natural selection's ability to act indirectly on two different aspects characteristic of a population of living organisms: the reproductive rate and adaptation. Kirkwood suggests that organisms have a limited amount of "energy" that can be divided between reproductive activities and the maintenance of individual life. If individual life cannot be maintained for a long time (as in the case of the mouse), then reproductive success is very important. Conversely, where life has to be maintained over a long period of time in order to reproduce itself, maintenance mechanisms (like DNA repair) are more important.[31] Yet, in this model, change in reproduction and maintenance are only promoted by genes, or pleiotropic effects of genes. Does it mean that there are some genes of age on which natural selection acts, without any consideration concerning the adaptability of a living organism?

Let me focus on a family of proteins studied in worms, in yeast and in drosophila: the Sirtuins. The ageing process is analysed by counting the number of cell daughters produced by an individual cell mother (replicative life span), or by measuring the survival time of populations of non-dividing cells (chronological life span). Sirtuins have an epigenetic activity. They act in yeast by removing histone acetyls groups in the presence of NAD^+. Thus, they are classified as "NAD^+-dependent deacetylases". An extra copy of Sir2 extends yeast replicative longevity by 40 percent. As a deacetylase, it prevents sterility, decreases the formation of rDNA circles and plays a crucial role in the formation of silent chromatin (heterochromatin). The starving of yeast cells also extends its life span. It increases cell respiration and the available amount of NAD^+. It has the potential to increase the activity of Sir2. The conclusion might be that the combination of starvation and the epigenetic mechanism mediated by Sir2 and NAD^+ explains the extent of the replicative life span. Experiments in Nematode and in the fruit fly seem to support those findings. Leonard Guarente thinks along those lines when he writes, "In our own studies, what seemed almost magical was that yeast genetics led us something that promoted survival. The 'SIR' troika works to counteract ageing".[32] This conclusion is debated today. It is well established that there is not just one pathway that determines the duration of life in yeast,[33] or in other animals, but this controversy doesn't affect my own analysis.

In my view, Guarente's conclusion simply means that ageing can be regulated through *systemic factors* in relation with environmental constraints, so that longevity can be extended (or not). Life span extension is not simply controlled by stochastic genetic mutation and/or natural selec-

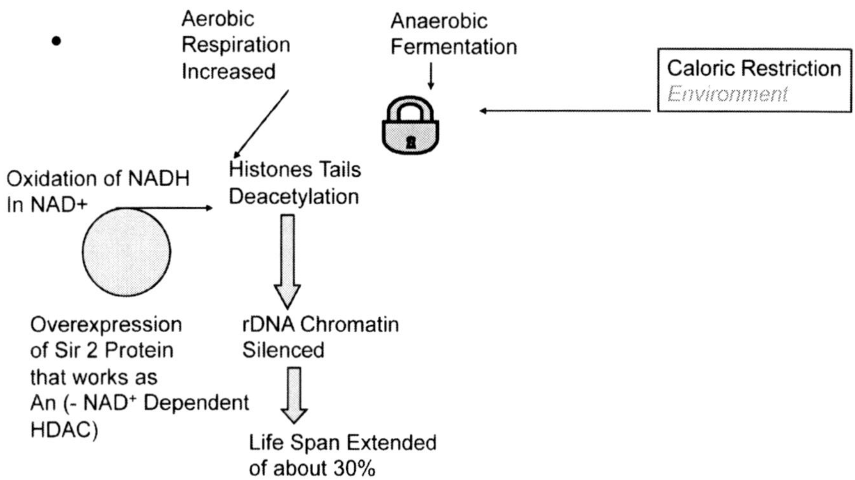

Figure 13.1. An Epigenentic Process of Ageing

tion, even if a mutation of the gene Sir2 can be involved in this case. Regulation of life span extension is a way for yeast to survive in the absence of food. It is a plastic norm that proves the ability of an organism to react to internal or external environmental input with a change in form, state, movement or rate of activity, and without any necessary change of genes, even if, in this instance, genes can also be involved.

I've just given an example of the impact of epigenetics on our understanding of ageing. Let me now develop it in a more speculative way. The French philosopher Henri Bergson[34] understood ageing as alteration and memory. I would add that longevity is nothing but "delay" in ageing, a delay that is not simply provided through heredity and natural selection, but also through the plasticity of the individual biological norms that characterize living organisms. For me, this means that adaptability must be understood as the limit between evolution and adaptation. Each organism is adapted by evolution through natural selection and other mechanisms. At the same time, however, each organism is able to adapt itself and change its norms by interacting with the environment. The more it is individuated, the more it is able to resist and counteract ageing. This resistance doesn't come simply from genetic and evolutionary factors. It also comes from its joint normative power of adaptation, which builds evolution at the same time that evolution builds it.

BIBLIOGRAPHY

Bergson, Henri. *Creative Evolution*. Translated by Arthur Mitchell. New York: Dover, 1998.

Bertalanffy, Ludwig von. "The Theory of Open Systems in Physics and Biology". *Science* 111 (1950): 23–29.

Canguilhem, Georges. *The Normal and the Pathological*. Translated by Carolyn R. Fawcett and Robert S. Cohen. New York: Zone Books, 1998.

Crick, Francis. "On Protein Synthesis". *The Symposia of the Society for Experimental Biology* 12 (1958): 138–63.

Denham, Harman. "Ageing: A Theory Based on Free Radical and Radiation Chemistry". *Journal of Gerontology* 11, no. 3 (1956): 298–300.

———. "Free Radical Theory of Ageing: Consequences of Mitochondrial Ageing". *AGE* 6, no. 3 (1972): 86–94.

Faury, Gabriel, et al. "Effect of Elastin Peptides on Vascular Tone". *Journal of Vascular Research* 32, no. 2 (1995): 112–19.

Friguet, Bertrand, et al. "Impairment of Proteasome Structure and Function in Ageing". *International Journal of Biochemistry and Cell Biology* 34, no. 11 (2002): 1461–74.

Guarente, Lenny. *Ageless Quest: One Scientist's Search for Genes That Prolong Youth*. New York: Cold Spring Harbor Press, 2003.

Hayflick, Leonard. "Entropy Explains Ageing, Genetic Determinism Explains Longevity, and Undefined Terminology Explains Misunderstanding". *PLoS Genetics* 3, no. 12 (2007): 2351–54.

Jenuwein, Thomas, and David Allis. "Translating the Histone Code". *Science* 293 (2001): 1074–79.

Kaeberlein, Matt, et al. "Sir2-Independent Life Span Extension by Caloric Restriction in Yeast". *PLoS Biology* 2 (2004).

Kauffman, Stuart. *Investigations*. Oxford: Oxford University Press, 2000.

Kirkwood, Thomas. "Evolution of Ageing". *Nature* 270 (1977).

Labat-Robert, Jacqueline, et al. "Age-Dependent Increase of Human Plasma Fibronectin". *Cellular Biology International* 5 (1981): 969–73.

Ladislas, Robert, et al., "Elastonectin and the Elastin Receptor". *Pathologie-Biologie* 37 (1989): 736–41.

Maffini, Marcel, et al. "The Stroma as a Crucial Target in Rat Mammary Gland Carcinogenesis". *Journal of Cell Science* 117 (2004): 1495–502.

Mayr, Ernst. "Cause and Effect in Biology". *Science* 134 (1961): 1501–5.

Monod, Jacques. *Le hasard et la nécessité*. Paris: Seuil, 1970.

Mossio, Matteo, and Alvaro Moreno. "Organisational Closure in Biological Organisms". *History and Philosophy of Life Sciences* 32 (2010): 269–88.

Prigogine, Ilya. "Symmetry Breaking Instabilities in Biological Systems". *Nature* 223 (1969): 913–16.

Prigogine, Ilya, and Gregor Nicolis. *Exploring Complexity*. New York: W. H. Freeman, 1989.

Rosen, Robert. "Some Relational Cell Models: The Metabolism-Repair System". In *Foundations of Mathematical Biology*, edited by Robert Rosen. New York: Academic Press, 1972.

Sanz Alberto, Pilar, et al. "Testing the Vicious Cycle Theory of Mitochondrial ROS Production: Effects of H_2O_2 and Cumene Hydroperoxide Treatment on Heart Mitochondria". *Journal of Bioenergetics and Biomembrane* (2006): 121–27.

Schrödinger, Erwin. *What Is Life?* Cambridge: Cambridge University Press, 1944.

Weismann, August. *Aufsätze über Vererbung und angewandet biologische Fragen*. Iena: Fischer, 1982.

NOTES

1. Thomas Kirkwood, "Evolution of Ageing", *Nature* 270 (1977): 301–4.

2. Denham Harman, "Ageing: A Theory Based on Free Radical and Radiation Chemistry", *Journal of Gerontology* 11, no. 3 (1956): 298–300.

3. Denham Harman, "Free Radical Theory of Ageing: Consequences of Mitochondrial Ageing", *AGE* 6, no. 3 (1972): 86–94.
4. Leonard Hayflick, "Entropy Explains Ageing, Genetic Determinism Explains Longevity, and Undefined Terminology Explains Misunderstanding Both", *PLoS Genetics* 3, no. 12 (2007): 2356–57.
5. Hayflick, "Entropy Explains Ageing".
6. Erwin Schrödinger, *What Is Life?* (Cambridge: Cambridge University Press, 1944).
7. L. von Bertalanffy, "The Theory of Open Systems in Physics and Biology", *Science* 111 (1950): 23–29.
8. Ilya Prigogine, "Symmetry Breaking Instabilities in Biological Systems", *Nature* 223 (1969): 913–16.
9. Jacques Monod, *Le hasard et la nécessité* (Paris: Seuil, 1970).
10. Francis Crick, "On Protein Synthesis", *The Symposia of the Society for Experimental Biology* 12 (1958): 138–63.
11. Robert Rosen, "Some Relational Cell Models: The Metabolism-Repair System", in *Foundations of Mathematical Biology*, ed. Robert Rosen (New York: Academic Press, 1972).
12. Matteo Mossio and Alvaro Moreno, "Organisational Closure in Biological Organisms", *History and Philosophy of Life Sciences* 32 (2010): 269–88.
13. Ernst Mayr, "Cause and Effect in Biology", *Science* 134 (1961): 1501–5.
14. Stuart Kauffman, *Investigations* (Oxford: Oxford University Press, 2000).
15. Kauffman, *Investigations*, 104.
16. Georges Canguilhem, *The Normal and the Pathological*, trans. Carolyn R. Fawcett and Robert S. Cohen (New York: Zone Books, 1998).
17. Canguilhem, *The Normal and the Pathological*, 206.
18. Canguilhem, *The Normal and the Pathological*, 206–7.
19. M. V. Maffini, A. M. Soto, J. M. Calabro, A. A. Ucci, and C. Sonnenschein, "The Stroma as a Crucial Target in Rat Mammary Gland Carcinogenesis", *Journal of Cell Science* 117 (2004): 1495–502.
20. Jacqueline Labat-Robert et al., "Age-Dependent Increase of Human Plasma Fibronectin", *Cellular Biology International* 5 (1981): 969–73.
21. A. Sanz, P. Caro, J. Gomez, and G. Baria, "Testing the Vicious Cycle Theory of Mitochondrial ROS Production: Effects of H2O2 and Cumene Hydroperoxide Treatment on Heart Mitochondria", *Journal of Bioenergetics and Biomembrane* (2006): 121–27.
22. B. Friguet et al., "Impairment of Proteasome Structure and Function in Ageing", *International Journal of Biochemistry and Cell Biology* 34, no. 11 (2002): 1461–74.
23. Jacqueline Labat-Robert et al., "Age-Dependent Increase of Human Plasma Fibronectin", *Cellular Biology* 5 (1981): 969–73.
24. Ladislas Robert et al., "Elastonectin and the Elastin Receptor", *Pathologie Biologie* 37 (1989): 736–41.
25. M. Maffini, A. Soto, J. M. Calabro, A. A. Ucci, and C. Sonnenschein, "The Stroma as a Crucial Target in Rat Mammary Gland Carcinogenesis", *Journal of Cell Science* 117 (2004): 1495–502.
26. Robert, "Elastonectin and the Elastin Receptor".
27. G. Faury, M. T. Ristori, J. Verdetti, M. P. Jacob, and L. Robert, "Effect of Elastin Peptides on Vascular Tone", *Journal of Vascular Research* 32, no. 2 (1995): 112–19.
28. Labat-Robert et al., "Age-Dependent Increase of Human Plasma Fibronectin".
29. Thomas Jenuwein and David Allis, "Translating the Histone Code", *Science* 293 (2001): 1074–79.
30. August Weismann, *Aufsätze über Vererbung und angewandet biologische Fragen* (Iena: Fischer, 1982).
31. Kirkwood, "Evolution of Ageing".
32. Lenny Guarente, *Ageless Quest: One Scientist's Search for Genes That Prolong Youth* (New York: Cold Spring Harbor Press, 2003), 32. See also Kirkwood, "Evolution of Ageing", 304.

33. Matt Kaeberlein et al., "Sir2-Independent Life Span Extension by Calorie Restriction in Yeast", *PLoS Biology* 2 (2004): 296.
34. Henri Bergson, *Creative Evolution*, trans. Arthur Mitchell (New York: Dover, 1998).

FOURTEEN

Generational Change as a Vehicle for Radical Conceptual Change: The Case for Periodic Rejuvenation

Steve Fuller

GENERATIONAL CHANGE AS A CENTRAL PROBLEM IN THE SOCIOLOGY OF KNOWLEDGE

Max Planck famously declared in his autobiography that significant scientific change requires generational change: you never change the minds of scientists who have spent their entire careers working within a certain research mindset. Only newcomers to the field, typically younger people, have the requisite open-mindedness to adopt an upstart perspective. In this respect, the public arguments that scientists periodically make to retain or shift their common research focus are made more for the gallery of posterity than for each other. Thomas Kuhn amplified Planck's observation into a general theory of scientific change, which helped to explain the "incommensurability" between paradigms that are dominant in a field before and after a "scientific revolution".[1]

When Kuhn first presented his general thesis in the 1960s, philosophers of science recoiled in horror at the prospect that scientists are not swayed by reason and evidence in their choice of theoretical direction. However, over the last fifty years, partly due to the rise of a more sociological understanding of science but also a greater sympathy for Darwinian "selectionist" explanations, the idea that "new science requires new scientists" perhaps no longer seems so radical. Nevertheless, as a normative proposition, it would suggest an argument for the planned obsolescence of personnel in science and perhaps other organized human activities that involve a regular transcendence of past practices in order to sustain their existence.[2] This proposition will be considered in relation to more respectful attitudes to the past (e.g., as "tradition", "wisdom" and arguably the "sanctity of life") as well as the current interest—motivated by both economic and transhumanist interests—in the indefinite extension of human longevity, not least in order to prolong one's working life.

Generational change, though relatively little studied today, was one of the original foundational topics in the sociology of knowledge, and Karl Mannheim's 1927 essay on the topic still sets the standard in terms of a conceptually sophisticated understanding of the phenomenon.[3] That this topic should have been taken up by this person at this time is understandable. The period 1870 to 1920—roughly two generations—marked the transition from the unification of the German-speaking states into an empire whose industrial and scientific base was the envy of the world until the Versailles settlement of the First World War brought the nation to its knees, resulting in Germany's first constitutional democracy, the Weimar Republic. Mannheim served as academic assistant to Alfred Weber, younger brother of Max, that iconic figure among young German academics. Max Weber's highly fluid, perspectival approach to social life, albeit stylistically congealed, reflected a life personally lived through these transitions, in which he sometimes had played a minor formative role.

However, Mannheim's nuanced account of generations as a principle of social structure brought together three distinct social relationships that are worth disaggregating: (a) people born within the same time range (i.e., a cohort); (b) people who orient their lives—at whatever age—to the same events (e.g., "Where were you when Kennedy was shot?"); (c) people who see themselves as bound in a common fate (e.g., everyone who plans to be alive in 2050). As Edmunds and Turner[4] observe, Mannheim himself was mainly concerned with (b), specifically how the response of people from different age groups to commonly experienced events in the recent German past explained the ideological spread of political parties in the newly established Weimar Republic. Thus, Mannheim believed that the Weimar political spectrum could be understood in terms of the causal narratives that people deployed to explain Germany's humiliation in the First World War. These narratives differed over the exact moment of "Original Sin" but typically it was located outside the battlefield, indexing events in distant places at distant times—as well as sustaining tendencies—that together provided a context for understanding the defeat as an eventuality of the various processes in play. To be sure, there is significant overlap in membership of the groups implied in the (a), (b) and (c) senses of "generation". Nevertheless, a disaggregation of Mannheim's complex notion of generation is in order, especially given our increasing capacity to know about the past as well as to extend our lives into the future. In this respect, age is indeed becoming simply a number.

THE GENERATIONAL SHIFT AGAINST NEWTON AND KANT AS COGNITIVE REJUVENATION

For purposes of this chapter, I wish to focus on (a), the cohort, the most "static" conception of generational change, in the sense that it aims to capture relationships between people simply by virtue of having been born at the same time. These relationships are usually understood in epistemic terms as complementary forms of blindness and insight. A younger generation is supposed to be more open-minded to new experience and hence more easily impressed by whatever happens, precisely because youth sets fewer prior expectations about what should happen. In sociological terms, the young are not so heavily invested in the past, even if they have been formally trained in established forms of knowledge, because they have not spent so much time routinely enacting their training as part of their self-presentation—and hence have not personally experienced the benefits of sticking to the old ways. For example, the classical (Newtonian) mechanics learned by the younger generation of physicists who eventually championed Einstein's relativity theory in the early twentieth century never acquired the ontological standing that it held for their elders. For the elders, classical mechanics embodied a rationalist worldview that, despite empirical shortcomings, had to be defended at any cost in the face of creeping cultural irrationalism.[5]

An interesting feature of this disparity in historical vision between older and younger cohorts is that by the time the generational conflict appears, the argument has become less about who has a better grasp of the ultimate truth than who is better adapted to survive in the future. This starting point already shows that the older cohort has conceded ground, as it shifts the stress from the realization of an overarching vision to the demonstration of a successful track record: a "progress *to*" is reduced to a "progress *from*". It is as if we had been lucky to get as far as we have, and it is only at our peril that we dare deviate from that path to whatever happens to be the fashionable alternative. In this respect, Kuhn's[6] notorious argument for the reluctantly revolutionary character of science may be seen as telling the entire history of science from the standpoint of a scientist in late middle age, someone like the physicist depicted in McCormmach.[7] The bottom line message is that it is too risky to behave counter-inductively.

Although Kant did not talk the language of risk, he introduced this defensive posture in the case of Newtonian mechanics by attempting to retro-fit Newton's fundamental conceptions of space, time and cause into the structure of the human mind, as it were, just waiting to be articulated by experience, which physics then aims to bring to full self-consciousness. This explains the theory's success, not only now but also in the future. But will this exfoliation of our innate cognitive powers actually result in the ultimate representation of reality? Kant himself was clear that this process may turn out to be no more than a projection of a self-generated "regulative ideal of reason". In any case, Kant's backwards-looking approach is a move to epistemic safer ground, reinforcing the foundations of what has been already shown to work (i.e., Newtonian mechanics in practice) rather than trying to leverage that empirical base into the completion of a specific worldview (i.e., Newtonian mechanics in promise).

As the nineteenth century wore on, Kant's cautious defence of the Newtonian orthodoxy seemed increasingly appropriate. It is worth recalling that by the dawn of the twentieth century, two centuries after Newtonian mechanics became the dominant scientific paradigm, the likelihood that it would resolve the anomalies surrounding the nature of light that had dogged the theory from its inception appeared small. Thus, in 1880, the pioneer neuroscientist Emil DuBois-Reymond invoked the slogan *ignoramus, ignorabimus* ("we do not know and we shall never know") to capture the humility of science. The slogan conveniently identified the limits of science with the limits of Newtonian mechanics, thereby obviating the logical possibility that perhaps Newton's theory needs to be replaced but science itself might continue apace.[8] Among these "conveniences" was an idea revived in recent years by Stephen Jay Gould:[9] namely, that religion begins where science can go no further. Moreover, even once the relativity and quantum revolutions had clearly taken hold in physics in the middle third of the twentieth century, it was common for both analytic and continental philosophers appealing, respectively, to ordinary language and phenomenology to argue that the great revolutions in physics, while positive developments in the history of science, implied that we were being taken away from our "natural" understanding of the world, on which any humanly intelligible mode of inquiry must be ultimately based.

I do not believe that these philosophical intuitions are as secure as they were, say, fifty years ago. Nevertheless, they remain as a vestige of the "older generation" sensibility, which is sometimes misidentified as "pre-theoretical". I say "misidentified" because the relevant understanding is not scientifically untutored. After all, as developmental psychologists from Jean Piaget to Michael McCloskey[10] have shown, the view of physical reality of the scientifically untutored child is closer to Aristotle's than Newton's. Instead "pre-theoretical" stands for someone tutored up to the level beyond which contemporary scientific research takes off. Thus, the old orthodoxy continues to furnish the "amateur" vision of the domain that the new orthodoxy now explores with authority, albeit to test and transcend its limits. In this context, it is interesting to observe the change in reputation that Ernst Mach underwent before and after the early twentieth century revolutions in physics.[11]

The prerevolutionary image of Mach was of a throwback to an earlier era, perhaps even back to Goethe, who still craved a human-centred worldview that sought phenomenological validity in our scientific understanding of the world. From this standpoint, Mach's compendium of conceptual and empirical objections to the Newtonian worldview, *The Science of Mechanics*, appeared as an exercise in resentment by someone who could not accept that a dehumanized "view from nowhere" had triumphed. To be sure, on the substantive scientific issues of his day, Mach was on the losing side of many debates, including the existence of atoms.[12] However, both the young Einstein and Bohr discerned in Mach's principled resistance to the Newtonian orthodoxy the seeds of a "relational" worldview that would come to be shared by relativity theory and the indeterminacy interpretation of quantum mechanics. Seen through the eyes of this younger generation, Mach looked like someone who was rethinking the foundations of physics by treating the officially forgotten objections as reminders of a truer path from which the field may have strayed because it responded too readily to research findings that promised to bear fruit in the short term.

The philosophical generation that tracked the experience of Einstein and Bohr—the logical positivists—should be understood in a similar light. (What is popularly called the "Vienna Circle" was originally named after Mach, who ended his career with the chair in the history of the inductive sciences at the University of Vienna.) Like that resolutely non-scientific thinker of the same generation, Martin Heidegger, the "positivists" now much derided "foundationalist epistemology" that would strip down the complexity of scientific discourse to observation statements and the logical operations performed upon them should be understood in the same spirit as Heidegger's much more seriously taken search for "the ground of being". Both projects should be understood in a rejuvenating spirit, whereby the recent difficulties encountered by the dominant "progressive" narratives are seen as symptomatic of what Heidegger would call a "forgetfulness of being" that requires a return to a more elemental way of understanding the world. Of course, Heidegger and Mach's Viennese followers interpreted this mandate rather differently and often in mutual antagonism. But common to the two enterprises was a Luther-like need to justify one's place in the world in terms directly available to oneself, not as a default inheritance from previous generations.[13]

However, it would be a mistake to see the breakdown of the older generation's Kant-inspired worldview purely in terms of the internal failure of classical mechanics. In addition, the nineteenth century witnessed a series of fundamental challenges to what might be called the "Adamic" premise shared by Newton and Kant. By this I mean the idea that humans, by virtue of having been created "in the image and likeness of God", are uniformly endowed with a conceptual apparatus that reproduced the blueprint of the divine plan. This helps to explain the ease with which epistemic finality was ascribed to classical mechanics. More to the point, it helps to explain why our predecessors should be taken as our epistemic superiors, if not our parents, in the sense to which Newton himself alluded when he said, "If I have seen as far as I have, it is because I have stood on the shoulder of giants": Those "giants" endowed us with the knowledge on the basis of which we go forward. (The Old Testament model may be Joshua vis à vis Moses.) The philosophical survival of this mentality is traceable through the signature Kantian "transcendental" style of argumentation.

While denying that the profound efficacy of the Newtonian worldview proved its ultimate truth, let alone its correspondence to "the mind of God", Kant nevertheless believed that it was impossible to think outside its parameters. From this innate limit to our imagination came the appearance of a divine hand. In any case, Kant's anthropic deity—real or imagined—was one who, as Leibniz had prescribed, chose to create the only world that would do justice to his divinity—namely, the best of all possible worlds, for which we are optimally equipped to understand. But these theological premises are not without their own problems: Might not this "best of all possible worlds" require humanity for its completion, so as to justify our existence as agents rather than as witnesses of creation? (Otherwise, we would seem to have inventive powers that are surplus to requirements.) In that case, as the Deists and Unitarians in Kant's day thought, we may mature as a species so as no longer to require parental oversight. Moreover, given that God created so many of us, might not this plurality suggest a need for different paths to the same ultimate goal? The questions raised by these complications to the image of God as universal optimizer spawned what by the early twentieth century would be a called a "relativistic" sensibility that accorded great epistemic significance to the historic survival of cultures that manifested radically different conceptions of space, time and cause. The philosopher who made the last great heroic attempt to reconcile the old Kantian vision with this superabundance of "ways of knowing" was Ernst Cassirer.

But the sharpest blow to the Kantian heartland was dealt perhaps unwittingly by art-historical scholarship, which suggested what amounted to a general cognitive rejuvenation strategy. Art historians showed that the so-called linear perspective formally developed out of Euclidean geometry that so clearly projected the worldview of classical mechanics was itself only a fifteenth century Florentine invention that was exported as an all-purpose construction principle to the rest of Europe during the Renaissance and then incorporated into the planning framework of colonial expansion. Despite its ancient Alexandrian provenance, Euclidean geometry lacked normative force as a generalized way of seeing the world until it was actively imposed on the layout of streets, the design of architecture, the art of painting and so on, after which it formed a "smart environment" that routinely conditioned ordinary people's sense of spatial (and, by analogy, temporal) perception. That Euclid's *Elements* had been "widely" studied in the interim must be understood in the context of the relatively low rates of schooling overall.

Nevertheless, by the end of the nineteenth century, partly reflecting classical mechanics' own persistent puzzles and partly inspired by the invention of non-Euclidean geometries, artists started to explore systematically the idea that the Euclidean space inhabited by classical mechanics is just one, perhaps overextended way of realizing of human cognitive potential, alternatives to which are realized in "oriental" and so-called primitive cultures or even an untutored child's perspective. Indeed, it might be possible to return to this epistemologically naïve state of mind. Thus, starting with Paul Cézanne and Vincent van Gogh, artists began to make a point of "unlearning" or "deconstructing" their craft in their practice, often resulting in the cultivation of a "hyperbolic" geometric perspective that envisages a horizon where there is no "vanishing point" but an open-endedness that encourages divergent resolutions of perceptual space on the part of the observer.[14]

THE POLITICAL CASE FOR COGNITIVE REJUVENATION IN SACRED AND SECULAR DEMOCRACIES

From the standpoint of the older generation, the epistemic horizons of the young veer dangerously close to an existential version of the equal-time doctrine of US broadcasting, whereby all sides of an argument are entitled to equivalent amounts of airtime, especially prior to an election. In effect, the charge is that the young do not give the track record of the past the weight it deserves while overestimating the promise of some unproven alternative. As the earlier allusion to Luther intimated, this problem has haunted the West from the dawn of Christianity. An especially potent site of contestation has been the status of infant baptism.

Generally speaking, the orthodox line followed by the Roman church has favoured a relatively automatic mechanism by which the new generation can immediately benefit from the decision of earlier generations to follow in the way of Christ. A secular descendant of this idea is that of citizenship as a right of birth. In contrast, heretics and later Protestants have called for baptism to take place later in life when one is in a position to make a free and rational choice. In this context, the person would be required to make a straight decision between being and not being a Christian, the result of which is binding on them but no one else. In the early days of the secular nation-state, followers of John Locke such as Thomas Jefferson argued for a collectivist version of this process, whereby each generation would be given the chance to ratify the social contract for themselves. Jefferson, writing while American ambassador to Paris in the period leading to the French Revolution, put the figure for decision at every nineteen years. He expressed this sentiment in a letter home that has been immortalized in the American libertarian slogan, though rarely quoted in full: "The tree of liberty must be refreshed from time to time with the blood of patriots and tyrants. It is its natural manure".[15]

Needless to say, this proposal would militate against the sorts of "in perpetuity" constitutions that characterize modern democracies—and to which Jefferson was opposed, in the case of the United States. A compromise solution was reached in the form of regular popular elections, often on a fixed-term basis, during which the standing government is forced to make explicit its plans to take its positive legacy into the future. Call it "planned political obsolescence". In several respects, this settlement has reduced politics to gamesmanship, with each election the functional equivalent of a match operating with a level playing field. The difference, of course, is that in politics, the winner is determined by the spectators whereas in sports it is determined by the players—in both cases subject to rules and refereeing. This is an interesting difference, the significance of which cannot be fully pursued here but is no doubt related to the fact that in politics the players are simply proxies for the spectators who are the real targets. (In this respect, the highly stylized character of modern warfare may be seen as a dialectical synthesis of democratic politics and competitive games.) In any case, if constitutions included operational definitions of their core values, then it would be easier to convert politics to a game format.

In a young person's ideal world, constitutions would not mime the same democratic platitudes, only to leave so much to the imagination—of the powerful—about how constitutional principles are implemented, as was the case of the United States and the Soviet Union during the Cold War. Rather, each nation-state would be constituted as an internationally recognized twenty-four-hour "rolling sport" (as in "rolling news"), which should be understood as slang for the pious phrase "living experiments in democracy". Thus, just as European football, American football and rugby are sports that aim to promote many of the same values but in significantly different ways, which are reflected in their respective rules of engagement, this is how we might imagine being a citizen of one or another nation. To be sure, this would leave open the question of how these "sporting" nation-states would relate to each other. But in a sense we already live in this world, as national spokespersons are quick to say whenever an electoral outcome does not meet with the rest of the world's approval: namely, that those who complain do not sufficiently recognize differences in the rules of the game—or the context of play.

From a Piagetian standpoint, such a sensibility marks the stage of "concrete operations" (or "conventional morality", in Lawrence Kohlberg's terms) during which adolescents refuse to let any further considerations override cognitive or moral judgements if they were not already formally available to the parties involved. Thus, even if judges function *ex post*, they are limited to spotting violations of the rules as implemented in recent memory rather than arguing that the "spirit" of the rules in a more general sense had been violated. They might reverse the outcome of a match but not the rules themselves. In other words, the judges would remain much more as "referees" than as members of a "supreme court". After all, an appeal to "spirit" would effectively grant the past greater license over the future, since persuasive appeals to *l'esprit de lois*, as Montesquieu put it, are difficult to make without according the past—if not the original lawmakers—considerable authority. What prevents such reversion to the past from inhibiting the prospect for legal innovation altogether is the ability to envisage what it would be like for the original lawmakers to be judging in today's world. In other words, we might counterfactually second-guess which aspects of their original judgements represented universal principles and which merely the exigency of the original circumstance. This speculative capacity corresponds to the highest stage of cognitive development, what Piaget called "formal operations" (and Kohlberg, "postconventional morality").

CONCLUSION: TRANSHUMANISM'S AMBIVALENCE TOWARDS COGNITIVE REJUVENATION

Contemporary transhumanism, despite its much vaunted "futurist" vision, remains largely oblivious to the intergenerational consequences of its quest for indefinite longevity. A step in the right direction is the grounds on which the European Union has been officially supportive of the "converging technologies" agenda that would provide incentives for the nano-, bio-, info- and cogno- sciences to join in research projects designed to "enhance" the human condition—namely, to extend the working life so that a pension can be legitimately drawn at an older age.[16] This would redress the balance between the producers and consumers of the welfare state's "universal benefits", thereby pushing back the spectre of the young having to subsidize their elders indefinitely with less for themselves once they retire. After all, Bismarck had designed the original social security system with a retirement age of sixty-five, when average life expectancy was just *below* that figure. Were he designing the same system today, the age would be eighty-five. However, it is equally clear that, partly due to the talismanic hold that "sixty-five" continues to have in our society, people are not merely allowed but also expected to enter into a period of protracted decrepitude after they leave the workforce. The remedy then would be to keep the body and mind fit—and hence keep people in post longer—through strategic regenerative interventions, as promised by the converging technologies agenda.

However welcomed, this strategy for enhancing ageing Europeans only scratches the surface of issues surrounding generational change. Perhaps the financial burden of the young is mitigated, but should the old really be expected to remain in post longer—or at least available as counsel to the young? Either prospect threatens the extent to which generational change might contribute to major socioeconomic transformation—"creative destruction", if you will. If new skills are needed for a new era, what, then, is the benefit of perpetuating older forms of labour and the mindsets associated with them? The reproductive processes of all societies are organized around age, which is to say, the time when people are expected to move from one role to another. The promise of indefinite longevity is poised to undermine the centrality of planned obsolescence to social succession—unless transhumanism's much-vaunted "black sky thinking" comes to the rescue and sends the able-bodied elderly to colonize other planets.[17]

Of course, one may wish to make a strong epistemic and economic case for the value of "wisdom" that older people possess that is detached from their specific work-related skills. But such arguments could easily veer towards turning older generations into second-order regulators of the younger generations who perform society's first-order labours. To be sure, this class of elders might serve to prevent harm in cases that readily recall past disasters—but it could equally, if not more likely, inhibit the sort of progress that comes from openness to novelty. Interestingly, there is a strand of Green thinking that associates such "wisdom" with a radically precautionary attitude to future generations.[18] For those of a more "proactionary" Enlightenment disposition, this attitude towards the young smacks of paternalism, even if it is vindicated in particular cases.[19]

In any case, it is striking that when, say, Aubrey de Grey[20] presents as attractive the prospect of our living several hundred years via various regenerative regimes, he presumes that however youthful our bodies may remain, our minds will carry forward the memories of the past to enrich our extended lives. Whatever else this utopia might be, it is *not* the future of, as Alexander Pope first said, "the eternal sunshine of the spotless mind", where forgetting—or simply not knowing—is considered liberating. For this reason, I have deemed de Grey's desire for indefinite longevity as a "middle youth" fantasy.[21] However, I mean this only as a normative judgement. I actually believe that a version of what transhumanists such as de Grey have proposed is likely to come to pass, but it will have enormous, largely unexpected consequences for how society is organized and how the value of life is understood more generally. In particular, the mindset of the younger generation—the sense of exhilaration that comes from acting in a bold yet naïve manner—may be threatened with extinction. When discussing the breakdown of the Newtonian worldview, I mentioned the role that artists such as Cézanne and van Gogh played in returning to a more open-ended way of seeing the world that approximated the standpoint of someone with a naïve sense of perspective, perhaps even a child. But how might one routinely simulate that mentality in a world where people are striving to live longer and continue to hold life's cumulative experience in high esteem?

Let me close by suggesting a role for virtual reality machines—a.k.a. video games—in addressing this question. The algorithms programmed into video games are designed to generate possible worlds from within a set of constraints, the nature of which the player comes to learn over the course of successive iterations of the game. Each iteration effectively returns the player to square one but now with an increased understanding of how the various possibilities permitted by the game emerge in various states of play, normally resulting in improved personal performance. The skills involved in this sphere of life, which is increasingly called "serious gaming", constitute a facility with counterfactual reasoning.[22] To be sure, such skills have been long cultivated with minimal technological interface—and perhaps more self-consciously—through exercises in modal logic, scenario building and "what if" historiography. However, for our purposes, the relevant feature of serious gaming is that it compels the gamer to pivot between the novel iteration that is the current game and her own experience of previous games. That novelty forces her to break any path-dependent-style expectations derived from the earlier games. Indeed, it may even make her a better temporal Gestalt-switcher: that is, someone who does not deny the past when envisaging the future but, on the contrary, reconfigures that very past into a quite different—and efficacious—future.[23] If one wanted a formula for cognitive rejuvenation, this would be it.

BIBLIOGRAPHY

Boyd, Jullien, et al., editors. *The Papers of Thomas Jefferson*. Vol. 12. Princeton, NJ: Princeton University of Press, 1950.

De Grey, Aubrey. *Ending Aging: The Rejuvenation Breakthroughs That Could Reverse Human Aging in Our Lifetime*. New York: St. Martin's Press, 2007.

Edmunds, June, and Bryan S. Turner. *Generations, Cultures and Society*. Milton Keynes: Open University Press, 2006.

Feuer, Lewis S. *Einstein and the Generations of Science*. New York: Basic Books, 1974.

Friedman, Michael. *A Parting of Ways: Carnap, Cassirer and Heidegger*. Peru, IL: Open Court Press, 2000.

Fuller, Steve. *Humanity 2.0: What It Means to Be Human Past, Present and Future*. London: Palgrave Macmillan, 2011.

———. *Knowledge: The Philosophical Quest in History*. London: Routledge, 2015.

———. *Kuhn vs. Popper: The Struggle for the Soul of Science*. Cambridge: Icon and Columbia University Press, 2003.

———. *Preparing for Life in Humanity 2.0*. London: Palgrave Macmillan, 2012.

———. *Thomas Kuhn: A Philosophical History for Our Times*. Chicago: University of Chicago Press, 2000.

Goodman, Nelson. *Fact, Fiction and Forecast*. Cambridge, MA: Harvard University Press, 1955.

Gould, Steven J. *Rocks of Ages*. New York: W. W. Norton, 1999.

Heelan, Patrick. *Space-Perception and the Philosophy of Science*. Berkeley: University of California Press, 1983.

Kuhn, Thomas. *The Structure of Scientific Revolutions* (1962). Chicago: University of Chicago Press, 1970.

Lipinska, Veronika, and Steve Fuller. *The Proactionary Imperative: A Foundation for Transhumanism.* London: Palgrave Macmillan, 2014.

Mannheim, Karl. "The Problem of Generations" (1927). In *Essays in the Sociology of Knowledge.* London: Routledge & Kegan Paul, 1952.

McCloskey, Michael. "Intuitive Physics". *Scientific American* 248, no. 4 (1983): 122–30.

McCormmach, Russell. *Night Thoughts of a Classical Physicist.* Cambridge, MA: Harvard University Press, 1982.

McGonigal, Jane. *Reality Is Broken: Why Games Make Us Better and How They Can Change the World.* London: Penguin, 2011.

Mean, Melissa, and James Wilsdon. *Masters of the Universe.* London: Demos, 2004.

Read, Rupert. *Guardians of the Future.* Weymouth, UK: Green House Publications, 2012.

Rescher, Nicholas. *The Limits of Science.* Pittsburgh: University of Pittsburgh Press, 1999.

NOTES

1. Thomas Kuhn, *The Structure of Scientific Revolutions* (Chicago: University of Chicago Press, 1970), 151.
2. Steve Fuller, *Kuhn vs. Popper: The Struggle for the Soul of Science* (Cambridge: Icon and Columbia University Press, 2003), chap. 3.
3. Karl Mannheim, "The Problem of Generations", in *Essays in the Sociology of Knowledge* (London: Routledge & Kegan Paul, 1952).
4. June Edmunds and Bryan S. Turner, *Generations, Cultures and Society* (Milton Keynes: Open University Press, 2002).
5. Russell McCormmach, *Night Thoughts of a Classical Physicist* (Cambridge, MA: Harvard University Press, 1982).
6. Kuhn, *The Structure of Scientific Revolutions.*
7. McCormmach, *Night Thoughts of a Classical Physicist.*
8. Nicholas Rescher, *The Limits of Science* (Pittsburgh: University of Pittsburgh Press, 1999).
9. Steven Jay Gould, *Rocks of Ages* (New York: W. W. Norton, 1999).
10. Michael McCloskey, "Intuitive Physics", *Scientific American* 248, no. 4 (1983): 122–30.
11. Lewis S. Feuer, *Einstein and the Generations of Science* (New York: Basic Books, 1974).
12. Steve Fuller, *Thomas Kuhn: A Philosophical History for Our Times* (Chicago: University of Chicago Press, 2000), chap. 2.
13. Michael Friedman, *A Parting of Ways: Carnap, Cassirer and Heidegger* (Peru, IL: Open Court Press, 2000).
14. Patrick Heelan, *Space-Perception and the Philosophy of Science* (Berkeley: University of California Press, 1983), esp. chap. 6.
15. *The Papers of Thomas Jefferson,* ed. Julian Parks Boyd et al., Vol. 12 (Princeton, NJ: Princeton University of Press, 1950), 356.
16. Steve Fuller, *Humanity 2.0: What It Means to Be Human Past, Present and Future* (London: Palgrave Macmillan, 2011), chap. 3.
17. James Wilsdon and Melissa Mean, *Masters of the Universe* (London: Demos, 2004).
18. Rupert Read, *Guardians of the Future* (Weymouth, UK: Green House Publications, 2012).
19. Steve Fuller and Veronika Lipinska, *The Proactionary Imperative: A Foundation for Transhumanism* (London: Palgrave Macmillan, 2014), chap. 1.

20. Aubrey de Grey, *Ending Aging: The Rejuvenation Breakthroughs that Could Reverse Human Aging in Our Lifetime* (New York: St. Martin's Press, 2007).
21. Steve Fuller, *Preparing for Life in Humanity 2.0* (London: Palgrave Macmillan, 2012).
22. Jane McGonigal, *Reality Is Broken: Why Games Make Us Better and How They Can Change the World* (London: Penguin, 2011).
23. Cf. Steve Fuller, *Knowledge: The Philosophical Quest in History* (London: Routledge, 2015), esp. chap. 6; and Nelson Goodman, *Fact, Fiction and Forecast* (Cambridge, MA: Harvard University Press, 1955).

Part IV

Desire and Pleasure

FIFTEEN

Sexuality and Liberalism

Patrick Singy

In the literature on the history of sexuality we often read that sexuality emerged within the historical period called "modernity". Since "modernity" usually implies, if only vaguely, the political context of "liberal democracies", the link between sexuality and liberalism would seem to be well established. Yet the exact nature of this link is usually not further explained and has all the appearance of a mere coincidence. For Michel Foucault, for instance, liberalism was not constitutive of sexuality: it was a way to administer sexuality and had only a relation of exteriority to it. As is well known, he traced the emergence of sexuality to the practice of post–Tridentine Catholic confession—hardly a liberal origin.[1] In this chapter I will try to tighten the connection between sexuality and liberalism and argue that sexuality is, in fact, a liberal concept.

LIBERALISM AND THE PROBLEM OF RADICAL EVIL

The label "liberalism" can apply to different approaches to governing, but the common thread is an emphasis on individual liberty: "everyone is the best judge of his/her interests and of the management of his/her life, is left as free as possible as long as he/she does not harm others and does not make an attempt on anyone's vital interests".[2] As Foucault and others have observed, the emphasis on liberty implies, in an apparent paradox, an emphasis on security and all the constraints on liberty that this implies.[3] What liberalism promotes is not liberty per se, but what has been called "ordered liberty".[4]

While security measures are unavoidable given that human beings are no angels, liberalism is possible only if human beings are not demons either. In *On Liberty*, John Stuart Mill made it clear that "those backward states of society" are not fit for liberalism: "Despotism is a legitimate mode of government in dealing with barbarians".[5] The liberal subject might be egoistic, but he or she is also reasonable, in all senses of the term. Catherine Audard traces liberalism's "resolutely optimistic conception of human nature"[6] to John Calvin and John Locke, who argued against Thomas Hobbes that human beings have a conscience, a "natural light" or reason, which guides them and makes it possible for them to be somewhat free without society falling into complete anarchy.[7] Liberalism contends that Hobbes had been overly pessimistic in believing that without an absolute monarch to keep people in check, "life of man [would be] solitary, poor, nasty, brutish, and short".[8]

Given the foundational optimism of liberalism, an obvious problem presents itself: how can the existence of radical evil be reconciled with a political theory that requires a fairly positive view of human nature? By "radical evil", I do not mean anything deeply philosophical: I only mean a form of evil that is not motivated by something other than itself—evil for the sake of evil. Obviously, radical evil is a problem in any type of society. But the existence of radical evil in a liberal society gives rise to two specific challenges.

First, there is a challenge to the functioning of the liberal criminal system. As Foucault described very well in *Surveiller et punir* and other works, in a liberal society both crimes and punishments involve a calculation of interests. An ordinary criminal has a reason to commit his crime (money, revenge, etc.), and the justice system needs to find a punishment that will be a reason for the criminal not to commit this crime.[9] But what can a liberal legal system do against a criminal who acts with nothing to gain, out of pure evil?

Second, there is a more profound challenge to the condition of possibility of liberalism described above. Doesn't the presence of radical evil demonstrate that human nature is so fundamentally evil that our faith in the possibility of liberalism is illusory at best, dangerous at worst? Wasn't Hobbes right after all?

Since the early nineteenth century, liberal societies have found an original solution to both challenges: they use psychiatry and other such sciences to define the perpetrators of radical evil out of humanity. The "degenerate" of the nineteenth century and the "sexual predator" of today are two notorious effects of liberalism. In response to the first challenge, the construction of an evil Other enables our modern societies to clear out a space within liberalism where punishment no longer needs to follow the rules of liberalism. For instance, in many Western democracies someone who has been deemed fundamentally evil by psychiatry can be confined *before* he commits a crime.[10] This preventive measure would clearly fly in the face of liberal reasoning if the person confined were thought to be a liberal subject. But his evil nature sets him apart from humanity, and we take care of him like we take care of a wild beast.

And of course, by giving a nature outside human nature to the perpetrators of radical evil, we can comfortably dismiss the second challenge as well: our faith in the reasonableness of human nature, this faith that is at the core of liberal thought, is not challenged by people who are not fully human. In the rest of this chapter I will show how sexuality emerged around the problem of radical evil and the challenges it poses to liberalism.

SEXUALITY AND LIBERALISM

"The facts that are being imputed to the defendant are so strange as to be difficult for the imagination to understand".[11] This is how the *Gazette des tribunaux* begins its report on a court martial that took place in Paris on 10 July 1849. The defendant, the Sergeant François Bertrand, is described as intelligent and of an ordinary size, with blond hair, clear blue eyes and a small and well-groomed moustache. He is accused of having desecrated corpses in a horrible fashion and of having had sex with some of them. I will return later to Bertrand's trial, which has left a long trail in the scientific and popular literature. It was discussed by many psychiatrists during the second half of the nineteenth century, including Krafft-Ebing, who used it as an example of sadism. For now, I want to focus on one specific exchange that took place during the trial, a short question followed by a short answer.

In the middle of the trial the presiding officer, Colonel Manselon, asks Bertrand the following question: "Have you ever wondered what was the point of destroying corpses that were already dead?" The absurdity of this question is what interests me. I do not mean "absurdity" in any kind of existentialist, Camusian sense, but in the straightforward dictionary sense of not being in harmony with reason, or, more precisely, of not being in harmony with our "style of reasoning".[12] What makes Manselon's question absurd is that, for us, an answer in terms of "what is the point" lies outside the realm of what is possible for us to think in a case like Bertrand's. We cannot imagine that rational motives would be behind repeated acts of mutilation and destruction of corpses. However, Bertrand's answer makes a lot of sense to us: "When my disease started, I felt, without being aware of it, this need to destroy".

At this moment in the trial, we might say that Manselon and Bertrand do not speak the same language. Manselon wants to hear about reasons and motives, and Bertrand answers in terms of need and disease. In this chapter I will locate my interpretation of the emergence of sexuality at this epistemological level. I want to offer a historical analysis that will show how Bertrand's answer came to be a possible candidate for an explanation of his behaviour. I will not be concerned primarily with the innumerable theoretical disputes or factual discoveries related to sexuality, but with the emergence of the concept of the sexual instinct, which is an "organizing concept", as Ian Hacking would say—that is, a concept that seems inescapable and that is used "for the intellectual and practical organization of a panoply of activities".[13] In a sense, the sexual instinct is what sexuality is all about: this is what Krafft-Ebing, Freud, Havelock Ellis and all the major players in the *scientia sexualis* are always talking about. It is what they explain, what they classify, what they cure—and it is what others liberate, celebrate and cultivate.

For reasons that I will explain next, the concept of the sexual instinct branched off from the earlier broader forensic concept of the instinct. The latter emerged with Philippe Pinel, who described in an article first published in 1798 a rather puzzling clinical case, which was later included in his path-breaking *Traité médico-philosophique sur l'aliénation mentale, ou la manie* of 1800. As far as I can tell, this is the first medical description of the kind of instinct that would become the focus of attention of so many nineteenth-century psychiatrists. Pinel's case is of a man who was taken by "the sudden invasion of a violent fury, which led him, with an irresistible propensity, to grasp an instrument or an offensive weapon in order to shed the blood of the first individual he would see". Most importantly, the man remained perfectly rational: "He said he felt a constant internal conflict between the ferocious impulse of a destructive instinct and the deep horror that the idea of a crime inspired in him. There was no sign of a lesion of memory, imagination or judgment".[14]

Pinel gave to this case of irresistible destructive instinct the diagnosis of "mania which consists only in a lesion of the will", or more simply of "mania without delirium". (Later it was also called "monomania".) This was a new kind of psychiatric disease, not a disease of the understanding but a disease of the instinct. Some defence attorneys were quick to use this disease to try to exonerate their clients, especially, as we will see, when the latter had committed horrible and senseless murders.[15] Among most legislators, however, the resistance to this type of defence was at first fierce.

Psychiatrists themselves were slow to engage in forensic debates, until Etienne-Jean Georget finally pushed psychiatry down the throat of legislators, with the publication of his *Examen médical des procès criminels des nommés Léger, Feldtmann, Lecouffe, Jean-Pierre et Papavoine* in 1825. In this short book, Georget looked back at five famous criminal cases and questioned the fairness of the sentences they had received. He argued that some of the defendants had been, precisely like Pinel's patient, the victims of an instinct they could not possibly have controlled; they should therefore not have been held legally responsible.

The difficulty for Georget was to distinguish as unambiguously as possible between an evil passion and a form of insanity characterized by a diseased instinct. While a passion might be an "excuse", in the legal sense of the term, and could mitigate the severity of punishments, it could not, unlike insanity, entirely exclude legal responsibility. This is why those in the legal profession who fought against Georget's effort to legitimize mania without delirium did so by stressing as much as possible the intimate relation between passion and insanity. The lawyer Collard de Martigny, for instance, published an essay in 1828 in which he claimed that there is "an exact similitude between homicidal monomania and any passion".[16]

It was therefore imperative for psychiatrists like Georget to sharply distinguish passion from mania without delirium: the fairness of justice as well as the legitimacy of forensic psychiatry rested on precisely this distinction. Unfortunately for Georget, there did not exist clear positive signs of mania without delirium: no specific physiology or anatomy, no pathognomonic symptom, no typical development of the disease. But a criterion was quickly found nonetheless: the absence of motives. Radical evil, such as killing for the sake of killing or stealing for the sake of stealing, became a disease.

"It is a truth that has been known for a long time", said Brierre de Boismont, "that all human actions have a motive, and that no individual commits a crime only for the pleasure of committing it".[17] Esquirol agreed: "The criminal always has a motive".[18] Any exceptions to this rule had to be explained away with a diagnosis of madness, especially when the crime was judged to be particularly gruesome. Antoine Léger, for instance, killed a young girl, opened up her body, drank her blood to satisfy his thirst, cut off her genital organs and finally ripped out her heart and ate it, or at least sucked on it (the facts in the report are not entirely consistent). The case of Léger was discussed by Georget, who noted that "anthropophagy is foreign to civilized people" and is not among "the ordinary motives of criminal actions", which are "cupidity, revenge, ambition, etc". Georget concluded that "Léger was therefore not pushed to crime by the passions that are its ordinary motives; his action does not have a motive that could be admitted by reason. He wanted *to drink blood! To eat human flesh!*"[19]

For Georget and others after him, it was not within the possibility of human nature to eat other people or to have sex with corpses, for instance. With one of the founders of modern liberalism, Jeremy Bentham, they clearly agreed that "the human heart does not have any absolutely evil passion".[20] "For the honour of the human species", asked Georget, "shouldn't the Caligulas, the Neros, the Louis XI be considered monomaniacs, who ordered to have unbelievable crimes committed with all the refinements of the most execrable cruelty?"[21] In 1816 André Matthey suggested that Pinel's expression "mania without delirium" should be called "tigridomanie"—that is, the disease of being ferocious like a tiger,[22] and today we talk about "sexual predators". These animal analogies are not insignificant: they signal that in our liberal age such criminals are no longer believed to be members of the human species.

By contrast, those arguing against Georget and his followers typically emphasized the depth and commonality of evil. Peyronnet, the prosecuting attorney in the famous 1825 case of Papavoine, a man who killed two children without any motive, anticipated the liberal objection to his argument against Papavoine: "You will tell us that human nature has never produced such monsters", by which he meant people who kill only "to spill human blood and satisfy a ferocious passion". But, Peyronnet went on, you should not trust "this first reaction of your heart". There were examples throughout history of individuals who enjoyed cruelty for its own sake, and Peyronnet referred his auditors to the Marquis de Sade's *Justine*.[23] In an 1830 book against monomania, Elias Regnault, after claiming that "no desire is foreign to human nature", had this dramatic expression that sums up quite well the position of Georget's adversaries: "There is then a demon in the heart of man; Hell resides in him".[24]

Although Regnault himself embraced some liberal causes—he fought for instance against the death penalty[25]—his belief in the evil side of human nature gives reason to Hobbes and weakens the possibility of a liberal society. Georget and his colleagues, on the other hand, were more profoundly following the liberal bent of the nineteenth century by insisting that the perpetrators of radical evil are no longer fully human. The monomaniac stands as the negative picture of the liberal subject.

Sergeant Bertrand was similar to famous monomaniacs like Léger, Papavoine and many others—save for the fact that in his case psychiatrists diagnosed a disease of the *sexual* instinct: "for any physician who has carefully studied mental alienation, this perversion of the genesic [= sexual] instinct is not more surprising than the suicidal, homicidal, or incendiary monomania".[26] It is true that the military court thought otherwise and condemned Bertrand to one year in prison, the maximum sentence for the crime of desecration according to article 360 of the penal code of 1810. But for the psychiatrists, this judgement only demonstrated the continuing lack of understanding of this type of insanity among judges. Articles were published in the most prestigious and influential journals of psychiatry, all arguing that the court had made a mistake and that Bertrand should not be held responsible since he suffered from a diseased sexual instinct. For the first time in history, a sexual "pervert" was being created.

This brief account of the emergence of the sexual instinct raises at least two questions. First, why did it take several decades for the *sexual* instinct to branch off from the forensic concept of the instinct? Indeed, in the first half of the nineteenth century libertinage was a cause of insanity, or it was caused by insanity, but it was not itself a type of insanity.[27] It is only after Bertrand that odd sexual behaviours came to be seen as the symptoms of a diseased sexual instinct. Why, then, is there a lag of half a century between Pinel's case and Bertrand? Second, is it a historical accident if the first pervert was a violent sexual criminal rather than a homosexual, a masochist or a fetishist, for instance? As I mentioned previously, later in the nineteenth century Krafft-Ebing coined the term "sadism" and used Bertrand as an example of it. Do sadism and other violent sexual perversions occupy a privileged place in the discourse of sexuality?

All these questions can be at least partly explained by the same two factors. First, as I have argued, in order for something to qualify as mania without delirium, both the lack of motive and a very deep sense of horror were crucial. But nonprocreative sex was seen as being usually very clearly motivated: people had nonprocreative sex because they were seeking pleasurable sensations. And while nonprocreative sex was officially condemned by most moral authorities, it was too widespread to elicit a real sense of horror, at least when compared to the one people felt when facing cases of cannibalism or infanticide.

Second, the concept of the instinct was created in the context of French forensic psychiatry. As it happens, in relatively liberal postrevolutionary France most sexual behaviours, while condemned by bourgeois morality and the Church, were legally irrelevant. Sodomy, in particular, had been decriminalized in France in 1791.[28] Only the few sexual acts that caused harm to others or constituted public offenses against decency had unequivocal legal implications, and therefore only such acts would be likely to raise the question of whether they had been committed freely or under the influence of an irresistible instinct.

These two complementary reasons for the delayed apparition of the sexual instinct amid instincts of murder or cannibalism also explain why the first significant case of sexual perversion was not a case of homosexuality, or fetishism or masochism, but rather was the case of Bertrand. Men who engaged in consensual nonprocreative sex would either not break the law or, if they did (for instance, by having sex in public), always be considered free liberal agents who had willingly turned away from normal, bourgeois sex in order to experience a forbidden kind of pleasure. In the first half of the nineteenth century, sodomites were still only sodomites, not perverts.[29] Bertrand, however, not only clearly broke the law but also did something that was seen as horrible beyond belief, so abominable that it could not possibly have been motivated by a quest for pleasure, especially since, as psychiatrists noted, he was a good-looking and intelligent fellow who could have satisfied his sexual desire with living women.[30] Only sexual behaviours that were at the same time criminal, horrible and unmotivated, necessitated a perverted sexual instinct as their principle of explanation, and only a violent sexual criminal, a "sadist", could meet all these early nineteenth-century criteria of perversion.

ON THE EPISTEMOLOGICAL INSTABILITY OF HOMOSEXUALITY

My analysis of the relation between the discourse of sexuality and liberalism invites us, against the current grain of the historiography of sexuality, to pay attention to the privileged role of violent types of sexuality, not only during but also after the first years of the discourse of sexuality. As anyone interested in the history of sexuality knows, most of the scholarship in this field is really about the history of homosexuality, while violent types of sexuality have generated comparatively very little interest. WorldCat yields about thirty times more entries with the key word "homosexuality" than with "sadism", for instance. This discursive imbalance would seem perfectly understandable: despite the importance of sadism and other violent types of sexuality in the first stage of the history of sexuality, homosexuality pushed aside the other perversions in the 1870s and quickly became the main obsession of psychiatrists.

No wonder, then, that scholars of all stripes, including historians, have responded in kind. But I think we should be careful not to take an avalanche of scholarship on homosexuality as a sign of its stability within the discourse of sexuality. I would argue precisely the opposite: one of the reasons why homosexuality has generated so much scholarship is because of its instability within the discourse of sexuality.

Homosexuality, but also playful S/M, fetishism, masochism and all the consensual sexual activities that do not violate the physical and emotional boundaries of those who engage in them, are legally innocuous within a modern liberal context. If I am correct to think that the sexual instinct, and therefore sexuality itself, emerged at the intersection of law and psychiatry in a liberal context, then all those perversions lack much *raison d'être* in a psychiatric classification. As a matter of fact, perversions other than the violent ones exist mostly because the rise of liberalism has never been more than incremental and partial, an ideal imposed by some and resisted by others. The concept of homosexuality, for instance, was created in Germany rather than in France, because the nineteenth-century German penal code was less liberal than the French one and enforced a sodomy law. This law was the motivation behind the pioneering works, in the 1860s, of the human rights campaigner Karl-Maria Kertbeny, who coined the word "homosexuality", and of the lawyer Karl Heinrich Ulrichs, who wrote tracts that directly influenced psychiatrists like Carl Westphal and Krafft-Ebing.

As a general rule, perversions other than the violent ones grew in the cracks of the liberal tradition, wherever and whenever the purpose of the law was not limited to defending individual liberty and was also used to enforce the moral views of society. And if some of these perversions survived, it is often because they have been repeatedly and deceptively associated with the desire to cause harm, thus being effectively transformed into varieties of sadism and other violent perversions.[31] It is this awkward and unstable position within the liberal tradition that makes nonviolent perversions prone to social resistance and vulnerable to academic critique, that puts some psychiatrists and politicians on the defensive and that inevitably leads to the spilling of much ink, and unfortunately of some blood as well.

By contrast, if violent perversions have generated less scholarship, it is not because they are fringe perversions, but because they are constitutive of sexuality, so deeply ingrained within it that to question their existence would undermine the entire system. Unlike the other perversions, their nosological legitimacy has never been seriously challenged. For instance, those who have lobbied to remove paedophilia from the *Diagnostic and Statistical Manual of Mental Disorders* (DSM), such as Richard Green or Charles Moser, have never been successful, and it is difficult to imagine them being successful in the near future, so entrenched is the belief in the radical Otherness of the paedophile.[32] More representative of our liberal age than Green and Moser is gay activist Charles Silverstein, who predicted recently that sexual perversions will disappear from the future editions of the DSM, but followed this statement with a qualification that guarantees the survival of the violent perversions: sexual perversions "will likely disappear from [the] DSM *for those who have consensual adult-adult sex*".[33] The new DSM-5, published in May 2013, agrees with Silverstein since it offers a definition of paraphilia that uses the legal concept of consent to demarcate paraphilias from other sexual interests.[34]

Given the historical relation that exists between sexuality and liberalism, what would a radical resistance to the discourse of sexuality look like? When we think of freedom and sexuality, we usually think of the sexual revolution and all the good things that followed, among which we count the decriminalization of sodomy in the United States in 2003 and the sprouting legalization of gay marriage in the Western world. One implication of my interpretation of sexuality as a liberal concept is that while we can make the discourse of sexuality gradually more tolerable by pursuing what the sexual revolution started, it will be difficult to escape fully from this discourse as long as liberalism remains the modern project of the Western world.

Homosexuality and other innocuous "perversions" are thankfully in the process of being dissolved into mere sexual quirks and erotic preferences. Without a doubt, the fight for gay rights has been and remains culturally important. But it is also epistemologically superficial. It is a side effect of the spread of liberalism in the Western world, and for this very reason, inasmuch as sexuality is a liberal concept, it can only reinforce the deeper structure of the discourse of sexuality itself. The resistance against the discourse of sexuality will require more of us than the recitation of our liberal principles. We will need to question the ontological separation that psychiatry has been building for almost two centuries between violent sexual criminals and the rest of us. We will need to open the gates of Hell, which Regnault claimed is within all of us—this Hell that saps our faith in the possibility of a liberal age.

BIBLIOGRAPHY

American Psychiatric Association. *Diagnostic and Statistical Manual of Mental Disorders: DSM-5.* Arlington, VA: American Psychiatric Association, 2013.

Anonymous. "Justice criminelle". *Gazette des tribunaux*. 11 July 1849.

Audard, Catherine. *Qu'est-ce que le libéralisme? Ethique, politique, société.* Paris: Gallimard, 2009.

Baillarger, Jules-Gabriel-François. "Cas remarquable de maladie mentale". *Annales médico-psychologiques* 4 (1858): 132–37.

Bentham, Jeremy. *Traités de législation civile et pénale.* 3 vols. Paris: Chez Bossange, Masson et Besson, 1802.

Brierre de Boismont and Alexandre Jacques François. "Observations médico-légales sur la monomanie homicide". *Revue médicale française et étrangère* 4 (1826): 225–44.

———. "Remarques médico-légales sur la perversion de l'instinct génésique". *Gazette médicale de Paris* 4 (1849): 555–64.

Carlier, Félix. *Les deux prostitutions.* Paris: E. Dentu, 1887.

Collard de Martigny, C. P. *Questions de jurisprudence médico-légale.* Paris: Chez Madame Auger-Méquignon, 1828.

Dagonet, Henri. *Considérations médico-légales sur l'aliénation mentale.* Paris: Rignoux, 1849.

Davidson, Arnold I. *The Emergence of Sexuality: Historical Epistemology and the Formation of Concepts.* Cambridge, MA: Harvard University Press, 2001.

Esquirol, Jean Etienne Dominique. "Note sur la monomanie-homicide". In *Médecine légale relative aux aliénés et aux sourds-muets, ou, Les lois appliquées aux désordres de l'intelligence,* by Johann Christoph Hoffbauer, 309–59. Paris: J.-B. Baillière, 1827.

Fodéré, François-Emmanuel. *Traité du délire.* 2 vols. Paris: Crapelet, 1816.

Foucault, Michel. *The Birth of Biopolitics. Lectures at Collège de France, 1978–1979.* Translated by Graham Burchell. New York: Palgrave Macmillan, 2008.

———. *The History of Sexuality Volume 1: An Introduction.* London: Allen Lane, 1979.

———. *Abnormal: Lectures at the Collège de France, 1974–1975.* London: Verso, 2003.

———. *Discipline and Punish: The Birth of the Prison.* Translated by Alan Sheridan. London: Allen Lane, Penguin, 1977.

Georget, Etienne-Jean. *De la folie.* Paris: Crevot, 1820.

———. *Examen médical des procès criminels des nommés Léger, Feldtmann, Lecouffe, Jean-Pierre et Papavoine.* Paris: Migneret, 1825.

Goldstein, Jan. *Console and Classify: The French Psychiatric Profession in the Nineteenth Century.* Chicago: University of Chicago Press, 2001.

Green, Richard. "Is Pedophilia a Mental Disorder?" *Archives of Sexual Behavior* 31, no. 6 (2002): 467–71.

Hacking, Ian. *Historical Ontology.* Cambridge, MA: Harvard University Press, 2002.

Hobbes, Thomas. *Leviathan.* Oxford: Oxford University Press, 1998.

Janus, Eric S. *Failure to Protect: America's Sexual Predator Laws and the Rise of the Preventive State.* Ithaca, NY: Cornell University Press, 2009.

Matthey, André. *Nouvelles recherches sur les maladies de l'esprit, précédées de considérations sur les difficultés de l'art de guérir.* Paris: J. J. Paschoud, 1816.

Mill, John Stuart. "On Liberty". In *On Liberty and Other Essays,* 1–128. Oxford: Oxford University Press, 1998.

Moser, Charles, and Peggy J. Kleinplatz. "DSM-IV-TR and the Paraphilias: An Argument for Removal". *Journal of Psychology and Human Sexuality* 17 (2005): 91–109.

Paillet, Alphonse Gabriel Victor. *Plaidoyers et discours*. Paris: Marchal, Billard et Cie, 1881.

Pinel, Philippe. *Traité médico-philosophique sur l'aliénation mentale, ou la manie*. Paris: Richard et al., 1800.

Regnault, Elias. *Du degré de compétence des médecins dans les questions judiciaires relatives aux aliénations mentales, et des théories physiologiques sur la monomanie*. Paris: J. B. Baillière, 1830.

Renneville, Marc. *Crime et folie. Deux siècles d'enquêtes médicales et judiciaires*. Paris: Fayard, 2003.

Sibalis, Michael David. "The Regulation of Male Homosexuality in Revolutionary and Napoleonic France, 1789–1815". In *Homosexuality in Modern France*, edited by Jeffrey Merrick and Bryant T. Ragan Jr., 80–101. New York: Oxford University Press, 1996.

Silverstein, Charles. "The Implications of Removing Homosexuality from the DSM as a Mental Disorder". *Archives of Sexual Behavior* 38 (2009): 161–63.

Singy, Patrick. "Danger and Difference: The Stakes of Hebephilia". In *The DSM-5 in Perspective: Philosophical Reflections on the Psychiatric Babel*. Springer, forthcoming.

———. "Gli 'stili di ragionamento' di Arnold Davidson". *Iride. Filosofia e discussione pubblica* 45 (2005): 437–42.

———. "L'ontologie, un problème historico-philosophique". *Agenda de la pensée contemporaine* 5 (2006): 97–108.

Valverde, Mariana. "'Despotism' and Ethical Liberal Governance". *Economy and Society* 25 (1996): 357–72.

NOTES

1. See Michel Foucault, *The History of Sexuality Volume 1: An Introduction*, trans. Graham Burchell (London: Allen Lane, 1979); Michel Foucault, *Abnormal: Lectures at the Collège de France, 1974–1975*, trans. Graham Burchell (London: Verso, 2003).

2. Catherine Audard, *Qu'est-ce que le libéralisme? Ethique, politique, société* (Paris: Gallimard, 2009), 10.

3. See Michel Foucault, *The Birth of Biopolitics. Lectures at Collège de France, 1978–1979*, trans. Graham Burchell (New York: Palgrave Macmillan, 2008), 65–67.

4. Among other places, this expression is used for instance in the important US Supreme Court case *Kansas v Hendricks*, 521 U.S. 346 (1997), 357.

5. John Stuart Mill, "On Liberty", in *On Liberty and Other Essays* (Oxford: Oxford University Press, 1998), 14–15. See Mariana Valverde, "'Despotism' and Ethical Liberal Governance", *Economy and Society* 25 (1996), 357–72.

6. Audard, *Qu'est-ce*, 176; see also 108.

7. Audard, *Qu'est-ce*, 41–59.

8. Thomas Hobbes, *Leviathan* (Oxford: Oxford University Press, 1998), 84.

9. Michel Foucault, *Discipline and Punish: The Birth of the Prison*, trans. Alan Sheridan (London: Allen Lane, Penguin, 1977). See, for instance, Jeremy Bentham, *Traités de législation civile et pénale* (Paris: Chez Bossange, Masson et Besson, 1802), 2:385–87.

10. See for instance the Sexually Violent Predator Laws in the USA. On the SVP laws, see Eric S. Janus, *Failure to Protect: America's Sexual Predator Laws and the Rise of the Preventive State* (Ithaca, NY: Cornell University Press, 2009). For a criticism of the US Sexually Violent Predator Laws in their relation with psychiatric categories, see Patrick Singy, "Danger and Difference: The Stakes of Hebephilia", in *The DSM-5 in Perspective: Philosophical Reflections on the Psychiatric Babel* (Springer, forthcoming).

11. Anonymous, "Justice criminelle", *Gazette des tribunaux*, 11 July 1849.

12. My understanding of "style of reasoning" is closer to Arnold Davidson's use of the expression than to Ian Hacking's. Cf. Arnold I. Davidson, *The Emergence of Sexuality: Historical Epistemology and the Formation of Concepts* (Cambridge, MA: Harvard University Press, 2001), chap. 5; Ian Hacking, *Historical Ontology* (Cambridge, MA: Harvard University Press, 2002), chap. 11 and 12. For an explanation of the differences

between Davidson's and Hacking's concepts of style of reasoning, see Patrick Singy, "Gli 'stili di ragionamento' di Arnold Davidson", *Iride. Filosofia e discussione pubblica* 45 (2005): 437–42; Patrick Singy, "L'ontologie, un problème historico-philosophique", *Agenda de la pensée contemporaine* 5 (2006): 97–108.

13. Hacking, *Historical Ontology*, 21.

14. Philippe Pinel, *Traité médico-philosophique sur l'aliénation mentale, ou la manie* (Paris: Richard et al., 1800), 81–82.

15. See for instance François-Emmanuel Fodéré, *Traité du délire* (Paris: Crapelet, 1816). See also Jan Goldstein, *Console and Classify: The French Psychiatric Profession in the Nineteenth Century* (Chicago: University of Chicago Press, 2001), 173.

16. C. P. Collard de Martigny, *Questions de jurisprudence médico-légale* (Paris: Chez Madame Auger-Méquignon, 1828), 74.

17. Alexandre Jacques François Brierre de Boismont, "Observations médico-légales sur La monomanie homicide", *Revue médicale française et étrangère* 4 (1826), 237.

18. Jean Etienne Dominique Esquirol, "Note sur la monomanie-homicide", in *Médecine légale relative aux aliénés et aux sourds-muets, ou, Les lois appliquées aux désordres de l'intelligence*, by Johann Christoph Hoffbauer (Paris: J.-B. Baillière, 1827), 353–54.

19. Etienne-Jean Georget, *Examen médical des procès criminels des nommés Léger, Feldtmann, Lecouffe, Jean-Pierre et Papavoine* (Paris: Migneret, 1825), 11.

20. Bentham, *Traités*, 3:27–28.

21. Etienne-Jean Georget, *De la folie* (Paris: Crevot, 1820), 75.

22. André Matthey, *Nouvelles recherches sur les maladies de l'esprit, précédées de considérations sur les difficultés de l'art de guérir* (Paris: J. J. Paschoud, 1816), 117. Antoine Léger was also described in the bill of indictment as a "tiger"; see Georget, *Examen*, 3.

23. Peyronnet, in Alphonse Gabriel Victor Paillet, *Plaidoyers et discours* (Paris: Marchal, Billard et Cie, 1881), 13–16.

24. Elias Regnault, *Du degré de compétence des médecins dans les questions judiciaires relatives aux aliénations mentales, et des théories physiologiques sur la monomanie* (Paris: J. B. Baillière, 1830), 207.

25. Marc Renneville, *Crime et folie. Deux siècles d'enquêtes médicales et judiciaires* (Paris: Fayard, 2003), 119–23.

26. Alexandre Jacques François Brierre de Boismont, "Remarques médico-légales sur la perversion de l'instinct génésique", *Gazette médicale de Paris* 4 (1849), 561.

27. See for instance Henri Dagonet, *Considérations médico-légales sur l'aliénation mentale* (Paris: Rignoux, 1849), 30.

28. Michael David Sibalis, "The Regulation of Male Homosexuality in Revolutionary and Napoleonic France, 1789–1815", in *Homosexuality in Modern France*, ed. Jeffrey Merrick and Bryant T. Ragan Jr. (New York: Oxford University Press, 1996), 80–101.

29. For instance, a psychiatrist like Johann Ludwig Casper, who is sometimes credited with having offered the first description of homosexuality in 1852, strongly and explicitly rejected his French colleagues' concept of perverted instinct, and so in my opinion falls clearly outside the discourse of sexuality, inasmuch as the concept of "sexuality" implies a clear distinction between the sexual instinct and mere sexual tastes or inclinations.

30. See for instance Brierre de Boismont, "Remarques médico-légales sur la perversion de l'instinct génésique", 559; Jules-Gabriel-François Baillarger, "Cas remarquable de maladie mentale", *Annales médico-psychologiques* 4 (1858), 134.

31. In the nineteenth century Félix Carlier for instance noted how "the ease with which pederasts [= male sodomites] shed blood is truly horrifying. . . . This development of the instincts of cruelty is certainly another cerebral consequence of habits of debauchery that are against nature"; Félix Carlier, *Les deux prostitutions* (Paris: E. Dentu, 1887), 469–70.

32. Richard Green, "Is Pedophilia a Mental Disorder?", *Archives of Sexual Behavior* 31:6 (2002): 467–71; Charles Moser and Peggy J. Kleinplatz, "DSM-IV-TR and the Paraphilias: An Argument for Removal", *Journal of Psychology & Human Sexuality* 17 (2005): 91–109.

33. Charles Silverstein, "The Implications of Removing Homosexuality from the DSM as a Mental Disorder", *Archives of Sexual Behavior* 38 (2009): 162. (Italics added.)

34. "The term *paraphilia* denotes any intense and persistent sexual interest other than sexual interest in genital stimulation or preparatory fondling with phenotypically normal, physically mature, consenting human partners"; American Psychiatric Association, *Diagnostic and Statistical Manual of Mental Disorders: DSM-5.* (Arlington, VA: American Psychiatric Association, 2013), 685.

SIXTEEN

Desire Within and Beyond Biopolitics

Miguel de Beistegui

This chapter seeks to explore the intersection between desire, as a fundamental feature of human nature, and biopolitics, as concerned with the conduct and government of life as a whole. Specifically, it is concerned with the point at which the problematic of desire (*eros, epithumia, concupiscentia, cupiditas*) comes into contact with biopolitics. Under what conditions and in what way does the question of desire become a *vital* problem, and then a *political* one? This, in turn, is tantamount to asking not only how desire becomes a problem for politics but also how politics transforms itself when it comes into contact with that question. With more time, we could show how, in Greek and Roman antiquity, but also in the Middle Ages, and even in the Renaissance, desire was a *constant* source of concern, precisely insofar as it led to all sorts of excesses or sins, and was in fact indicative of an imperfect, fallen or at least troubled human nature. As such, though, it belonged to a regime of discourse that was moral and religious, as well as political, and was inscribed in a problematic of hubris and *hamartia*, of transgression and excess. It was the object of a certain concern and a certain practice—a concern with a life of temperance, and a practice of control and mastery. In order to govern—whether oneself or others—and to govern *well*, one needed to *dominate* one's own desires. The ability to hold one's desires in check was the sign of a strong and good nature. Why? Because of the very structure or morphology of desire, which remains in place to this day, and consists of two main features: on the one hand, desire presupposes the experience of a lack as its origin; on the other hand, desire is oriented towards its own satisfaction, in which it comes to an end. Yet because the lack in question isn't temporary or accidental, but structural, and because the satisfaction is itself only temporary and accidental, the life of desire is necessarily unfulfilled and unhappy. And that is precisely why human nature, *qua* desire, requires a certain care or technology of the self, a certain shaping and moulding of life, one that could de described as spiritual and therapeutic in the broadest sense of the term.

THE ECONOMIC REGIME OF DESIRE

What, if any, is the place of desire in politics today? Does it still belong to the problematic of care, or does it fall within a technology of a different kind? Following Foucault, we could offer a twofold response to those questions, and identify a crucial transformation that began to take place towards the end of the eighteenth and the beginning of the nineteenth century. Firstly, we need to acknowledge the production of a new object—namely, sexuality—which coincided with the emergence of a new discourse (the *scientia sexualis*) and a new form of power, exercised over the body as well as the mind. As a phenomenon, sexuality unfolds at the junction of a discursive regime and a configuration of power, which creates its own norms and normative processes. It signals a physiology of desire framed by a new scientific discourse about, and a new power over, life itself. Secondly, and more importantly for the problem I am concerned with here, we need to acknowledge a transformation of the meaning and role of desire itself, or a displacement of what I would call its register, which began at roughly the same time, but experienced a remarkable radicalization in the twentieth century. This development, which we could refer to as a new *regime* of desire, is made all the more interesting by the fact that Foucault himself recognized it, but didn't dwell on it. The transformation in question presupposed not a new philosophical anthropology as such, but a new perspective on a classical one, and a new way of relating to one's desires and nature. From an object of mastery and domination, desire progressively became the necessary mechanism for the production of the greater good. This means that the hitherto dominant *ascetic* regime of desire was itself transformed radically, and replaced by a very different one, the extreme limits and natural conclusion of which became apparent in the last thirty years, under the neoliberal form of governmentality.

Before this shift, one governed oneself *against* one's desires, or at least in spite of them. After it, one governs oneself (and others) *with* desire and *for* it. As such, it amounts to a rehabilitation of desire as a *natural* or *vital* feature. What is remarkable about this conversion or displacement is that it is accompanied by a profound transformation of power itself, which shifts the locus of sovereignty from the state (or the Prince) to the individual. And the space where this rehabilitated desire exists in its free state, and expresses its sovereignty, is the market, precisely as a counter-power to that of the state. Now where there is power—in this instance, that of the individual in its relation to the state—there is also a specific type of knowing, or discourse. The specific form of knowing that this reversal takes is political economy. This is the second point of contact between desire and life, or the second articulation between life, on the one hand, and power and knowing, on the other. Desire is now integrated in an *economic* system.

Now before we go any further, we need to emphasize the exceptional, even violent nature of such an association of terms, one that, because it has become so familiar to us in the course of the last two hundred years, we're no longer able to recognize. For a very long period of time, politics was indeed thought to have a strong connection with life. But that was in the sense of the *bios politikos*, which was clearly distinguished from the life and science of the *oikos*.[1] "Economics" was concerned with the administration and management of bare life (*zoe*), which was seen as qualitatively different from the life of the *polis*. It was limited to the domestic sphere, and remained outside the (exclusively masculine) logic of freedom and sovereignty. By becoming political in the eighteenth century, economics introduced *zoe* within *bios*, or bare life within political life. But what's remarkable is that by making its way into politics, *zoe* immediately became *bios*, and thus abolished the distinction. In yet more concrete terms, the question of needs, as relating to vital *needs*, became a question for the public power itself. At the same time, the question that traditionally belonged to it and was its prerogative—namely, the question of sovereignty—became a question for the individual subject, now envisaged as a subject of desires and interests. The hypothesis I'd like to develop in this chapter is that desire stands precisely at the junction of this double movement: it signals the presence of sovereignty at the level of life, or the vital form that the new, so-called liberal governmentality takes. Indeed, it's not so much political economy as such (that of Marx, for example) which sees desire as the engine of the new sovereignty and the progress of history, but that of Adam Smith and Bentham and, more explicitly still, and closer to us, that of the Vienna and Chicago schools of economics.

From a threat to the *bios politikos*, and a tendency that did not find its place in economics understood as the management of the needs and the affairs of the *oikos*, desire, as the irreducible trait of human life, becomes the very engine or energy of political life, the instrument of a new sovereignty and the object of a new political science. Once associated with a dimension of human life that needed to be controlled, dominated or even erased through a variety of techniques or technologies of the self, and thus become the object of an ethics or a spirituality, desire eventually came to be associated with the development and flourishing of life in a political sense, thus altering the meaning of desire as well as those of politics. It's through this process of economization that desire became a biopolitical phenomenon.

A third and possibly final point of entry into this question would be through the immunological paradigm, as developed by Roberto Esposito especially.[2] Were we to relate the development in question to the question of biopolitics as immunology, we could say this: beyond the immunitary mechanism of sovereignty in the Hobbesian sense—the Leviathan, which enables life and its natural tendency to assert its own power (this is known as natural right) to preserve or save itself (*conatus sese praeservandi*) by giving up something that is integral to itself—yet extending and reinforcing that mechanism, and forcing the liberal tradition of political thought into a new direction, there is the mechanism of the market, and the science of political economy.

How does this mechanism operate? By creating a space or an object (the market), the expansion of which is, in principle, without limits, because it is itself the limit of sovereign or state power—and thus the manner in which the power in question immunizes itself against its own pathologies—as well as the vehicle through which desires are expressed freely, and ultimately satisfied. But this new space is justified only to the extent that it is presented and seen as a place of veridiction—that is, as sustained or underpinned by a hidden rationality: the "invisible hand". In other words, insofar as the free play of desires in the market place is shown or believed to be the principal generator of the well-being and happiness of the majority, it no longer requires the intervention of an external force, nor the therapeutic procedures of the Ancient world. It is a self-regulating, self-immunizing system generating its own, spontaneous order. The creation of the market as a place of veridiction, governed by laws akin to those of nature, and not by the will of the sovereign power or the Prince, is what allows human life to express itself freely—that is, according to its natural inclination (*conatus*). The "economization" of life is the mechanism that allows it to preserve and assert itself as a whole. It is the space in which human beings—now referred to as "individuals"—will be able to define and reach the ultimate goal of life itself, which is pleasure. In the end, desire is recognized as the basis for a new immunological strategy, which we could label "epithumology". My thesis is that this specific immunological strategy is one that, in the end, turns life (and desire) against itself, and as such can be seen as a specifically modern type of self-infection.

DESIRE AND THE BIRTH OF LIBERALISM

The philosophical groundwork of this decisive shift is carried out largely by British philosophers of the eighteenth and nineteenth century. One of its first formulations can be found in Locke's *An Essay Concerning Human Understanding*.[3] In that book, Locke defines desire as "an *uneasiness* of the mind for want of some absent good", and thus as involving the idea of pain, which we naturally seek to avoid.[4] "Life itself", he continues, "is a burden cannot be borne under the everlasting and unremoved pressure of such an *uneasiness*".[5] As such, desire is what motivates the *will* to act, and is the main, if not the only engine of human action. In that respect, desire and the will need to be distinguished very clearly: it is through the will that we act, but desire, and not the good, is what *determines* the will: "*good*, the *greater good*, though apprehended and judged to be so, does not determine the *will*, until our desire, raised proportionably to it, make us *uneasy* in the want of it".[6] So long as some sort of want, privation and uneasiness isn't felt, there is no reason to act: "Good and Evil, present and absent, 'tis true, work upon the mind: but that which immediately determines the *will*, from time to time, to every voluntary action, is the *uneasiness* of *desire*, fixed on some absent good, either negative, as indolence to one in pain; or positive, as enjoyment of pleasure".[7]

Desire, in that respect, is "the spring of action". This "law" applies to natural needs, such as the desire to satisfy one's thirst, hunger or one's sexual appetites, which work towards the preservation of ourselves and the continuation of the species. But it applies equally to moral principles (a man may be "persuaded of the advantages of virtue", yet "till he 'hungers and thirsts after righteousness' [Matthew 5.5]" and "feels an *uneasiness* in the want of it, his *will* will not be determined to any action in pursuit of this greater good"[8]) or "habits acquired by fashion, example, and education", such as "the itch after *honour*, *power*, or *riches*, *etc*.".[9] Ultimately, desire is the only power that moves us, as it allows us to experience the pain and want that we seek to remove as an obstacle towards the achievement of happiness, or pleasure, which is the ultimate goal and highest good for man. What we desire is happiness, or at least its lowest degree, which is the absence of pain. This, however, does not mean that we are led by our desires as if blindly, nor that we should seek to pursue each and every one of them, or even as they emerge, for it is in our power—a power of the *mind*—to "*suspend* the execution and satisfaction" of any of our desires and to "examine them on all sides, and weigh them with others" with a view to *judging* "the good or evil of what we are going to do" and so avoid a "variety of mistakes, errors and faults which we run into, in the conduct of our lives, and our endeavours after happiness".[10] When we have done this, "we have done our duty, and all that is in our power".[11]

This process of rational deliberation, reminiscent of the Aristotelian *bouleusis*, which keeps desire at bay, or temporarily suspended, is precisely the exercise and "source of all *liberty*".[12] Far from being a restraint or limit of freedom, such a process of examination and judgement is the "end" and "use of our *liberty*", as well as its "very improvement and benefit".[13] In addition, "the further we are removed from such a determination, the nearer we are to misery and slavery".[14] Consequently, we need to examine what may have looked like an unconditional rehabilitation of desire as the source of action, and come to the conclusion that whilst no action and no happiness is possible without desire, our liberty consists in our ability to examine each and every one of our desires and decide which ones to prioritize in our pursuit of happiness. It is only when judging the good or evil of an action that we are genuinely free—that is, self-determined or autonomous.

In that respect, Locke's account continues to echo the therapeutic strategies of desire put in place in the ancient and medieval worlds. But there is another, unspoken reason, for which desire cannot be entirely rehabilitated. To be sure, desire is necessary in the pursuit of happiness, and thus in action. Yet insofar as it is always accompanied by pain, what it really seeks is the end of desire, or a state of permanent happiness. But that is precisely what it will never be able to achieve: the human condition is that of a desiring being who desires only one thing, and that is to no longer desire. He or she may achieve temporary happiness. But, Locke is forced to admit, "as soon as any new uneasiness comes in, this happiness is disturbed, and we are set afresh on work in the pursuit of happiness".[15]

Hume takes up and develops further the shift that begins to take place in Locke. "The chief spring or actuating principle of the human mind", he famously writes, "is pleasure or pain".[16] Reason alone, then, is not enough to motivate the will, and any form of action, including the virtuous kind, requires passion.[17] Hume goes as far as to say that "reason is, and ought only to be the slave of passions, and can never pretend to any other office than to serve and obey them".[18] Why? By virtue of the same principle—pleasure and pain—that allowed Locke to rehabilitate desire, and whose terminology Hume is content to repeat: "'Tis obvious, that when we have the prospect of pain and pleasure from any object, we feel a consequent emotion of aversion or propensity [which Hume also calls *desire* in 2.3.9], and are carry'd to avoid or embrace what will give us this uneasiness or satisfaction. . . . 'Tis from the prospect of pain or pleasure that the aversion or propensity arises towards any object".[19]

Regarding the just or morally good action, and its origin, Hume rejects the idea that it is derived from nature: it arises artificially, from education and human conventions (3.2.1), and requires some *motive* to be carried out: "*no action can be virtuous, or morally good, unless there be in human nature some motive to produce it, distinct from the sense of its morality*".[20] What is the motive in question? Hume rejects the thought that it be a concern for the common good or interest: "men, in the ordinary conduct of life, look not so far as the public interest, when they pay their creditors, perform their promises, and abstain from theft, and robbery, and injustice of every kind. That is a motive too remote and too sublime to affect the generality of mankind".[21]

There is, Hume goes on to write, "no such passion in human minds, as the love of mankind, merely as such, independent of personal qualities, of services, or of relation to ourselves".[22] There is no love, and no positive feeling towards others, which is not rooted in the pleasure "that any action, sentiment or character gives us by the mere view and contemplation" of it.[23] There are no exceptions to the rule of pleasure and pain as the principal motivation for human action and tendencies. More specifically, there is a clear *natural* advantage for man to behave according to the rules and laws of society, including from the point of view of sexual reproduction. It is as if nature, which has exercised considerable cruelty towards human beings by loading them with countless needs and necessities, and extremely limited means to relieve them of such necessities, had encouraged human beings to enter into society. It is through society alone that human beings are able to compensate for their natural disadvantage and acquire superiority over their fellow creatures. Since human beings are *naturally* governed by interest, and that "even when they extend their concern beyond themselves, 'tis not to any great distance",[24] it would be unwise, if not altogether foolish (for simply ineffective), to govern them any differently than according to their own interest and relative selfishness, especially regarding private property and the acquisition of riches: human beings "establish government, as a new invention to attain their [natural] ends, and preserve the old [society]", the primary motive of the invention being "nothing but self-interest".[25]

The science of political economy, which emerges in the eighteenth century, capitalizes on such views regarding human nature. It legitimizes and formalizes the shift of the locus of power, from the sovereign to the individual; it defines desire and interest no longer only in relation to the sovereign, but in relation to this new political entity, this new political subject, known as the individual. But for that, it needs to create a new space, the *market*, in which desires will be able to express themselves freely, and be satisfied. Let me be more specific. Markets, which naturally existed before the emergence of political economy, are, as Foucault emphasizes, subjected to an epistemological transformation as a result of the emergence of that discipline.[26] From a place of "jurisdiction", which bore the mark of the sovereign and expressed his Law, the market becomes a place of "veridiction", with laws that are now ascribed to human nature and to the market as a quasi-physical field, governed by human desires and interests: "Just as the physical world is ruled by the laws of movement", Helvétius writes, "no less is the moral universe ruled by the laws of interest".[27]

It would be unwise, therefore, as well as vain, to seek to govern (whether oneself or others) by going against the laws of human nature. What is required, rather, is a proper and complete understanding of such laws, which alone can decide what will constitute good and bad government. Quite logically, good government will be seen as allowing the maximum amount of space for the free expression of those laws, which themselves, insofar they are likened to laws of nature, spontaneously tend to produce a state of balance, equilibrium or happiness. And the market is precisely seen as the space in which this spontaneous order can unfold, and human nature flourish.

This is how, in the words of Adam Smith, and once the idea of the "invisible hand" (or "Providence") has been adopted, it is possible to affirm that even the "natural selfishness and rapacity of the rich", with their "most frivolous desires", "their own vain and insatiable desires", actually contribute to the common good.[28] Having said that, and simply by way of further clarification, it's important to emphasize that, even in Adam Smith, who is often singled out as the market advocate *par excellence*, there is a certain tension between, on the hand, the affirmation of "self-love" and the "insatiable desires" of human beings as the necessary conditions of economic activity and well-being, and, on the other hand, "sympathy" as the basic mechanism of moral action—a tension that can perhaps be attributed to a residual Christian spirituality.

It's really with the birth of utilitarianism that this tension is resolved and that the market is unequivocally asserted as the place or space in which the natural tendency of human beings to seek and maximize their own pleasure can be realized. Insofar as "pain and pleasure" are the "two sovereign masters" that govern human nature: that is, "govern us in all we do, in all we say, in all we think",[29] the principle of good government can only derive from such a human nature. In other words, it can no longer be a question of governing oneself in spite of, or even *against* one's desires, but *with* them, or according to them. From an essentially therapeutic strategy of domination and control we have moved to a strategy of enhancement and maximization—that is, of *management*. From an essentially *ascetic* regime of desire, which dominated the ethical and political ideal of the West for centuries, we have moved to an *economic* regime and a *libidinal* economy. The question is no longer one of knowing what it is legitimate (or not) to desire, but what can generate the highest degree of satisfaction for any *individual*.

The question has become one of knowing how to best govern individuals who are naturally governed by their own desires, and who recognize as their true "sovereign" the principles of pleasure and pain. The problem of governmentality has become an economic problem, where the "science" of economics and the object it seeks to understand and predict—namely, the market—define the solution to that problem. Unlike political economy in the Marxist sense, which, as we saw, aims to neutralize desires and address the problem of needs, political economy in the liberal sense is the science of needs and *especially* of desires. Where Marx problematized the biopolitics of needs, I would say the *zoepolitics* of bare life, liberalism problematizes the biopolitics of desires, and sees the market as the space of their resolution.

Liberalism claims that there can be a genuine *politics* of desire, which is irreducible to a morality, spirituality, religion or philosophy of desire.

It rehabilitates desire as the very spring of economic and political action, and turns it into a new sovereignty. In that respect, it turns desire into an object of life *beyond bare life*. This shift regarding the locus of sovereignty, from the state or the Prince to the individual as a subject of desire transforms the question of government, and the place and role of the state: "The great object, the great *desideratum*", writes Bentham, "is to know what ought and what ought not be done by government" once it has been established—I would say posited axiomatically—that the market is the space in which desires can express themselves freely.[30] Freedom—this is yet another decisive development—is itself defined as the freedom to pursue such desires and seek their satisfaction.

DESIRE IN THE NEOLIBERAL CONTEXT

This further shift was carried out by the founding fathers of neoliberalism, and expressed as their own *credo*. The most radical conception of freedom thus understood is perhaps that of the Chicago School of economics, which understands it, in the words of Rob van Horn and Philip Mirowski, as "the capacity for self-realization attained through individual striving for a set of necessarily unexplained (and usually interpersonally ineffable) prior wants and desires".[31] It overlaps significantly with the following definition of the science of economics, which seems unaware of its tautological and ironic nature: economics, we are told, studies "how people choose to use *limited* or *scarce* resources in attempting to satisfy their unlimited wants".[32] Precisely to the extent that it is now invested with an efficiency and a rationality that is carried out, paradoxically, by individual interests, desires and passions, the market is seen as the principle, the model and the form of *good* "governmentality", and of the state itself. The latter is now seen as governing *for* the market—that is, with a view to its maximal efficiency. It is wrong, therefore, as Foucault emphasizes, to believe that in the neoliberal paradigm, the state has no role, or is reduced to one that is increasingly insignificant.[33] Insofar as it puts in place and guarantees the necessary conditions for the existence and growth of the market, and of competition in particular, the state is an irreducible part of the market operation. In that respect, the status of the market changes somewhat from its liberal conception: it's no longer seen as a mechanical system, governed by immutable laws, the model of which is gravity as expressed by Newton's laws of motion, but as an open, dynamic and especially *thermodynamic* system, the prime model of which is *life*, both in the sense of an organism that needs to be sustained in a milieu and in the evolutionary sense of a system that evolves through natural selection and competition.

As a result of the infinite expansion (at least *de jure*) of the market, and competition as the new paradigm, or essential feature, of the market, a new type of man emerges, the *homo economicus*. Now the term *homo economicus* dates back to the end of the nineteenth century, and was coined as a critical response to Mill's work on political economy.[34] It would be more apt, therefore, to speak of the *new* homo economicus. What, exactly, defines this new subject? The new *homo economicus* is no longer the partner of an exchange relating to a problematic of needs, on which a utility is founded, and leading to the process of exchange. He is no longer defined by a system of exchange, value and even consumption, at least understood in the traditional sense—that is, as the instance that exchanges money for goods. Rather, he has internalized the value of competition, to the point of making it a principle of conduct of life itself, of his *own* life. In short, he has become the *entrepreneur* of his own self, or the self that produces itself through entrepreneurial techniques.

The aim of neoliberalism is to allow each and every one of us, every individual, to recognize and experience him or herself as an entrepreneur, albeit of him or herself, of his own home, property, family, body and mind. As Foucault says, we have witnessed the de-multiplication of the model of enterprise within the social body.[35] The worker is no longer defined by his labour force—although one could of course argue that, across the globe, the number of proletarians in the Marxist sense is actually increasing—but by his or her "skills" and "human" capital, which now includes one's genetic inheritance ("genetic capital"), cultural background and education ("cultural capital") and even looks ("erotic capital").[36] The idea of a labour force, which needed to sell itself at the market price to a capital that would be invested in a firm, has been replaced by the idea of skills *as* capital, which receives an income in return for its services.

Through the figure of the entrepreneur, and the theory of human capital, it's precisely the difference between labour and capital that's erased. And, to quote a commentator, "the opposition between capitalist and worker had been effaced not by a transformation of the mode of production and distribution of wealth, but by the mode of subjection, a new production of subjectivity".[37] The worker is no longer compensated for a quantum of force that he or she expresses, but for an (essentially libidinal) investment that he or she made, and continues to make—for example, in education, now a service industry selling skills that are negotiable in the market economy, and in need of regular updating and upgrading. There is no longer anything like a pure salary: salaries themselves are viewed as income, and by that we need to understand a return on investment in human capital broadly defined. And insofar as the investor-consumer generates his or her own satisfaction or utility in that way, he or she is also a *producer*. Human capital, Schultz writes, is "*human* because it is embodied in man, and *capital* because it is a source of future satisfactions, or of future earnings, or of both".[38] In other words, he is the producer of his own enjoyment. Every worker is an agent or subject engaged in the same activity, that of the maximization of the utility function, and in that respect equivalent to any other activity. As Miller and Rose put it, the worker is "no longer considered as a social creature seeking satisfaction of his or her need for security, solidarity, and welfare, but as an individual actively seeking to shape and manage his or her own life in order to maximize its returns in terms of success and achievements".[39] The worker is now a (seemingly) self-owned enterprise requiring constant investment and improvement so as to perform in the best possible way in a competitive environment. The latter has become increasingly competitive and invaded all spheres of life.

As I was suggesting a moment ago, the energy on which this new system runs is no longer simply *physical*, whether we understand it as the energy that's extracted from the proletarian or from nature's resources. It is also, and increasingly, *libidinal*, and attached to a new type of subject, the entrepreneur-consumer. The market is itself no longer sustained by the *physical* or *calorific* energy of the proletarian (that energy has not disappeared, but rather has been largely "outsourced"), and by the mechanisms of exchange, but by the *libidinal* energy of the entrepreneur. What has taken place, then, is a shift in the type of energetics defining economic and social relations, and the emergence of a new form of energy.[40] Out of all energy sources energies, the libidinal one is by far the cheapest and the most renewable. It is also unlimited and, as such, endlessly exploitable. Capitalism has proved remarkably adept at creating techniques and technologies to capture, channel, package and sell that energy. One thinks of marketing, communication and advertising, of course, which are all technologies of the self and of life in general. This new approach was once very candidly, or perhaps cynically expressed by Paul Mazur of Lehman Brothers in an article from 1927, published in the *Harvard Business Review*. He writes that "we must shift America from a needs- to a desires-culture. People must be trained to desire, to want new things, even before the old have been entirely consumed. . . . Man's desires must overshadow his needs".[41]

Advertising was a major tool in enacting that shift, in that, according to a specialist and former director of the General Motors Research Lab, it is nothing other than "the organized creation of dissatisfaction".[42] But one also increasingly thinks of the computer technology which uses and capitalizes on the extraordinary development of social networks, online videos, tweets, clickstreams and other "unstructured sources" by gathering, analysing and ultimately selling to other companies what's referred to as "big data", and which a recent advertisement by IBM characterizes as the data of "desire".[43] Yet if, through this new technology, firms are able to understand, predict and anticipate the desires of their (actual or potential) clients, desire also constitutes its internal mode of organization. It radiates through the firm as a whole, from its lowest echelons to its highest peak, and through the creation of new hierarchies and grades between those extreme poles (middle management, back office, intermediaries, etc.) corresponding to a quasi-infinite list of titles (director, vice president, president, CEO, CFO, etc.).

As systems of desire, companies—and, increasingly, universities—also require the assistance of various techniques of "motivation" (such as seminars, conferences, trips and social gatherings aimed at encouraging and consolidating the corporate ethos) as well as "reflection" (such as coaching, performance evaluations, self-evaluations and targets), which aim to align all desires with the meta-desire of capital.[44] The bipolarity of the old schema has been replaced by the infinitely more nuanced and wide spectrum of a single Desire, by a series of stages or steps that one climbs patiently, by the ladder of the unifying Desire—the desire to maximize one's potential, or to obtain a maximal return on one's investment. Finally, and as we already suggested, the model of the enterprise has been internalized and applied to life itself and as a whole: we are encouraged to comport and govern ourselves as units of capital, for which we are responsible, and which require a never ending cycle of investment and return. Capital now defines the very *being* of the *human* being; it is the new anthropological paradigm that claims to speak the truth regarding human life as a whole.

To be sure, such techniques of subjectivation are different from the disciplinary techniques of, say, the military, the prison or even the school. In a sense, they are more effective—that is, more productive and "rational"—precisely to the extent that they achieve their goals through consent and a softer *dressage*. But let us not forget that, ultimately, it is a question of *dressage*—that is, of making the multitude behave in a certain way, or of conducting its conduct. Specifically, it is a matter of producing "individuals" through the realization and maximization of their capital, of generating skilled subjects able to compete on the global marketplace. On the surface, and through the market, it seems that desire was freed, and that the market is precisely the expression of the multiplicity, the infinity, even of human desires. But it is of the utmost importance that those desires all work in the same direction, that each step or stage becomes a cog of the same mechanism, the desire of a unique, infinitely differentiated Desire—the Capital-Desire. This is how, already in 1972, Deleuze and Guattari summarized it: "The wage earner's desire, the capitalist's desire, everything moves to the rhythm of one and the same desire, founded on the differential relation of flows having no assignable exterior limit, and where capitalism reproduces its immanent limits on an ever widening and more comprehensive scale".[45]

In that respect, capitalism can be seen as the greatest apparatus of capture of desire ever invented, the greatest (and constantly evolving) force to have aligned the multiplicity of desires on a meta-desire. It's an apparatus that, following F. Lordon, we could characterize as "epithumosynthetic", in that it manages to gather, federate and organize the majority of desires. But insofar as it also generates or produces its desires, it is also "epithumogenetic". At once federator and generator of desires, postindustrial capitalism has become something like the World Organisation of Desire (WOD).

DESIRE BEYOND NEOLIBERALISM: A NEW SOVEREIGNTY

Aristotle had already warned his readers against such a desire, and all the desires that follow from it: if economic exchange, and the exchange of money in particular, consists *in theory* of a just and equitable operation between equal citizens, it is in fact the constant object of an unlimited desire (*epithumia*) that threatens the social link (*philia*) that brings together free, equal and sovereign citizens. In the *Politics*, Aristotle warns us against this particular *techne* that he calls "chrematistics" and which consists in the accumulation of goods and riches for purely personal aims. He distinguishes it very clearly from economics in the strict sense of the term—that is, from the *techne* oriented towards the natural needs of life (*zoe*) and the home (*oikos*) or the estate. The following two passages from the *Politics* are, in that respect, illuminating:

> As in the art of medicine there is no limit to the pursuit of health, and as in the other arts there is no limit to the pursuit of their several ends, for they aim at accomplishing their ends to the uttermost (but of the means there is a limit, for the end is always a limit), so, too, in this art of wealth-getting [*chrematistike*] there is no limit of the end, which is riches of the spurious kind, and the acquisition of wealth. But the art of wealth-getting which consists in household management [*he oikonomia*], on the other hand, has a limit; the unlimited acquisition of wealth is not its business. And, therefore, in one point of view, all riches must have a limit; nevertheless, as a matter of fact, we find the opposite to be the case; for all getters of wealth increase their hoard of coin without limit.[46]

> Hence some persons are led to believe that getting wealth is the object of household management, and the whole idea of their lives is that they ought either to increase their money without limit, or at any rate not to lose it. The origin of this disposition in men is that they are intent upon *living only*, and not upon *living well* [*to eu zen*]; and, as their desires [*epithumia*] are unlimited, they also desire that the means of gratifying them should be without limit.[47]

Needless to say, such a "system" amounts to the production and organization of lack, and is oriented towards an always increasing, yet endless consumption, known today as "growth". But if, as neoliberalism suggests, our desires are necessarily unlimited, are we not necessarily condemned to dearth—and endless dissatisfaction, however great the resources may be? Furthermore, are we not forced to increasing those resources *ad infinitum*, in what amounts to an ultimately pointless and destructive exercise? Paradoxically, we arrive at a situation that is the exact opposite of the one that is explicitly mentioned as the natural outcome of the market economy: the market, we recall, is supposed to be the place where pleasure and happiness are maximized. The question, then, becomes one of knowing to what extent the market economy, especially as it has developed in the last thirty years, and generated its own doxa and dogma, isn't in the end the agent of its own toxicity, and thus ultimately self-destruction.

The question is one of knowing whether, precisely to the extent that they offer a solution to the problem of desire, rather than that of needs (which I believe they *can* give), the market, and the "science" that defines it, don't offer a false, illusory and ultimately harmful solution. To be more specific: from a market *economy*, the aim of which is to produce a situation of abundance from which new social relations and a life of *otium* could emerge, we moved progressively, and I would say catastrophically, to a market society and a life of *negotium*, the ultimate but necessarily unachievable goal of which is growth itself. What was once considered a means has become an end in itself. Growth requires the limitless conquest and creation of new markets, as well as an increasingly invasive consumption, to the point that the space separating life, and the question of life, from consumption, becomes increasingly difficult to define. It's not in reality that growth is unlimited. The recent financial and economic crises are there to remind us that growth is *actually* limited.

Such crises are bound to become more regular, and could in fact signal, in what would amount to a paradoxical reversal, the collapse of the system as a whole, precisely to the extent that it relies on the infinity of a form of desire that sees a way forward only through its perpetuation and precipitation, or its catastrophe in the consumerist and speculative frenzy. It's a system that has quite literally become catastrophic. So, if growth is unlimited, it's not in practice or in reality, but rather in theory or in principle: it is, and needs to be, without horizon or end. It is, to use a Hegelian expression, a bad infinite, which constantly needs to invent new mechanisms in order to perpetuate itself—mechanisms that aim to capture, channel or funnel desires, but also, and as their necessary corollary, to get hold of natural resources, including by force, to seek cheap labour and so on.

Were there to be a failure of desire, or a shortage of libidinal investment in the economic system, it's the system as a whole that would be threatened. At the same time, we need to wonder whether the constant investment of desire, one which, as we saw, is structurally and necessarily unsatisfactory, doesn't or won't lead to the same dead end, and to further bubbles, further crises, which risk bringing down the system as a whole. It's as if desire had become the condition of possibility and impossibility of our socioeconomic system, that without which it cannot function, and that which threatens it like no external force or will. But it is also through desire, its mutation or transformation, or the creation of a new morphology of desire, that the system can be neutralized or, better still, ignored. In that respect, desire is pharmacological: at once poison and remedy. It's all a question of the type of investment, of dosage and direction. In the end, might not biopolitics also be concerned, and even coincide, with that question?

As the current economic crisis tolls the bell of the growth and consumerist model, the vast majority of politicians and elected representatives are unable to recognize it as such. On the contrary, all the solutions and measures that are adopted or envisaged confirm the same trend, which inevitably translates into an increase and an exponential acceleration of speculation, that is, in the end, of a short-term vision that has become the toxic, possibly lethal agent generated by the capitalist body itself. It's as if it too were prone to a form of auto-immunization—or so it believes—that is actually one of auto-infection. What the economic regime of desire, especially in its neoliberal and hyper consumerist phase, cannot contemplate is the possibility of a libidinal *disinvestment*. By that, I don't mean a neutralization or eradication of desire *as such*. Rather, I mean a type of desire that would be both unproductive (I would say "improductive") *and* vital.

Insofar as it has entirely embraced the cause of productivity and the principle of maximum yield, this regime, which dominates life today, is no longer able to envisage a form of life that would be free of all calculation or economic rationality, all strategies of investment and return—free, in short, of the logic of interest. Such a shift would have nothing to do with the so-called moralization of markets and capitalism that's often called for, simply because one couldn't possibly ask economic agents to act out of disinterestedness. This would simply be contradictory.

But there is also an economic, or perhaps *aneconomic* response to the economic regime of desire, one that would nonetheless escape the logic of interest and investment of capitalism. It's the desire of what Bataille calls free expenditure, and which he equates with a "glorious operation". In the capitalist economy, he says, "every expenditure, even if it is unproductive, is subordinated to the acquisition of new forces of production". To this economy of investment and entrepreneurship, "which absorbs the excess of forces with a view to the unlimited development of wealth",[48] of which we saw that it has penetrated all spheres of life (*zoe* and *bios*), subsuming the latter under its imperatives, Bataille opposes an economy of "sacrifice" and "consummation", which mobilizes a very different type of desire—I would go as far as to say a different morphology of desire—one that is not bound to the bad infinite and the lack normally associated with desire, but to a purely positive and excessive economy.

Beyond calculation and utility, the "sovereign" desire that's in question here counts for "nothing" and cannot be counted. Yet it is *intimately* related to life, and forces one to reconsider the relation between life and the economy. This is the sense in which it designates the *limit* of the principle of utility that governs us today, and with which we govern ourselves—a principle that Bataille doesn't hesitate to qualify as a "ridiculous stupidity". It is a question, therefore, of identifying a *limit* of the regime of utility—yet one that does not signal a way out or an outside of economy. Rather, it's a matter of recognizing such a limit as an economic limit, or a self-limitation of the economy, which signals the very meaning of the life of the subject.

This is the meaning of life that I would characterize as more than bare life, beyond and after the meaning of life as survival (*survie*), which economics is *initially* and *necessarily* concerned with. The problem with the economy of unlimited growth and expansion in which we live is that it treats desires as if they were needs, and thus cannot see the specificity of the sense of life as sovereign life, or *care*: "Further than need", Bataille writes in *The Accursed Share*, and "beyond the necessities defined by suffering" lies the realm of desire, the true object of which is "sovereign life".[49]

This means that whereas the sphere of needs and the necessities of life are a matter for the economy, and economics, the sense of life as sovereign life, or as desire, exceeds the boundaries of the market, and the problems it can solve. Whereas the first refer to the (vital) problem of abundance, the second refer to the (equally vital) problem of sovereignty. Sovereign life, he adds, "begins when, once the necessary has been taken care of, the possibility of life opens up, in a way that is unlimited". But this limitlessness isn't the toxic limitlessness of consumption and growth. Rather, it's the healing infinity of consummation, or *sacrifice*: "we call sovereign the enjoyment [*jouissance*] of possibilities that utility doesn't justify (utility: that which has productive activity as its end). What lies beyond utility is the domain of sovereignty".[50]

Enjoyment (*jouissance*) is to be distinguished from any form of satisfaction in the utilitarian (and even Freudian) sense, insofar as it bypasses the economy of investment and return, as well as the symbolic economy of sublimation. It operates within an economy of expenditure—that is, *as* and *at* the very limit of that which can be retained. This means that if one of the key, if not the central, problems for our postindustrial market economy is that of what David Harvey calls the "capital surplus absorption problem", which forces capitalists to recapitalize and reinvest in expansion a proportion of the surplus they produce, and generate a situation of over-accumulation, then, consummation, rather than consumption, could begin to address that systemic problem by inserting a vital limit within what is in effect a bad (or toxic) infinite.[51] Against liberal economics, which poses the question of the modes of production and accumulation, and constructs its rationality accordingly, identifying desire as its necessary form of energy in the process, Bataille's general economy recognizes the existence of a form of desire that is directly connected to a surplus of vital energy, and which can be realized in consummation only. It follows that the mechanism of production and exchange are themselves subordinated to that other, "glorious" economy, within which the question of how we can take care of ourselves and of others begins to acquire a different signification.

BIBLIOGRAPHY

"From Details to Desire: The Power of Big Data". *Financial Times*, 4/5 May 2013.

Agamben, Giorgio. *Homo Sacer: Sovereign Power and Bare Life*. Translated by Daniel Heller-Roazen. Stanford, CA: Stanford University Press, 1998.

———. *Remnants of Auschwitz: The Witness and the Archive*. Translated by Daniel Heller-Roazen. New York: Zone Books, 1999.

———. *The Open: Man and Animal*. Translated by Kevin Attell. Stanford, CA: Stanford University Press, 2004.

Arendt, Hannah. *The Human Condition*. Chicago: University of Chicago Press, 1958.

Bataille, Georges. *The Accursed Share: An Essay on General Economy*. Translated by Robert Hurley. New York: Zone Books, 1991.

Becker, Gary. *Human Capital*. Chicago: University of Chicago Press, 1993.

Bentham, John. *An Introduction to the Principles of Morals and Legislation* (1823). Mineola, NY: Dover Publications, 2007.

Bentham, John. *Manual of Political Economy*. In *Jeremy's Bentham's Economic Writings*, edited by W. Stark, Vol. 1. London: Allen & Unwin, 1952; Routledge, 2004.

Deleuze, Gilles, and Félix Guattari. *Anti-Oedipus*. Translated by Robert Hurley, Mark Seem and Helen R. Lane. London: Athlone Press, 1984.

Derrida, Jacques. "Autoimmunity: Real and Symbolic Suicides". In *Philosophy in a Time of Terror: Dialogues with Jürgen Habermas and Jacques Derrida*, edited by Giovanna Borradori. Chicago: University of Chicago Press, 2003.

———. *Rogues: Two Essays on Reason*. Translated by Pascale-Anne Brault and Michael Naas. Stanford, CA: Stanford University Press, 2005.

Douglas, Niall, and Norbert Häring. *Economists and the Powerful: Convenient Theories, Distorted Facts, Ample Rewards*. London: Anthem Press, 2012.

Esposito, Roberto. *Bios: Biopolitics and Philosophy*. Translated by Timothy C. Campbell. Minneapolis: Minnesota University Press, 2008.

———. *Immunitas: The Protection and Negation of Life*. Translated by Zakyia Hanafi. London: Polity Press, 2011.

Foucault, Michel. *The Birth of Biopolitics: Lectures at the Collège de France*. Translated by Graham Burchell. New York: Palgrave Macmillan, 2008.

Helvétius, Claude Adrien. *De l'Esprit* (1758–1759). Paris: Diffusion Inter-forum, 1973.

Hume, David. *A Treatise of Human Nature*. Edited by David Fate Norton and Mary J. Norton. Oxford: Oxford University Press, 2000.

Lordon, Frédéric. *Capitalisme, désir et servitude*. Paris: La Fabrique, 2010.

Lyotard, Jean-François. *The Postmodern Condition*. Translated by Geoffrey Bennington and Brian Massumi. Manchester: Manchester University Press, 1984.

Mazur, Paul M. *American Prosperity: Its Causes and Consequences*. New York: Viking Press, 1928.

Miller, Peter, and Nikolas Rose. *Governing the Present*. Cambridge: Polity, 2008.

Persky, Joseph. "The Ethology of Homo Economicus". *Journal of Economic Perspectives* 9, no. 2 (1995): 221–31.

Read, Jason. "A Genealogy of Homo Economicus". *Foucault Studies* 6 (2009): 25–36.

Schor, Juliet. *The Overworked American: The Unexpected Decline of Leisure*. New York: Basic Books, 1991.

Schultz, Theodore W. *Human Resources.* New York: National Bureau of Economic Research, 1972.

———. *Investment in Human Capital: The Role of Education and of Research.* New York: Free Press, 1971.

Skidelsky, Eward, and Robert Skidelsky. *How Much Money Is Enough? The Love of Money, and the Case for the Good Life.* London: Allen Lane, 2012.

Smith, Adam. *The Theory of Moral Sentiments.* Edited by David Daiches Raphael and Alec Lawrence. Oxford: Oxford University Press, 1976.

Steuart, James. *An Inquiry into the Principles of Political Economy.* Dublin, 1770.

van Horn, Robert, and Philip Mirowski. *The Road from Mont Pèlerin: The Making of the Neoliberal Thought Collective.* Edited by Philip Mirowski and D. Plehwe. Cambridge, MA: Harvard University Press, 2009.

NOTES

1. See Hannah Arendt, *The Human Condition* (Chicago: University of Chicago Press, 1958), chap. 2; Giorgio Agamben, *Homo Sacer: Sovereign Power and Bare Life,* trans. Daniel Heller-Roazen (Stanford, CA: Stanford University Press, 1998), 1; *Remnants of Auschwitz: The Witness and the Archive,* trans. Daniel Heller-Roazen (New York: Zone Books, 1999); and *The Open: Man and Animal,* trans. Kevin Attell (Stanford, CA: Stanford University Press, 2004).

2. Roberto Esposito, *Bios: Biopolitics and Philosophy,* trans. Timothy C. Campbell (Minneapolis: Minnesota University Press, 2008); *Immunitas: The Protection and Negation of Life,* trans. Zakyia Hanafi (London: Polity Press, 2011). See also Jacques Derrida, "Autoimmunity: Real and Symbolic Suicides", in *Philosophy in a Time of Terror: Dialogues with Jürgen Habermas and Jacques Derrida,* ed. Giovanna Borradori (Chicago: University of Chicago Press, 2003), and *Rogues: Two Essays on Reason,* trans. Pascale-Anne Brault and Michael Naas (Stanford, CA: Stanford University Press, 2005).

3. John Locke, *An Essay Concerning Human Understanding* (London: Penguin Classics, 1997).

4. Locke, *An Essay Concerning Human Understanding,* 233.

5. Locke, *An Essay Concerning Human Understanding,* 234.

6. Locke, *An Essay Concerning Human Understanding,* 234.

7. Locke, *An Essay Concerning Human Understanding,* 234.

8. Locke, *An Essay Concerning Human Understanding,* 235.

9. Locke, *An Essay Concerning Human Understanding,* 241. As James Steuart would argue eighty years later, it's desire that pushes man to work and prosper, and sets him on the path to freedom: James Steuart, *An Inquiry into the Principles of Political Economy* (Dublin, 1770), Vol. I., Book II, x, chap. 1, 166.

10. Locke, *An Essay Concerning Human Understanding,* 242.

11. Locke, *An Essay Concerning Human Understanding,* 243.

12. Locke, *An Essay Concerning Human Understanding,* 242.

13. Locke, *An Essay Concerning Human Understanding,* 243.

14. Locke, *An Essay Concerning Human Understanding,* 243.

15. Locke, *An Essay Concerning Human Understanding,* 250.

16. David Hume, *A Treatise of Human Nature,* edited by David Fate Norton and Mary J. Norton (Oxford: Oxford University Press, 2000), 367.

17. "Reason alone can never be a motive to any action of the will" (Hume, *A Treatise of Human Nature,* 265).

18. Hume, *A Treatise of Human Nature,* 266.

19. Hume, *A Treatise of Human Nature,* 266.

20. Hume, *A Treatise of Human Nature,* 308.

21. Hume, *A Treatise of Human Nature,* 309.

22. Hume, *A Treatise of Human Nature,* 309.

23. Hume, *A Treatise of Human Nature,* 305.

24. Hume, *A Treatise of Human Nature*, 342.
25. Hume, *A Treatise of Human Nature*, 348.
26. Michel Foucault, *The Birth of Biopolitics: Lectures at the Collège de France*, trans. Graham Burchell (New York: Palgrave Macmillan, 2008), 30–31.
27. Claude Adrien Helvétius, *De l'Esprit* (1758–1759) (Paris: Diffusion Inter-forum, 1973), 60.
28. Adam Smith, *The Theory of Moral Sentiments*, ed. D. D. Raphael and A. L. Macfie (Oxford: Oxford University Press, 1976), IV.1: 9 and 10, respectively.
29. John Bentham, *An Introduction to the Principles of Morals and Legislation* (1823) (Mineola, NY: Dover Publications, 2007), 1.
30. John Bentham, *Manual of Political Economy*, in *Jeremy's Bentham's Economic Writings*, ed. W. Stark, Vol. 1 (London: Allen & Unwin, 1952; Routledge, 2004), 224.
31. Rob van Horn and Philip Mirowski, *The Road from Mont Pèlerin: The Making of the Neoliberal Thought Collective*, ed. P. Mirowski and D. Plehwe (Cambridge, MA: Harvard University Press, 2009), 162.
32. IMSciences.net. Cited by Robert and Edward Skidelsky in *How Much Money Is Enough? The Love of Money, and the Case for the Good Life* (London: Allen Lane, 2012), 12.
33. Foucault, *The Birth of Biopolitics*, 133–46.
34. See Joseph Persky, "The Ethology of Homo Economicus", *Journal of Economic Perspectives* 9, 2 (Spring 1995), 221–31.
35. See Foucault, *The Birth of Biopolitics*, 14 February 1979.
36. Although the idea of "human capital" first appears in 1902 in Gabriel Tarde's sociology, it is usually attributed to Theodore W. Schultz. See, for example, his *Investment in Human Capital: The Role of Education and of Research* (New York: Free Press, 1971) and *Human Resources* (New York: National Bureau of Economic Research, 1972). Another strong proponent of the theory of human capital, also from the Chicago School, is Gary Becker, *Human Capital* (Chicago: University of Chicago Press, 1993).
37. Jason Read, "A Genealogy of Homo Economicus", *Foucault Studies* 6 (February 2009), 33.
38. T. W. Schultz, *Investment in Human Capital*, 48.
39. Peter Miller and Nikolas Rose, *Governing the Present* (Cambridge: Polity, 2008), 49.
40. In *The Postmodern Condition* (1979), Lyotard had already insisted on this significant shift, with specific reference to the future of higher education. The explicit aim of the latter, he explains, will increasingly be the production of skills as negotiable currency in the marketplace. See Jean-François Lyotard, *The Postmodern Condition*, trans. Geoffrey Bennington and Brian Massumi (Manchester: Manchester University Press, 1984).
41. Paul Mazur in Norbert Häring and Niall Douglas, *Economists and the Powerful: Convenient Theories, Distorted Facts, Ample Rewards* (London: Anthem Press, 2012), 17. In a book published a year after his article, he expressed the same idea, before adding that human nature "very conveniently presents a variety of strings upon which an appreciative sales manager can play fortissimo"—strings such as "threats, fear, beauty, sparkle" (Paul M. Mazur, *American Prosperity: Its Causes and Consequences* [New York: Viking Press, 1928], 44, 47).
42. See Juliet Schor, *The Overworked American: The Unexpected Decline of Leisure* (New York: Basic Books, 1991), 120.
43. "From Details to Desire: The Power of Big Data", *Financial Times*, 4/5 May 2013.
44. For a more detailed description of those techniques, see Frédéric Lordon, *Capitalisme, désir et servitude* (Paris: La Fabrique, 2010), 127–30.
45. Gilles Deleuze and Félix Guattari, *Anti-Oedipus*, trans. Robert Hurley, Mark Seem, and Helen R. Lane (London: Athlone Press, 1984), 259.
46. Aristotle, *Politics*, 1257b25–34.
47. Aristotle, *Politics*, 1257b38–1258a2.
48. Georges Bataille, *The Accursed Share: An Essay on General Economy*, trans. Robert Hurley, Vol. 1 (New York: Zone Books, 1991), 59.

49. Bataille, *The Accursed Share*, Vol. 3: *Sovereignty*, 199.
50. Bataille, *Sovereignty*, 198.
51. See David Harvey, *The Enigma of Capital: And the Crises of Capitalism* (London: Profile Books, 2010).

SEVENTEEN

The Pragmatics of Desire and Pleasure

Rethinking Somatic Powers with Foucault, Deleuze and Guattari

Marjorie Gracieuse and Nicolae Morar

By privileging a critical problematization of ourselves over a set of ethical and political prescriptions, Foucault, Deleuze and Guattari have not only invited us to abandon our habitual conception of power but also made philosophy and psychoanalysis fall off their traditional pedestal, depriving them of their alleged epistemic superiority and presumptuousness "in speaking in the name of others". This radical shift consists both in purging what is commonly called "desire" and "pleasure" of their historical and moral determinations and, also, in revealing the contingent nature of our dominant schemes of intelligibility and social paradigms, while pointing us towards more creative, nuanced and self-aware forms of life. The goal of this chapter is not to provide a comparative study, but to move beyond Foucault and Deleuze and Guattari's theoretical differences in order to explore the conditions under which desire and pleasure can be conceived as decisive powers of transformation and creation, and therefore be employed as powerful forces of resistance against the economy of servitude and affective misery that characterizes the cultural commodification and political economy of our present.

In order to do so, we would like to make three distinct but interrelated points. First, Foucault, Deleuze and Guattari emphasize the material conditions of power, where the source and the very object it exerts its effects on are located not in abstract social formations primarily, but rather in somatic relations. Second, Deleuze and Guattari's account of power not only shows the ways in which power can produce effects of repression and generate reactive desires but also sheds light on the conditions under which the vital power of desire can become a force of empowerment and self-transformation. Third, Foucault is suspicious of the *traditional* notion of desire since its so-called scientific, religious and political character promotes a normalizing politics of sex, and ends up being co-opted in an entire series of biopolitical mechanism regulating our own (sexual) conduct.

THE POWERS OF BODIES AND THE MATERIAL IMMANENCE OF SOCIAL RELATIONS

Deleuze and Guattari's concept of desire and Foucault's notion of pleasure are essentially problematic notions. Whenever they appear in their writings, they signal a kind of blind spot or an empty space that cannot and, more importantly, *should not* be occupied by any ultimate signifier. In fact, desire and pleasure function as strategic and semiotic operators; they aim to create, within thought and through discourse, an interval or void making possible the critical assessment of what we are presently becoming. Indeed, Deleuze, Guattari and Foucault purposefully maintain the relative indeterminacy of these notions, which, instead of legitimating what we already know, makes us feel and think differently about what we are and what we do.

It should thus come as no surprise when Foucault says in an interview that he does not know what pleasure is.[1] It is a "virgin territory" almost devoid of meaning since it signals an event "outside the subject, at the limit of the subject or between two subjects".[2] Deleuze and Guattari use a similar argumentative technique when they claim that desire has neither a fixed subject nor a preexisting object, but is rather a vital movement susceptible of different uses or regimes, and an informal energy with which we collectively produce the texture of our sociocultural reality and the sense and value of our various *modus vivendi*.

It is important to acknowledge the extent to which desire and pleasure are not merely concepts, but rather modes of our vitality, such as infra-discursive and material processes that are directly related to the vicissitudes of our somatic life. Certainly, Foucault's notion of utopian body, "always elsewhere than in the world", could stand as counterevidence. And yet the body is "the principal actor of all utopias" that does not cease to turn its utopian power against itself, "allowing all the space of the religious and the sacred, all the space of the other world, all the space of the counter world, to enter into the space that is reserved for it. So the body, then, *in its materiality* . . . would be like the product of its own phantasms".[3]

Deleuze, in turn, describes the body as a network of material forces in tension with one another, which themselves constitute the informal, unconscious and intensive life of thought. The power of thought lies indeed in its virtual reality, in its potential to construct the tools to overcome its organic and discursive delimitations through transformative experiences and artifices of writing. For this reason, in his joint work with Felix Guattari, Deleuze presents the body as a "desiring machine" or "body without organs",[4] whose spiritual vitality (*hubris* or desire) is eminently creative. It continuously generates new perspectives on life corresponding to unequal levels of perception and singular stylizations of existence.

In a world in which the traditional belief in a transcendent foundation and justification of power has more or less vanished, we are forced to abandon the old model of sovereignty and think of political power in a pluralist, materialist and decentred way. Thus, power is constituted by the mobile forces of our somatic and collective life. What is left, Deleuze asks, once we start thinking of bodies as "forces, *and* nothing but forces"? What is left, if force no longer refers to a centre or is simply meant to confront a set of obstacles? If force only confronts other forces, it refers to other forces only insofar it affects them or is affected by them, then a new conception of power emerges. "Power . . . is this power to affect and be affected, this relation between one force and others".[5]

What Deleuze calls "power" or "desire" or "life" are the transcendental forces of matter itself: they constitute the "plane of nature" as plane of intersections of singular vectors of power. Following Lucretius's materialistic vision of the world, Deleuze even asserts that "nature, to be precise, is power"[6] (*la puissance*). From this perspective, each individual body is now conceived as intrinsically open to its "outside", entering in relation of composition and disjunction, of attraction and repulsion, of love and hatred with other bodies, and continuously actualizing and affirming its power up to this or that degree.

Thus, in a philosophy of pure immanence, bodies, through their combinations and struggles, are the productive forces of reality. This is a reciprocal process of production. Bodies appear as purely relational entities continuously shaped, united and divided by power relations since bodies are both the agents and the vectors of power, and also that onto which power exerts and inscribes itself. They constitute human history as the history of desire or power, without ever reaching a permanent or final state that could constitute the end of history or could provide us with an ultimate essence of the human. This is also the sense in which nature and culture become indiscernible.

This conception of bodies as networks of forces or powers, inherited from Nietzsche, is common to Foucault, Deleuze and Guattari. It enables them to posit the problem of power in purely materialistic and etho-political terms, without appealing to traditional categories of ontology. Since power does not refer to an essence, it is not of the order of being, but it amounts to what a body can do and to its variation of vitality. This is why Foucault claims that "power produces reality". Since power only exists in action and "affects the body",[7] it gives rise to new thoughts, new capacities and new conducts. Similarly to Foucault, Deleuze defines power as an "affection of desire".[8] Sometimes, Deleuze needs to distinguish between power (*le pouvoir*) and power (*la puissance*) or potency of desire, the former being supposed to account for what Foucault calls "a state of domination" as "the terminal form of power" (the moment when power reaches its limit and ossifies into the monopoly of legitimate violence).

A spectrum view of power, which extends from hatred to love, faces the challenge of accounting for conditions under which power can constitute a force of empowerment or, on the contrary, can take the form of a coercive and disciplinary instance. It demands from us to think beyond the "Power principle"—that is, beyond (or, more precisely, beneath) this abstraction called Power (*le pouvoir*). Indeed, Foucault, Deleuze and Guattari think of this abstraction as a mystification thanks to which micro-relations of socioeconomical domination and exploitation can maintain themselves (economical exploitation is based on cultural domination and not the reverse).

If bodily forces can be schematized and channelled, if the process of affecting certain forces by other forces can be implemented and reproduced, it is because power as domination functions only thanks to what Foucault calls power/knowledge "*dispositifs*" and Deleuze "social institutions" or "desiring assemblages". Those are discursive and nondiscursive infrastructures that give power its local concreteness, material repeatability and efficiency. In that sense, power (*le pouvoir*) does not refer to a particular institution, such as the State or the Church. Rather, it consists in a cultural dominant way of knowing and perceiving the world that generates the illusory impression of a fixed and stable arrangement of social reality. This particularly dominant schema or abstract diagram of action and perception nonetheless produces real effects upon bodies, not simply through discursive practices, but also through somatic disciplines. Power makes bodies act and speak, since it incites to specific conducts and utterances in a nonviolent way. Power also uses the perspective of violence and punishment to influence or prevent possible actions.

While bodies are the genetic and transformative forces of power apparatuses, the latter can become separated from the bodies, becoming and progressively solidified into laws of domination and oppression. For Deleuze, the process of separation of institutions from the social forces or singularities defines the alienation of bodies' power to laws or policies. This alienation mechanism no longer uses the cultural power of institutions to facilitate rational actions but directly intervenes upon bodies through a system of prohibitions and sanctions. In this context, the political problem is that of disclosing the ways in which bodily actions and practices produce a variety of models of sociality or institutions, which in turn react upon bodies and can hinder their power as much as enable them to think and act differently.

If institutions are, as Deleuze notes, means to satisfy instincts and tendencies and, as such, precious schemata of actions and intelligibility, they only do so by integrating tendencies into a system of anticipation. This regulation system imposes a predetermined set of rules of actions upon them. "Every institution imposes a series of models on our bodies, even in its involuntary structures, and offers our intelligence a sort of knowledge, a possibility of foresight as project. We come to the following conclusion: humans have no instincts, they build institutions. The human is an animal decimating its species".[9]

If so, the human animal constantly remakes its human nature. By creating social institutions, we endlessly overcome our species through a culturally creative activity, whose ethical and political value can be measured according to the degree of freedom and plasticity that it allows in its institutions and in its subjects. "Such a theory will afford us the following political criteria: tyranny is a regime in which there are many laws and few institutions; democracy is a regime in which there are many institutions, and few laws. Oppression becomes apparent when laws bear directly on people, and not on the prior institutions that protect them".[10]

However, Foucault and Deleuze and Guattari do not think of power simply as an institutional instance of domination and oppression. Rather, they highlight the way power is always dispersed in the social field and thus always multipolar and infrastructural. By insisting on the fact that one needs to elaborate a micro-physics of power (Foucault) or a micropolitical analysis of desire (Deleuze and Guattari), they show that the apparent "transcendence" of power is *always* an "effect of surface" or "product" of social immanence. This radical shift redefines the fundamental problem of political philosophy. The task does not consist anymore in defining the conditions for certain political institutions (i.e., state, church) but rather in showing how power can be incorporated into bodies to such an extent that it no longer merely educates our sensibility, but takes the form of a self-subjectivation process, which in turn generates a fascination for domination and a profound bitterness that makes apparatuses of domination prosper: "the strategic adversary is fascism . . . the fascism in us all, in our heads and in our everyday behaviour, the fascism that causes us to love power, to desire the very thing that dominates and exploits us".[11]

FROM THE POWER OF DESIRE TO THE FASCINATION FOR POWER: DELEUZE AND GUATTARI

In *Nietzsche and Philosophy*,[12] we find Deleuze's most full-fledged theory of power. This view inspires his entire *corpus*, and more particularly his conception of desire, as an unconscious and vital activity, of which knowledge as "will to truth" and mastery is only the lowest degree. The theoretical challenge is not to oppose power and desire, but to analyse what Deleuze calls "the becoming-reactive of desire", as a particularly weak mode of exercising one's singular power. It consists in becoming incapable of opening and connecting oneself to other forces without negating or seeking to dominate them by enforcing new types of identity. If the immanent power of bodies can be subjected to training, maturation and self-overcoming and, as such, can constitute a critical force of self-liberation, it can also be subjugated by a variety of external and internalized constraints.

By incorporating an external norm or belief, a body can indeed become the very agent of its own subjection. We can make sense of this process of self-subjection *only* insofar as we recognize the two ways in which human power comes into existence and affirms itself: either by borrowing the authority of established values and complying with preexisting norms (reactive mode of desire) or by actualizing its immanent and normative potential, and thus by introducing in the existing game of social forces a new way of living and thinking (active mode of desire). Thus, "active" or "reactive" forces do not designate the nature of bodily forces *abstractly*, but refer to concrete assemblages of forces, as institutions or ways of thinking and perceiving power, which can only be identified and evaluated once they have reached a dominant status and relative consistency.

When thought is deprived of the cultural and political conditions necessary to exercise its immanent power of invention and self-fashioning, it satisfies itself with ready-made opinions and representations. Instead of being able to learn how to combine itself with other forces and develop its faculties and capacities, thought remains governed by a passional regime of fear and superstition, which prevents the development of its creative and critical powers. Deleuze claims that the exercise of thought is, in this case, reduced to the regime of "organic representation", which is a particularly reactive regime of thought that judges the real according to abstract categories and imposes upon its material becoming an ideal "plane of organization" as a measure of judgement, thanks to which one continuously judges oneself and other beings.

This stratification of the human's thinking powers into a veritable "system of judgement" constitutes the operation through which the creative and critical forces of thought are captured by static forms of knowledge. Once captured, these critical forces end up contributing to the reinforcement of established states of domination. Even worse, this (semiotic) process of subjection enacts in each being, through social emulation, "a desire for recognition", which, Deleuze claims, converts the liberating and transformative virtue of knowledge into a principle of social distinction and social reproduction. This reactive process polarizes the forces of desire and reduces it to its most reactionary and autarchic mode: "Our security, the great molar organization that sustains us, the arborescences we cling to, the binary machines that give us well-defined status, the resonances we enter into, the system of overcoding that dominate us—we desire all that".[13]

Thus, the enslaving power of what Deleuze calls, in his early work, "representation"[14] is not simply epistemological, but thoroughly political since it constitutes a dogmatic image of thought and a codified regime of discourse that "separate beings from what they can do and think". Although we are encouraged to "represent" ourselves and speak in our name, the key political problem is that a large number of human beings are deprived of means of expression that would allow them to formulate their social problems in terms that do not merely reinforce the dominant use of language. A power dictum would encourage us to "be ourselves—being understood that this self must be that of others. As if we would not remain slaves so long as we do not control the problems themselves, so long as we do not possess a right to access and participate in problems".[15]

For this reason, Deleuze argues that contradiction is "not the weapon of the proletariat but, rather, the manner in which the bourgeoisie defends and preserves itself, the shadow behind which it maintains its claims to decide what the problems are".[16] What can the "masses" do when they are culturally separated from their critical power to evaluate the very schema that make them think and act? Can they do anything else than turning their power against themselves or become reactive? With the becoming-reactive of the human's active forces, the vital potency of desire becomes a contagious will to judge and a fascination for mastery and power. From a force of empowerment and self-transformation, desire progressively transforms itself merely into a will to gain or maintain power over others (*le pouvoir* as domination).

Power, along with its fascination effects leading to domination, is for Deleuze the lowest degree of desire and the symptom of its disempowerment. It signals an exhaustion of vital force, which can no longer act its organic reactions. It thus folds back onto itself, constituting a fictitious interiority that can only oppose itself to the other forces it encounters. This "reactive type" of desire is, strictly speaking, a will to dominate others that fails to dominate itself. Given its powerlessness, this reactive type often finds refuge in the cogs of established power and in the stratifications of a dominant way of speaking and acting. In each case, the human desire is sick of its own temporal, social, political and economical codifications: the reactive man is a sad, dissatisfied man whose passive affects in turn reinforce its servitude to dominant and oppressive instances. These various stasis of desire constitute the Modern illness of desire, as a form of desire fostering the conservation of social status and static hierarchies and the thirst for power and security. This is why, ultimately, Deleuze and Guattari affirm that

> there isn't a desire for power; *it is power itself that is desire*. Not a desire-lack, but desire as plenitude, exercise, and functioning, even in the most subalterns workers. Being an assemblage, desire is precisely one with the gears and components of the machine. And the desire that someone has for power is only his fascination for these gears, his desire to make certain of these gears go into operation, to be himself one of these gears.[17]

CAPITALISM AND NIHILISM: "ACCELERATE THE PROCESS"

This analytic of desire and genealogy of power would be of little importance if it did not allow us to make a parallel between Deleuze's notions of nihilism (coming from Nietzsche) and capitalism (from his writings with Guattari). And when, Deleuze and Guattari encourage us to "go still further in the movement of the market . . . not to withdraw from the process",[18] they actually invoke Nietzsche and his will to "accelerate the process" of nihilism in order for it to be ultimately vanquished by this very mechanisms. But how is that possible? And what forms can this self-dissolution of capitalist and nihilist subjectivity take?

"Going further with the movement of the market" does not amount to an apology of economical liberalism. According to Deleuze and Guattari, we must realize that the living basis of capitalism is precisely the promotion of an indifferent liberation of all flows of desires in order for them to be immediately recuperated and channelled by the interests of the market. "Capitalism liberates the flows of desire but under the social conditions that define its limit and the possibility of its own dissolution, so that it is constantly opposing with all its exasperated strength the movement that drives it toward this limit".[19] Our very affectivity becomes dependent on external causes as source of its arousal and this capitalist mechanism constitutes a regime of passionate servitude and recurrent dissatisfaction. The more we possess things, the more we are possessed by them.[20] This polarization of affectivity facilitates the emergence of "passive joys" as evanescent moments of satisfaction. Following Spinoza, Deleuze and Guattari show not only that we are not the direct cause of such events but also, and more importantly, that our desiring energy is captured and polarized in objects of consumption.

The promotion of a hedonistic way of life and the incessant encouragement of a form of liberation of all desires are the two ways in which capitalism secretes, in the bourgeoisie and in the masses, an indefinite quest for an impossible object of enjoyment. Through this, it generates an imaginary lack that provokes a general feeling of frustration. We are always running after something we do not possess. And, because our desires are infinite in their variation since they can always latch onto a new object we lack, we live under the impression of an infinite debt towards the capitalist system itself. This process entails an infinite deferral of *jouissance* and, as such, produces an accumulation of goods and a constant reinvestment: "the bourgeois sets the example", Deleuze and Guattari claim, since "he absorbs surplus value for ends that, taken as a whole, have nothing to do with his own enjoyment: more utterly enslaved than the lowest slaves, he is the first servant of the ravenous machine, the beast of the reproduction of the capital, internalization of the infinite debt. 'I too am a slave-these are the new words spoken by the master".[21]

However, Deleuze and Guattari do not believe that the capitalist form of subjectivity is a universal, invariant and necessary structure. Capitalism is a contingent, cultural organization, rather than the natural outcome of processes of fulfilment of our (alleged) human essence. In fact, the main thesis of *Anti-Oedipus* is that even if capitalism is the dominant formation of our present society, other formations and modes of desire are possible. And, such modes of desire are already developing *within* capitalism itself. The capitalist pursuit of happiness and intense stimulations is progressively turning the human body into a quasi-anesthetized body insofar it creates the demand for an ever-stronger stimulation in order to meet an ever-rising sensory threshold. The economical cult for enjoyment and pleasure is not meant to develop our sensibility but to saturate it. Paradoxically, this is the place where one can observe one of the limits of capitalism. The limit lies in the specific type of body it produces as a nervous and hypersensitive body-being constantly traversed by waves of precarious pains and pleasures. Such a body could be conceived either as a "remainder" or as a "symptom" of the economical process of production.

This body, developed at the core of the capitalist production, does not always recognize itself in the forms of identity and ready-made pleasures that submerge it, and it is in this inadequacy between the forces of the body and their modes of captures that Deleuze and Guattari locate the breach within the system itself: "At the very heart of this production . . . the body suffers from being organised in this way, from not having some other sort of organisation, or no organisation at all".[22] For Deleuze and Guattari, the "body-without-organs" is incapable of *counter-cathexis*, since it remains reduced to its mere materiality in a sterile and nonproductive stasis. It is incapable of reinvesting its mental or emotional energy into something other than an active *repulsion* of all these apparatuses of power and enjoyment that persecute it. This evanescent stage or state of the body, which Deleuze and Guattari name "anti-production", does not necessarily end up in the clinical figure of the schizophrenic. It equally constitutes a potential, present within each one of us, to *become* revolutionary. It characterizes a transcendental experience of the loss of the Ego and thus an active mode of desire, capable of struggling against its own habitual and memorial tendencies and of disinvesting itself from the various social machines that usually regulate its functioning.

It comes as no surprise that the intensive and material force of the "body" without organs, which Deleuze and Guattari assimilates to an egg, also designates the swarming vitality of the brain[23] itself, as a volume traversed by unbound energies capable of creating new connections between ideas, and of redetermining its habitual associations of ideas. Hence, a universal schizophrenia: the brain constitutes the possibility of an internal self-distancing and self-differentiality, through which we can become able to evaluate and refuse what we are becoming. Among the many powers our brain grants us with (habit, memory, etc.), "the power of decision"[24] is the highest for Deleuze. It is the power of thought to abstract itself from its habitual sociohistorical determinations and to redetermine the sense and value of its becoming.

While capitalism does not cease to push its limits further, implementing more policies and forms of control upon bodies and imposing a purely *economical* form of time on our lives, we see that the real political problem consists in reconquering an uneconomical and unproductive time of desire. This time is neither the time of leisure nor that of pleasure, but the necessary interval of time that thought needs in order to construct "a body-without-organs" for itself. This is what Deleuze calls "a plane of immanence" as a critical plane of thought and a starting point for a becoming-revolutionary, or what Spinoza calls "Reason". It makes possible a break from the dictatorship of passions. This plane is precisely what we are *lacking*.[25] We are lacking a common plane that would allow us to resist the present established values and the stereotypical behaviours that are imposed on our lives.

This plane needs to be constructed both individually and collectively as a plane of cultural analysis and experimentation since it offers us the necessary distance and the material means to analyse critically our institutions. In Foucault, this critical and political task takes the form of a new practice of History, while in Deleuze and Guattari it is that of a hierarchical classification of desiring productions. All this forms a "war machine", a series of direct actions upon the order of discourse, which for these thinkers constitutes our "cultural unconscious" or a semiotic power apparatus of which we must be aware of *if* we are to turn it against itself. "If a struggle can be led against the capitalist system, it can only be done . . . through combining a struggle –with visible, external objectives—against the power of the bourgeoisie, against its institutions and systems of exploitation, with a thorough understanding of all the semiotic infiltrations on which capital is based".[26]

FOUCAULT'S CRITIQUE OF THE BIOPOLITICAL IMAGE OF DESIRE

While Deleuze and Guattari adopt a constructivist approach that consists in creating an entirely new metaphysical concept of power of desire, Foucault takes a different path[27] towards a kind of historical nominalism. Foucault is not so much concerned with "desire" as a physical and relational process, but rather with its most common abstract representations, which, in one way or another, assimilate desire to an obscure psychological instance structured by an ontological lack. Foucault is interested in understanding the historical process by which this interpretation of desire became a cultural paradigm since it fostered and continues to foster a certain discourse about sexuality that enables a series of apparatus of control and normalization over bodily conducts and feelings. "Desire" is above all a word. It is a cultural production whose history we should reconstruct in order to understand how such an abstraction has nonetheless exerted real transformative effects on our worldviews and conducts.

In *The Will to Knowledge*, Foucault highlights a historically constituted conception of desire as the main contributor to the development of system of normalization of human conduct. His trajectory consists in unearthing the historical constitution and development, and thus the contingency of this particularly moral concept of desire in order "to think of sex without the law and power without the king".[28] The stakes are high. Foucault targets the very dismantling of an enduring myth of Western thought. Since Plato, we continue to somehow believe the myth of a purity and neutrality of scientific and philosophical knowledge *vis à vis* religious and political powers.

Foucault's critique of the bourgeois image of desire emphasizes the "*unreality*" of "sex", conceiving it as an abstract instance or "imaginary point", around which a particular system of power/knowledge revolves and thanks to which it functions. The "sex-desire" *dispositif*, as both a power apparatus and a historical discursive formation, demonstrates how "*scientia sexualis*" has come to constitute bodies as "subjects of desire" and to subject them to the fiction of sexual identity. The deployment of sexuality as a power *dispositif* is the expression of bourgeois hegemony, which has progressively imposed its "austere monarchy of sex" and its identitarian model of subjectivity to the rest of society. Sexuality has progressively become not simply an object of fascination but also an abstract value and a criterion of reference of socialization. This *dispositif* has enacted new forms of sexual identities and has generalized an obsession in talking about sex as a way to uncover some "truth" about sexuality.

Thus, the sex-desire *dispositif* cannot be assimilated to the repression or inhibition of our (natural) sexuality. Rather, it implies a multiplication of sexualities, an increase of their medical supervision and, ultimately, a proliferation of institutional, scientific but also individual discourses about sex. Far from repressing sexual practices, the operation of the *dispositif* of sexuality consists in naming, in classifying and in rendering visible certain conducts which existed previously more or less silently. The multiplication of new truths or moral norms about sexuality coincides with the development of new pathological concerns of the social body. Mental alienation is intrinsically correlated to economical alienation.

> This valorisation of the body on a level that is not simply moral but political and economical, has been one of the key features of the Occident. And it is curious that this political and economical valorisation of the body, this importance that was attached to the body, was accompanied by an increasing moral devaluation. . . . This led to a kind of dissociation, of disjunction, that has certainly been the origin of many individual psychological disturbances, maybe also of much larger collective and cultural disturbances: an economically overvalued and a morally undervalued body.[29]

Once the strategies of the *dispositif of sexuality* are made explicit, Foucault emphasizes that "the counterattack against the deployment of sexuality ought *not* to be sex-desire, but bodies and pleasures".[30] Foucault rejects desire to the extent that it functions as the very vehicle of control (*scientia sexualis*). And, he affirms bodies and pleasures (*ars erotica*) since the notion of pleasure, unlike the notion of desire, "avoids the entire psychological and medical armature that was built into the traditional notion of desire".[31]

Moreover, Foucault claims that the multiplication of norms about sexuality coincides not only with the economical expansion of capitalism but also with the emergence of a new form of power: biopower. This concept is supposed to capture the ways in which power operates at the level of population, including a series of everyday life practices (biopolitics of the population) and through which each one of us is persuaded to engage in processes of self-surveillance and self-discipline (anatomo-politics of the body). The emergence of social medicine is thus inseparable from the development of capitalism:

> capitalism, which developed from the end of the eighteenth century to the beginning of the nineteenth century, started by socializing a first object, the body, as a factor of productive force, of labour power. Society's control over individuals was accomplished not only through consciousness or ideology but also in the body and with the body. For capitalist society, it was biopolitics, the biological, the somatic, the corporeal that mattered more than anything else. The body is a biopolitical reality, medicine is a biopolitical strategy.[32]

However, the fact that biopower has become a "power over life" should not prevent us from reclaiming "a right to desire"[33] in order to reappropriate the forces of our vital and productive body. If it is true that the exercise of power creates a certain type of pleasure in the various effects of knowledge it produces, this pleasure of knowing pleasure and of exerting power should not prevent us from thinking that we can experience pleasures and desires out of the *dispositifs* of power. To experiment with our bodies, with our pleasures and desires, does not entail a return to an alleged "pure experience". Rather, it requires resisting any self-enclosed form of subjectivity by a constant work of self-fashioning, which does not function without a meticulous work of self-alteration, as active refusal of one's own reactive and conservative tendencies.

This process of desubjectivation is per se an act of resistance against one's own fascination for power and will to judge (Deleuze). It implies a perpetual combat against oneself and against one's reactive becoming. It enables us to constitute liberating and empowering compounds of power and to develop our affective capacities, and the vitality of our desires and pleasures. Similarly, for Foucault, pleasures are not simply what we feel, but also what we can *use* to explore our capacities, transform our sensibility and open our subjectivity to change. The "care of self", central to Foucault's later work, involves a series of transformative operations one can exert onto oneself in order to become a truly ethical subject—that is, a subject capable of using knowledge not as a power of social distinction, but rather as a tool for collective and self-transformation. In this context, ethical subjectivity does not preexist the practice of ethics. Rather, it is the product of a constant process of self-experimentation, which not only takes the form of a progressive self-mastery but also provides us with a kind of sobriety and joy of self-creation. Resisting to established power relations means creating new uses of our power in order to transform the very elements (reactive desires and artificial needs) that disempower us.

To the techniques of domination (power as action over others' actions), Foucault opposes the pragmatics of the self, as active and deliberate cultivation of one's capacity to govern oneself (action upon one's power to act).

> I tried to see how and through what concrete forms of the relation to self the individual was called upon to constitute him or herself as moral subject of his or her sexual conduct. In other words, once again this involved bringing about a shift from the question of the subject to the analysis of forms of subjectivation, and to the analysis of these forms of subjectivation through the techniques/technologies of the relation to self, or, if you like, through what could be called *the pragmatics of self*.[34]

In conclusion, the cultural politics of counter-conducts and resistance as promoted by Foucault, Deleuze and Guattari do not consist in an indifferent liberation of desires. Rather, it requires a careful and constant analytical work on our desires, pleasures and institutions, in order to raise a kind of collective form of intelligence and empowerment, capable of competing with the culture of ignorance and servitude through which a dominant class maintains its economical hegemony and cultural monopoly. As Felix Guattari justly notes:

> it would be absurd to oppose desire and power. Desire is power; power is desire. What is at issue is what type of politics is pursued with regard to different linguistic arrangement that exist . . . capitalist and socialist bureaucratic power infiltrate and intervene in all modes of individual semiotization today, they proceed more through semiotic subjugation than through direct subjugation by the police or by explicit use of physical pressure.[35]

BIBLIOGRAPHY

Deleuze, Gilles. *Cinema 2: Tthe Time-Image*. Translated by Hugh Tomlinson and Robert Galeta. Minneapolis: University of Minnesota Press, 1989.

———. "Desire and Pleasure". In *Two Regimes of Madness. Texts and Interviews, 1975–1995*, 122–34. Edited by David Lapoujade, translated by Ames Hodges and Mike Taormina. New York: Semiotext(e), 2005.

———. *Difference and Repetition*. Translated by Paul Patton. New York: Columbia University Press, 1994.

———. "Gilles Deleuze in conversation with Antonio Negri". In *Negotiations*, translated by Martin Joughin. New York: Columbia University Press, 1997.

———. *Francis Bacon: The Logic of Sensation*. Translated by Dan W. Smith. London: Continuum, 2003.

———. "Instincts and Institutions". In *Desert Islands and Other Texts (1953–1974)*. Translated by Mike Taormina. New York: *Semiotext(e)*, 2004.

———. *The Logic of Sense*. Translated by Mark Lester and Charles Stivale, edited by Constantin V. Boundas. New York: Columbia University Press, 1990.

———. *Nietzsche and Philosophy*. Translated by Hugh Tomlinson. New York: Columbia University Press, 2006.

Deleuze, Gilles, and Felix Guattari. *Anti-Oedipus*. Translated by Robert Hurley, Mark Seem and Helen R. Lane. London: Athlone Press, 1984.

———. *A Thousand Plateaus: Capitalism and Schizophrenia II*. Translated by Brian Massumi. Minneapolis: University of Minnesota Press, 1987.

———. *Kafka: Towards a Minor Literature*. Translated by Dana Polan. Minneapolis: University of Minnesota Press, 1986.

———. *What Is Philosophy?* Translated by Janis Tomlinson and Graham Burchell. New York: Columbia University Press, 1996.

Deleuze, Gilles, and Claire Parnet. *Dialogues*. Translated by Barbara Habberjam, Eliot Ross Albert and Hugh Tomlinson. New York: Columbia University Press, 2002.

Foucault, Michel. "The Birth of Social Medicine". In *Power*. Edited Paul Rabinow, trans. Robert Hurley, 139–42. New York: New Press, 2000.

———. *The Foucault Reader: An Introduction to Foucault's Thought*. Edited by Paul Rabinow, Penguin Social Sciences. London: Penguin, 1984.

———. "The Gay Science: An Interview with Foucault". Translated by Nicolae Morar and Daniel W. Smith. *Critical Inquiry* 37 (2011): 385–403.

———. *The History of Sexuality: An Introduction*. Vol. 1. Translated by Robert Hurley. London: Penguin, 2000.

———. "Power Affects the Body". Translated by John Johnston. In *Foucault Live: Collected Interviews, 1961–1984*, edited by Sylvère Lotringer. New York: Semiotext(e), 1989.

———. "The Minimalist Self". In *Michel Foucault: Politics Philosophy, Culture*. Translated by Alan Sharidan. London: Routledge, 1988.

———. "Preface to *Anti-Oedipus* by Gilles Deleuze and Felix Guattari". Translated *Capitalism and Schizophrenia: Anti-Oedipus*. New York: Viking Press, 1977.

———. "The Stage of Philosophy". Interview with Moriaki Watanabe, 22 April 1978. *New York Magazine of Contemporary Art and Philosophy*, Issue 1.5.

———. "The Utopian Body". Translated by Lucia Allais. In *Sensorium*, edited by Caroline A. Jones, 229–34. Cambridge, MA: MIT Press, 2006.

Grace, Wendy. "Faux Amis: Foucault and Deleuze on Sexuality and Desire". *Critical Inquiry* 36, no. 1 (2009): 52–75.

Guattari, Felix. "The Cinema of Desire". In *Chaosophy: Texts and Interviews, 1972–1977*. Translated and edited by Sylvère Lotringer. New York: Semiotext(e), 2008.

Spinoza, Baruch. *Treatise of the Emendation of the Intellect*. Translated by Samuel Shirley. In *The Essential Spinoza: Ethics and Related Writings*, edited by Michael L. Morgan. Indianapolis: Hackett, 2006.

NOTES

1. "Acts are not very important, and pleasure—nobody knows what it is", in Michel Foucault, *The Foucault Reader: An Introduction to Foucault's Thought*, ed. Paul Rabinow (London: Penguin, 1984), 359.

2. Michel Foucault, "The Gay Science: An Interview with Foucault", trans. Nicolae Morar and Daniel W. Smith, *Critical Inquiry* 37 (2011): 389. See also "The Minimalist Self", in *Michel Foucault: Politics Philosophy, Culture* (London: Routledge, 1988), where Foucault confesses, "I think I have real difficulty in experiencing pleasure. I think that pleasure is a very difficult behaviour. It is not as simple as that (*Laughter*) to enjoy oneself. And I must say that's my dream. I would like and I hope I'll die of an overdose (*Laughter*) of pleasure of any kind . . . because I think it's really difficult and I always have the feeling that I do not feel *the* pleasure, the complete total pleasure and, for me, it is related to death".

3. Michel Foucault, "The Utopian Body", trans. Lucia Allais, in Caroline A. Jones, ed., *Sensorium* (Cambridge, MA: MIT Press, 2006), 229–34. Our emphasis.

4. "The body without organs is flesh and nerves; a wave flows through it and traces levels upon it; a sensation is produced when the wave encounters the forces acting on the body . . . it elevates mechanical forces to sensible intuitions, it works through violent movements. . . . It also attests for a high spirituality, since what leads it to seek the elementary forces beyond the organic is a spiritual will. But this spirituality is a spirituality of the body: the body is the spirit itself, the body without organs". Gilles Deleuze, *Francis Bacon: The Logic of Sensation*, trans. Dan W. Smith (London: Continuum, 2003), 47.

5. Gilles Deleuze, *Cinema 2: The Time-Image*, trans. Hugh Tomlinson and Robert Galeta (Minneapolis: University of Minnesota Press, 1989), 139.

6. Deleuze, *The Logic of Sense*, trans . Mark Lester and Charles Stivale, ed. Constantin V. Boundas (New York: Columbia University Press, 1990), 267.

7. Michel Foucault, "Power Affects the Body", in *Foucault Live: Collected Interviews, 1961–1984*, trans. John Johnston, ed. Sylvère Lotringer (New York: Semiotext(e), 1989), 214.

8. Gilles Deleuze, "Desire and Pleasure", in *Two Regimes of Madness: Texts and Interviews, 1975–1995*, ed. David Lapoujade, trans. Ames Hodges and Mike Taormina (New York: Semiotext(e), 2005), 122–34.

9. Deleuze, "Instincts and Institutions", *Desert Islands and Other Texts (1953–1974)*, trans. Mike Taormina (New York: Semiotext(e), 2003), 20.

10. Deleuze, "Instincts and Institutions", 21.

11. Foucault's preface to Deleuze and Guattari's book *Anti-Oedipus: Capitalism and Schizophrenia*, trans. Robert Hurley, Mark Seem and Helen R. Lane (New York: Viking Press, 1977), xi–xiv.

12. Gilles Deleuze, *Nietzsche and Philosophy*, trans. Hugh Tomlinson (New York: Columbia University Press, 2006).

13. Gilles Deleuze and Felix Guattari, *A Thousand Plateaus: Capitalism and Schizophrenia II*, trans. Brian Massumi (Minneapolis: University of Minessota Press, 1987), 219.

14. "The slave only conceives of power as the object of a recognition, the concept of a representation, the stake in a competition, and therefore makes it depend, at the end of a fight, on a simple attribution of established values". See Gilles Deleuze, *Nietzsche and Philosophy*, 10. On page 81 of the same source, Deleuze goes on to say that "the notion of representation poisons philosophy: it is the direct product of the slave and of the relations between slaves, it constitutes the worst, most mediocre and most base interpretation of power".

15. Gilles Deleuze, *Difference and Repetition*, trans. Paul Patton (New York: Columbia University Press, 1994), 171.

16. Deleuze, *Difference and Repetition*, 268.

17. Deleuze and Guattari, *Kafka: Towards a Minor Literature*, trans. Dana Polan (Minneapolis: University of Minnesota Press, 1986), 56. Our emphasis.

18. Deleuze and Guattari, *Anti-Oedipus*, 240.

19. Deleuze and Guattari, *Anti-Oedipus*, 140.

20. On this problem of affective dependency and desire's subjugation to external objects, see Baruch Spinoza, *Treatise of the Emendation of the Intellect*, in *The Essential Spinoza: Ethics and Related Writings*, ed. Michael L. Morgan, trans. Samuel Shirley (Indianapolis: Hackett, 2006), 165.

21. Deleuze and Guattari, *Anti-Oedipus*, 254.

22. Deleuze and Guattari, *Anti-Oedipus*, 8.

23. "This is nothing to do with going back to 'the subject', that is, to something invested with duties, power, and knowledge. One might equally well speak of new kinds of event, rather than processes of subjectification: events that can't be explained by the situations that give rise to them, or into which they lead. They appear for a moment, and it's that moment that matters, it's the chance we must seize. Or we can simply talk about the brain: the brain's precisely this boundary of a continuous two-way movement between an Inside and Outside, this membrane between them. New cerebral pathways, new ways of thinking, aren't explicable in terms of microsurgery; it's for science, rather, to try and discover what might have happened in the brain for one to start thinking this way or that. I think subjectification, events, and brains are more or less the same thing". "Gilles Deleuze in conversation with Antonio Negri", an interview published in Gilles Deleuze, *Negotiations*, trans. Martin Joughin (New York: Columbia University Press, 1997), 176.

24. See Deleuze, *Difference and Repetition*, 268: "the most important task, that of determining problems and realising in them our power of creation and decision".

25. "Those who lack is real have no possible plane of consistence which would allow them to desire. There are prevented from doing this in a thousand ways. And as soon as they construct one, they lack nothing on this plane, and from this starting-point they set off victoriously towards that which they lack outside. Lack refers to a positivity of desire and not the desire to a negativity of lack". Gilles Deleuze and Claire Parnet, *Dialogues*, trans. Barbara Habberjam, Eliot Ross Albert and Hugh Tomlinson (New York: Columbia University Press, 2002), 91. See also Deleuze and Guattari, *What Is Philosophy?*, trans. Janis Tomlinson and Graham Burchell (New York: Columbia University Press, 1996), 101 ("We lack a genuine plane, misled as we are by Christian transcendence") and 108 ("we do not lack communication. On the contrary, we have too much of it. We lack creation. We lack resistance to the present").

26. Félix Guattari, "The Cinema of Desire", *Chaosophy: Texts and Interviews, 1972–1977*, ed. Sylvère Lotringer (Cambridge, MA: MIT Press, Semiotext(e), 2008), 276.

27. We further explore this difference in our essay "The Desire-Pleasure Problem", in *Between Deleuze and Foucault*, eds. Nicolae Morar, Thomas Nail and Dan Smith (under review with University of Minnesota Press). There, we defend an argument against the *incompatibility thesis* as developed by Wendy Grace (2009).

28. Foucault, *The History of Sexuality: The Will to Knowledge*, vol. 1, trans. R. Hurley (London: Penguin, 2000), 128.
29. "The Stage of Philosophy" first published as "La Scène de la Philosophie", in Michel Foucault, *Dits et Écrits*, Vol. 3 (1976–1979), ed. Daniel Defert and François Ewald (Paris: Gallimard, 1994), 571–95.
30. Foucault, *The History of Sexuality*, Vol. 1, 157.
31. Foucault, "The Gay Science: An Interview with Foucault", 390.
32. The term "biopolitics" first appeared in October 1974 in Foucault's conference titled "The Birth of Social Medicine", delivered at the Institute of Social Medicine of the University of Rio de Janeiro, published in *Power*, ed. James D. Faubion, trans. Robert Hurley et al., vol. 3 of *Essential Works of Foucault 1954–1984*, series ed. Paul Rabinow (New York: New Press, 2000), 139–42.
33. Deleuze and Parnet, *Dialogues*, 147.
34. Foucault, *The Government of Self and Others: Lectures at The Collège de France 1982–1983*, ed. Frédéric Gros, English series editor Arnold I. Davidson, trans. Graham Burchell (New York: Palgrave Macmillan, 2010), 5. Our emphasis.
35. Guattari, *Chaosophy*, 287.

EIGHTEEN

Revenge of the Tender Pervert: On a Troubling Concept That Refuses to Go Away

Hector Kollias

Why does perversion *still* get such bad press? There was a time in the early 1990s, before queer theory became entrenched as the dominant discursive posture for all those who wish to speak about, for, or on behalf of aberrant sexualities, when Mandy Merck could happily entitle a collection of essays *Perversions*, and just as happily discuss the meanings that made her choose that title, including "the broader opposition to what is expected or accepted (e.g. 'You're just being perverse')", later subtly—politically—modified as "opposition to 'what is right, reasonable, or accepted'"—with, of course, the inevitable etymological twist: "(from the Latin *perversus*, 'turned the wrong way, awry')—on the other side"; a time when Teresa de Lauretis could scarcely hope to hide, under the less contentious title *The Practice of Love*, the thorough-going and immensely productive reimagining of the idea of perversion brimming from every page even though muffled in the subtitle "Lesbian Sexuality and Perverse Desire"; a time also, on the other side of the Atlantic, when Jonathan Dollimore's remarkably prescient elaboration, queer (just) *avant la lettre*, of what he termed *Sexual Dissidence* was entirely based upon a reworking of perversion.[1]

Nowadays the very word "perversion" is absent from the vocabulary of those to whom the baton was passed by the queer pioneers, suggesting that the more hegemonic queer became, in the academy no less than in the general intellectual landscape, the more perversion had to be renounced, already found wanting by its radical questioning as a category by Saint Foucault, the founding father. Queer became, in some respects, the slogan whose reference covers what used to be the field of perversions but without the latter's normalizing baggage, perceived as inevitable and regrettable. That was down to Foucault. Perversion's fate was sealed in the following citation, which more than merits the epithet "classic":

> The growth of perversions is not a moralizing theme that obsessed the scrupulous minds of the Victorians. It is the real product of the encroachment of a type of power on bodies and their pleasures. It is possible that the West has not been capable of inventing any new pleasures, and it has doubtless not discovered any original vices. But it has defined new rules for the game of powers and pleasures. The frozen countenance of the perversions is a fixture of this game.[2]

This countenance though turned out to be anything but frozen, at least if the vicissitudes of perversion that I aim to trace in this chapter are anything to go by. And before even delineating the path my argument will take, I feel compelled to note the pregnant irony in the fact that this "classic" citation is commandeered both by Tim Dean, who places it as epigraph to his excoriating queer critique of the continuing application of the category "perversion" in Lacanian psychoanalysis, and also by Élisabeth Roudinesco, who, in a book that is both history of and *apologia* for perversion from a psychoanalytic perspective, has no qualms in following it with this statement: "It would be difficult to put it better".[3] The irony does not just bear on Foucault's legacy, on a project which, having been received as a decisive repudiation of psychoanalysis, can also be appropriated by it when psychoanalysts find it "difficult to put it better". More importantly, I take the irony to be constitutive of the notion and the destiny of perversion itself. "Perversion" appears to be both: useful and useless; problematic and indispensable; sexual and nonsexual, to wit social and/or political; intrinsically conservative and inherently radical; historically limited or contingent and synonymous with the human condition; a source of indignation and one of admiration.

If I stop short of concluding "all things to all men", it is because, precisely, perversion is *not* that: what "thing" it might be is entirely contingent on the "man" observing it, but the variations are *not* infinite. The irony through which I felt compelled to begin discussing it persists throughout this chapter that is, at least in part, devoted to understanding perversion (dare I say *logically* or even *ontologically*) as a category that, even when described as a "transhistorical and structural phenomenon that is present in all human societies", still demands to be historicized.[4] The apparent contradiction here should be maintained as an ironic marker, a knowing wink, a sign of dissatisfaction. *It won't do* to theorize perversion as "our dark side", a quasi-mystical entity serving to define, as yet more negative theology, "what it means to be human" or some such vapid abstraction, just as *it won't do* to assume an extreme nominalist position holding that the historically contingent advent of multiple perversions in the service of a new regime of biopolitical power invalidates *ipso facto* the reach and force, in other historical contexts, of what the notion designates.[5]

My contention is that perversion operates *logically*, if not *ontologically*, in a way that profoundly unsettles calls for its "raising to the dignity of (part of) the human condition" just as much as it calls for its complete erasure because of the historical contingency of its determination.[6] I turn first to this history and to perversion's own Thucydides (or perhaps its Gibbon): Foucault.

A COMPLETE HISTORY OF PERVERSION (PARTS 17–24)[7]

If in the *History of Sexuality* Foucault is content with establishing the framework of what he calls "the perverse implantation", in the series of Lectures at the Collège de France devoted to the "Abnormal" in 1975, he gives a much more thorough, historically precise account of the genesis of the concept of perversion, and of its "implantation" in the then-nascent fields of psychiatry and criminology which are in effect born out of this implantation, out of the transformation of the erstwhile "monster" of the times when the Law was divine and aberrations were sins, to the garden-variety little pervert of the Victorians, apprehended within a juridical/medical discourse whose parameters are no longer god, sovereignty and sin but norms, illness and public hygiene. More specifically, Foucault postulates that what used to be the "mutual exclusion of medical and juridical discourses" becomes, in the course of the nineteenth century, a more complex "game of dual, medical and juridical, qualification" organizing "the realm of that very strange notion 'perversity'", a notion therefore seen as the *product* of the "game" mingling the juridical and the medical order of discourse.[8] Only a little later, in a move typical of his thought process, Foucault now places "perversion" alongside "danger" as the ostensible *agents*, "enabling", in the first case, "the series of medical concepts and the series of juridical concepts to be stitched together", and, in the second, "the justification and the theoretical foundation of an uninterrupted chain of medico-judicial institutions".[9]

Already there is a series of duplications and reciprocities forming around perversion in its very notional genesis: it marks the domain where the *dual* game of juridical and medical discourses unfolds; it is *produced* by this dual game just as much as it *enables* the "stitching together" of the two series of elements making up the game's discourses; and it is *coupled* with the also emerging concept of "danger", together forming "the essential theoretical core" for the dual—but henceforth *united*—discourses.[10] As Foucault goes on to explain, "danger", better specified as "the dangerous individual", is an essential ingredient for the advent of the change most memorably encapsulated in the famous quip about the sodomite and the homosexual in the *History of Sexuality*—namely, the conversion of a crime, an *act* into a character type, a *person*. He suggests, referring specifically to "special courts for children", that these new legal processes bear much more on the context of the individual's existence, life and discipline than on the act for which he has been brought before the children's court. The child is brought before a court of perversity and danger rather than before a criminal court.[11] The constitution of perversion and of the dangerous individual, alongside the "dual game" of juridical and medical discourses form what Foucault calls "the techniques of normalization of sexuality".[12] Unsurprisingly, normalization itself enters into a reciprocal or "dual" game of cause and effect, condition and goal. Foucault can claim to be looking at "the mechanics of the disciplinary apparatus" for "their *effects* of normalization", and, less than a page later, he can equally write, "The norm is not simply and not even a principle of intelligibility; it is an element *on the basis of which* a certain exercise of power is founded and legitimized".[13]

While the polemical focus on debunking "the repressive hypothesis" in *A History of Sexuality* has always been read as concomitant with that on "the multiple implantation of perversions", I believe the more closely observed historical inflections of the Lectures on the "Abnormal" allow for a more nuanced reading of what Foucault was investigating.[14] Thus, after he has introduced perversion and the dangerous individual, and the norm against which they are discursively mobilized, Foucault goes in search of key moments in the transformation of previous medical and juridical discursive regimes and of "the monster" or the madman peopling them, into the domain of the abnormal. Here, however, he does not propose a multiple generation for the perversions later to be multiple in their implantation. Instead, in his examination of the discourses of medical scientists like Esquirol and Marc, acting, respectively, for the prosecution and the defence in the seminal case of Henriette Cornier, Foucault presents an essentially *dual* argument focusing on subtle, yet consequential, modulations on the values of presence and absence of the concepts hitherto available to such discourses in the judgement of a crime such as Cornier's.

The scandal of her crime, the fundamental problem it raised, was the combination of lucidity and premeditation suggesting she was sane and fit for the scaffold, together with a stupefying lack of motive or justification for her crime—suggesting the exact opposite. Proceeding from what Foucault had already established as the shift of focus from the crime to the criminal, a justifiable diagnosis of Cornier is as psychopath, and a lucid, nondelirious one at that: in other words—a pervert. With such terminology not yet available, the prosecution opted to argue that the *absence of reasoning*, of justification for her crime, was overruled by *the presence of rationality*, of reason as mental capacity, but the defence countered with the opposite argument: the *absence of reasoning* implied a *presence of madness*. For Foucault, this becomes the pivotal moment when the older rationalist view of the mind was given a fatal blow. If Cornier had no motive or reason for her crime, she was *compelled* to commit it, for she was perfectly lucid (rational) in her understanding that it would cost her her life: since she was aware of this and *yet* acted against it, this must mean that she was led to the crime not by reason but by something else, by a compulsion. Foucault finally offers: *an instinct*.

It is this unprecedented appeal to the instincts which will eventually allow for the proliferation of psychopathies and perversions—that much is made quickly evident. But this crucial moment when the focus switches "from the motiveless act to the instinctive act" does not merely provide the necessary conditions for the birth and incredible fruition of sexology, and thence of psychoanalysis itself, as a supposed science *of the instinct*.[15] As Foucault observes, "Basing itself on the instincts, nineteenth-century psychiatry is able to bring into the ambit of illness and mental illness all the serious disorders and little irregularities of conduct that are not, strictly speaking, due to madness". This is what effectuates "the transition from the great monster to the little pervert".[16] I believe it is no great extrapolation to surmise that perversion *as such*, as it is engendered (in Foucault's genealogy) in the nineteenth century, comes thus to be placed as a *third term*, a term mediating between, on the one hand, the out-and-out, delirious madness that will still be associated with the constitutive phenomenal presence of delirium and, on the other hand, the calculated, "justified" (in the sense of "motivated") criminal behaviour of a sane and, for lack of a better word, "normal" criminal. The *ab*normality of the pervert will be a different kind of abnormality from that of the "mad" person, or of the psychotic in psychoanalytic language—just as it would be different from the normality of criminality or neurosis.

I do not want to make grandstanding claims aligning Foucault's gradual elaboration of the "multiple implantation of perversions" with the "rigid" structure of the subject found in Lacanian (particularly) psychoanalysis. But I do want to suggest that, at least in the earlier trajectory of Foucault's thinking as presented in the Lectures on the "Abnormal", we find a conceptual framework that has certainly not solidified into the "three structures" of Lacanian lore, but that, being crucially derived from a double demarcation from madness *and also* from normality, belies any immediate "multiple" operation. At this point at least, it is more accurate to see perversion as a *tiers parti*, a third term emerging out of the opposition between reason and madness. Thus it *is* possible, despite Foucault's own reticence, to conceptualize instead of perversions, perversion—a discrete category that gathers, under the aegis of the first conceptualizations of instinct, all the subsequently "multiple implantations" Foucault's analysis draws on. And it may be surprising to discover that this possibility becomes most eminent at a juncture that Foucault's later, more polemical position would have made seem unlikely. It is in his startling analysis of the first *Psychopathia Sexualis*, published by Heinrich Kaan some forty years before the more (in)famous tome by Kraft-Ebbing, that Foucault first discovers what will become synonymous with the "perverse implantation": the proliferation of sexual aberrations and the classification compulsion germane to sexology.

There he also discovers a concept of the sexual instinct that "begins before and goes beyond copulation".[17] But he also discovers that "the set of these both natural and abnormal aberrations [that] constitute the domain of *psychopathia sexualis* [also] . . . constitute a *unified* domain", unified, moreover, by "the imagination, what he [Kaan] calls *phantasia*".[18] This may not prefigure Lacan's subjective structure of perversion—but it *does*, unmistakably, prefigure the Freudian elaboration of the drive as itself going "beyond copulation", as tending *by nature* to abnormality, since the "goal" of copulation is far too restricted for its multifarious activities; just as the stress on *phantasia* also prefigures Freud's focus on fantasy. Foucault does not comment on this (he merely offers a mordant joke at the expense of Kraft-Ebbing's taxonomical vocation), and no doubt *instinct* remains, in the 1840s, *Instinkt* and not *Trieb*, just as Foucault's gloss of "imagination" for Kaan's *phantasia* no doubt relates it more to the Romantic or perhaps the Schopenhauerian conceptions of the imagination. But surely the Freud of the "Three Essays on Sexuality" and of "Instincts and their Vicissitudes" would have recognized an ancestral presence in this older *Psychopathia Sexualis*.

THE ANGELS ARE PARAPHILIACS[19]

It is not unknown for psychoanalysts or psychiatrists to have their own historicist conversions; at their best, these can provide a framework in which Foucault's historicism can itself be historicized. Even though he only cites Foucault for the earlier works on madness, the role of Lacanian Thucydides goes to the Belgian psychoanalyst Paul Verhaeghe and to his monumental attempt at creating "a manual for clinical psychodiagnostics".[20] It should be instructive, for those committed to the denigration of psychoanalytic approaches to sexuality, to learn that Verhaeghe's effort to raise his diagnostics on secure foundations involves a thorough debunking of the notions of normality and abnormality. And the first principle Verhaeghe uses to reject normality as diagnostic guide is precisely the blatantly contingent proliferation of abnormalities (or of "perversions")—only, intriguingly, here it is not, as in Foucault's account of sexology, the identification of multiple perversions that shores up a protected zone of normality, but the other way round. Quoting perhaps the most trenchant, most eloquent and least remembered of the anti-psychiatrists, Verhaeghe offers the following: "Every rule or norm of psychological health generates a new category of mental illness".[21]

We encounter again, as with Foucault's formulations, this odd reciprocity between cause and effect, suggesting that what is problematic about "the abnormal" is that one quickly discovers that its identification, and the concomitant drawing of the boundaries constituting the norm, is a hopelessly contingent event and yet, as Foucault and anti-psychiatry alike most fervently protested, it has real, material effects on the bodies and minds of patients as well as on the culture and society in which they live. The question "who is a pervert?" is a remarkably difficult one to answer satisfactorily, but every *unsatisfactory* answer has profound effects on the very context in which, the very position from which, the question is asked.

Verhaeghe seems to be fully espousing Foucault's judgements when he contends that the adoption of norms as criteria for the diagnosis of mental illness or of psychosexual abnormality soon becomes mired in inescapable paradox:

> first, psychopathology and criminality become so interwoven that the knife begins to cut both ways. Criminal behaviour becomes acceptable because it is explicable; psychopathological behaviour becomes punishable because it is deviant. . . . Second, psychiatry and clinical psychology become the judges of the social order, with the prosecutors of the pathological society at one extreme and the guardians of the public good on the other.[22]

From this, it takes but a breath to reach the inevitable conclusion that what has occurred *in reaction to* this indictment of Verhaeghe's own profession is, on the one hand, a result of the influence of people like Szasz and Foucault, but, on the other, even *less* satisfactory. Foucault's genealogical work and the anti-psychiatrists' emphasis on the pernicious role of categorizations of abnormality can both be seen as giving rise to what is, by now, widely understood as a bankrupt new system of classifications: the DSM, the *Diagnostic and Statistical Manual of Mental Disorders* issued by the American Psychiatric Association.[23] It should be obvious that the DSM makes no mention of perversions, singular or plural. Instead, since the DSM-III of 1980, it has adopted the term "paraphilia", to which it attaches subcategories that map onto already named perversions such as fetishism, sadism, voyeurism and the like (but famously not, since 1974, homosexuality). Out of Verhaeghe's many objections to the DSM, two are most consequential for the notion of perversion/paraphilia, and both of them point to a loop that has taken psychiatry back to the conceptual regime that Foucault was analysing historically, even though, and this is the paradox, the move towards the DSM was *occasioned* by critiques (such as Foucault's) of these earlier regimes.

First, even though different versions of the DSM offer a different focus on the constitutive elements of paraphilias, the very adoption of the term, with its Greek rather than Latin derivation, changes perversion only to the degree that it specifies, in *philia*, and even then highly ambiguously, that what is being designated is to do with sex.[24] The change of one prefix to another, however, is a safe indication that the term really means exactly what perversion always did. Verhaeghe's sarcasm shines through: "Indeed, the pervert—excuse me, the paraphiliac—is someone who transgresses the norm, be it customary, religious, or legal. We are back at the beginnings again, with sinners and criminals".[25] There is *nothing* in the cosmetic makeover of perversion into paraphilia to answer the thoroughgoing critique of normalization which ostensibly led to it. Second, and worse, not directly linked to the adoption of the term "paraphilia" (even though with the DSM-5's further distinction between paraphilias and paraphilic *disorders* we are getting dangerously close), the empirical, observational perspective that successive editions of the DSM proceed from appears unmistakably to harken back to the ghosts of sexology and to Kraft-Ebbing's taxonomical excesses, to say nothing of Szasz's indictment. Verhaeghe's sarcasm is again apposite: "soon all the antiquated psychiatric entities will have found their contemporary equivalents, albeit with the suffix 'personality disorder'".[26]

Foucault and others like him sought to redress what they perceived as the problematic normative assumptions of psychoanalysis by providing the historical genealogy of its concepts and therefore of its normative assumptions. It is highly ironic, then, that in helping to discredit Freudian theories on perversion, they have unintentionally contributed to a most retrograde and most imbecilic renunciation of theory, of what Verhaeghe calls, after Freud, "metapsychology", in favour of the observational grids that lead back to the naïve empirical taxonomies initially denounced. If it be objected that Foucault's aim would have been nearer to rejecting the very idea of perversion/paraphilia, it should just as readily be admitted that such rejection has taken place or is taking place, albeit in piecemeal fashion: homosexuality may have been dropped from the DSM, but the criteria used for classifying paraphilias as either potential disorders or otherwise mere harmless kinks *remain*, and even gain credence (in some circles) by not being directly referred to a norm. In other words, denouncing the "multiple implantation of perversions" has done strictly *nothing* to change the way in which what was being implanted, regardless of what it was called or how it was described or given content, was defined as abnormal, as against the norm. And that's not all. In a double, if not duplicitous, move that Foucault would have been perhaps the first to point the finger at, the very gesture of progressively "dropping" more and more "harmless kinks" from the paraphilia list is counterbalanced by the creation, from the DSM-IV-TR onward, of the ominous category of NOS (not otherwise specified) paraphilias, where it would not take a paranoid sensibility to predict seeing any and every type of sexuality that has yet to become acceptable as kink.

When no one is named a pervert anymore, it seems that potentially *everyone* could be a paraphiliac. Verhaeghe laments having "lost sight" of a crucial "motivation" in anti-psychiatry—namely, the critique of "authoritarianism" in the exercise of power by psychiatrists.[27] One could be more precise and say that the power matrices Foucault identified in the concurrent emergence of psychiatry and criminology in the nineteenth century have both co-opted his critique of them and used it to their benefit. Not only has "the multiple implantation of perversions" continued its forward march, but it has also shored itself up against critique by dressing up normative criteria into observational grids and perversion into paraphilia, to such an extent that a diagnosis such as the following by Roudinesco, highly disturbing from a Foucauldian perspective, has begun to be perceived as unproblematically true:

> The more psychiatric discourse replaces the word "perversion" with "paraphilia" in the belief that it can do away with the implicit reference to God, good, evil, the Law and transgression of the Law, or even to *jouissance* and desire, the more it becomes synonymous with "perversity" in civil society.[28]

Before examining this highly dubious statement further, a less grandiose and more careful observation concerning the "loss" of perversion as a concept is necessary, if only as a reminder that Foucault's project (and in certain respects certainly also that of anti-psychiatry) has suffered, not only from being tacitly co-opted by the very discourses it sought to uncover but crucially *also* from its most vociferous defendants and epigones. Surely the enthusiastic and overzealous embracing, not only within queer theory but perhaps there particularly, of an extreme nominalist interpretation of the polemics of *A History of Sexuality* has meant that the more subtle, conceptually more accommodating history of the emergence of "that odd concept 'perversity'" in the Lectures on the "Abnormal" has been brushed aside. Equally—and with what irony—it led to the alignment, in the eyes of a certain type of psychoanalytically inspired cultural and political critique, no longer of queers with perverts, but rather of queer, and (not just by extension) of Foucault's ideas themselves, with perversion.

I WAS A MARXIST/LENINIST/STALINIST/ MAOIST/BADIOUIST INTELLECTUAL

So why are both "civil society" today and queer theory associated with "perversity" in Roudinesco's terms? In the space afforded me here I can only provide the sketch of an answer. The proper and complete answer would require a thorough-going exploration of what perversion has meant and what it has come to mean in the psychoanalytic lexicon where it has found its safest conceptual haven. It would require an understanding of the concept's history from Freud's earliest ruminations, through to the very influential and, surprisingly, not always deleterious displacement it underwent with object-relations theory, when it became soldered to aggression and to the death drive; most importantly, most frustratingly, most inevitably caught in labyrinthine complexity, it would also require a reckoning with Lacan's cruelly under-theorized but much-adopted (and adapted) interpretation.[29] It would also require a questioning of one of the staple psychoanalytic responses to the problematic of perversion, one that has contributed the most to making perversion one of psychoanalysis' most reliable resources in its attempts to present itself as cultural theory: namely, once again using Roudinesco as cipher, the understanding of perversion as a privileged short-circuit between what is most abject in sexuality, in the drives, and what, in civilization (in art, ethics and science), is most sublime: "Perversion fascinates us precisely because it can sometimes be sublime and sometimes abject".[30]

Curiously, it is this last incarnation of perversion as both sublime and abject that tends to get swept under the carpet when theorists fly the banner stigmatizing today's "civil society" or "late capitalism"—which amounts to no more than the proverbial half-a-dozen—as perverse. Roudinesco will now sound rather like a meek avant-garde, castigating the DSM as "a perverse classification of perversions" that "realizes, in a deadly way, Sade's great social utopia [where] differences are abolished [and] subjects are reduced to objects under surveillance".[31] Take away the reference to perversion and perhaps also that to Sade, and this is a statement that Foucault himself might have made, or Deleuze—on the contrary, I am quite uncertain as to whether *Lacan* might have made it. Nor do I think there is ultimately much of Lacan left in one of the key positions refined and reconfigured in over twenty years of publications by Slavoj Žižek, who represents the artillery in this offensive. What there is, in the obdurate reference to perversion anchoring political critique, is an undoubtedly forceful but unsubtle extrapolation of selected points in Lacanian theory, often violently extracted from the clinical contexts of their primary application.

The full deployment of a point-by-point dispute with Žižek would also be impossible within the limits of this chapter, but the essential rudimentary steps can be traced. The *incipit* is axiomatic: ours is the "era of the 'decline of Oedipus', when the paradigmatic mode of subjectivity is . . . the 'polymorphously perverse' subject following the superego injunction to enjoy".[32] Bypassing yet another slide between different modalities of perversion (in this case between perversion in the Lacanian sense of "mode of subjectivity" and Freud's "polymorphously perverse" assignation, reserved for *infantile* sexuality), what Žižek expresses here is a mostly unchallenged *topos* of contemporary left critique: late capitalism is a perverse socioeconomic system whose organizing principle is the imperative compelling subjects towards *jouissance*. The salient connection here is the one Žižek makes between certain "radical" thinkers, notably Deleuze and Foucault, and the perversion this system embodies. In Foucault's case, this amounts to a reading of his conception of power as formulated in the 1970s which may happily praise it as *description* of the status quo (as when Foucault, "a perverse philosopher if ever there was one", is praised for "the strength of his forceful argumentation [that] lies in his claim that resistances to power are generated by the vary matrix they seem to oppose") but which also resolutely *denies* that Foucault permits any space for contesting or resisting the status quo: "There is nothing more misguided than to argue that Foucault . . . opens up the way for individuals to rearticulate-resignify-displace the power mechanisms they are caught in".[33] Since Žižek emphatically denies any possibility that Foucault's historicist circumscription of perversion as one of the many indices of a biopolitics of power could have any emancipatory or politically salutary application, so, and this (especially after the well-documented souring of his dialogue with Judith Butler) should surprise no one, Žižek *voids* queer theory, its radical credentials and progressive goals altogether.

I should never have bemoaned the bad press perversion gets, then. I should never have thought that, never mind the rather partially consumed polemics of Foucault's *History of Sexuality*, there may be something in perversion as a concept that troubles by its very logic the attempts made to implant, taxonomize or domesticate it. No, perversion was evil all along, aligned with the pernicious processes of capitalism intent not only on forcing me to "enjoy" to the limits of my capacity to do so, but also on dressing up the very perverse conceptual mechanisms that ensure my enslavement in the queerest, most radical colours ostensibly aimed at debunking them. Foucault, whose analyses I wanted to reconstruct beyond the standard reading obfuscating the emergence of perversion as a "third category", a *tiers exclu* whose disappearance from contemporary thinking about sexual (and social) aberrations is a regrettable error, was really the Sadean figure who, hiding all along behind the rhetoric *against* the "multiple implantation of perversions", in effect *propagated* that multiple implantation itself.

All of a sudden, I realize: forget Freud, Lacan or psychoanalysis as we know it; never mind the crystal-clear sentence: "in psychoanalysis we cannot speak of perversion except in relation to sexuality".[34] Perversion is *not sexual*. Rather, it is by *extrapolating from* the sexual, from Lacan's dogged insistence on identifying "subjective structures" according to their mode of relation to the phallus, or the Other, or perhaps the mother, and castration evidently, and sexual difference, let's not forget that . . . it is from all this—oh, plus the four discourses, *obviously*—that Žižek is able to extract these "antagonistic" essences, the Pervert and the Hysteric, mapping them onto "attitudes" with political import.[35] Because, is it not clear? "Following Freud, Lacan *repeatedly insisted* that perversion is always a socially constructive attitude, while hysteria is much more subversive and threatening to the predominant hegemony".[36] I guess one of those instances where we see Lacan *repeatedly insisting* might be the time in his sixth Seminar when he defines perversion as "that which, in the human being, resists any normalisation".[37] But normalization is not the issue here. Not when the world needs to be carved up between "subversive" hysterics and "reactionary", or "socially constructive" perverts who only ever serve the political status quo—perverts like Foucault, of course, or like Deleuze, who is "the model of false subversive radicalisation that fits the existing power constellation perfectly".[38]

Another strange loop is thus looped. When perversion is no longer the outcome of a multiple implantation but the very structural/social/subjective position from which such an implantation is both executed *and* denounced, there is no longer any sense in either defending whatever sexual or other aberrations come to be gathered under perversion's banner or rejecting the banner and the gathering of aberrations altogether, since each of these alternatives will have *always already* been denounced as perverse—by whom? Žižek and his acolytes demand only one answer to this question: by hysterics. The irony I insisted discourses on perversion were marked with from the beginning of this chapter is back once again—welcome! If hysteria is subversive and perversion a mask of radicalization concealing strict adherence to the status quo, exactly what, in the "existing power constellation" that is unable to shake off the persistence of normativity in sexuality, does Žižek's position subvert? Does it subvert that very persistence of normativity? If it did, would it not thereby come to rejoin those perverts it aimed to denounce? And if it is not normativity, but *the lack thereof* that is subverted, would that not return us to *another form of the norm*, or of the Law? Aren't hysterics really just crying out for a New Master? Žižek is very careful, for instance, to demarcate a category of *perverse* homosexuality from the category of homosexuality *as such*—to me that sounds already very close to the panicked capitulations to normalization animating the DSM.

Who is a pervert, then? Surely, if no one is, then everyone could be, or so the bizarre logic of repudiation goes, the logic behind both the knee-jerk rejection of the term (in some incarnations of queer theory, for example) and its unquestioning adoption, by Žižek and others, when this adoption proclaims to have *nothing to do* with notions of normativity, appealing instead to half-digested crumbs of Lacanian structuralism. But the absurdity of such logic is evident. It is never everyone, nor no one, it is always *someone*. It is always that someone who, willingly or not, consciously or not (and these alternatives obviously have implications), takes on the mantle of "our dark side" and wears it like a tattoo on the forehead, so that the rest of us will know we march in light. Or, to be less portentous, the pervert is that someone who *both subverts and upholds* the projected norm that makes them a pervert. The pervert is like the subject of the following verses, which speak *not* of "the homosexual" as category, as one of the "multiple implantations" of perversion, but of "the homosexual" as the name of a perverse strategy of subversion and revenge, as the thrilling and destructive, vengeful *jouissance* located precisely *in the name "pervert"*, the name "homosexual", the name for the abnormal that every "normal" subject desperately clings to:

> "The Homosexual" they call me
> It's all the same to me
> That spectre they projected I will now pretend to be
> Since their neurosis is what passes for normality
> It's okay with me if I'm queer
> Since their tone-deafness is called the love of music
> I won't disabuse them
> I'll make love with their women
> I'll make them sing notes of pleasure
> Their husbands will never hear.[39]

It is tempting to try to map the proper names that have appeared in this chapter (but why end there?) onto the general "they" these viciously clever song lines target. That would be the final ironic twist by which perversion holds us (and "them") in its grip, for however "they" fit the schemes of the song's subject, "they" are united in believing that "their neurosis passes for normality". This is the point, after all, of Žižek's passionate attachment to hysteria. So if, in the end, the subject of the song remains elusive, assuming only "the spectre they projected" and refusing to "disabuse them", and if, as I maintain, that subject *is* "the pervert", I hope I am allowed, if only by means of a projection that hopes to be different from "theirs", to bask a little in the reflection of a light as strong as it is gleeful. The pervert, thankfully, is the one who has the last laugh.

BIBLIOGRAPHY

Dean, Tim. "The Frozen Countenance of the Perversions". *Parallax* 46 (2008): 93–114.

Dollimore, Jonathan. *Sexual Dissidence: Augustine to Wilde, Freud to Foucault*. Oxford: Oxford University Press, 1991.

Foucault, Michel. *The History of Sexuality Volume 1: An Introduction*. London: Allen Lane, 1979.

———. *Abnormal: Lectures at the Collège de France, 1974–1975*. London: Verso, 2003.

Halperin, David. *One Hundred Years of Homosexuality*. New York/London: Routledge, 1990.

Kollias, Hector. "'Another Man Is An Other': Impersonal Narcissism, Male Homosexuality, Perversion". *Textual Practice* 27, no. 6 (2013): 991–1011.

Lacan, Jacques. *The Ethics of Psychoanalysis, 1959–1960. The Seminar of Jacques Lacan, Book VII*, edited by Jacques-Alain Miller. New York: Routledge, 1992.

———. *Le Séminaire. Livre VIII. Le Transfert.* New edition by Jacques-Alain Miller. Paris: Seuil, 2001.

———. *Le Séminaire, Livre VI: Le Désir et son interpretation*, edited by Jacques-Alain Miller. Paris: La Martinière, 2013.

de Lauretis, Teresa. *The Practice of Love: Lesbian Sexuality and Perverse Desire.* Indianapolis: Indiana University Press, 1994.

Merck, Mandy. *Perversions: Deviant Readings.* London: Virago, 1993.

Roudinesco, Élisabeth. *Our Dark Side: A History of Perversion.* Cambridge: Polity, 2009.

Stoller, Robert J. *Perversion: The Erotic Form of Hatred.* New York: Pantheon, 1975.

Vadolas, Antonios. *Perversions of Fascism.* London: Karnac, 2009.

Verhaeghe, Paul. *On Being Normal and Other Disorders: A Manual for Clinical Psychodiagnostics.* London: Karnac, 2008.

Žižek, Slavoj. *The Ticklish Subject: The Absent Centre of Political Ontology.* London: Verso, 2000.

NOTES

1. Mandy Merck, *Perversions: Deviant Readings* (London: Virago, 1993), 2, 6, 10; Teresa de Lauretis, *The Practice of Love: Lesbian Sexuality and Perverse Desire* (Indianapolis: Indiana University Press, 1994); Jonathan Dollimore, *Sexual Dissidence: Augustine to Wilde, Freud to Foucault* (Oxford: Oxford University Press, 1991).

2. Michel Foucault, *The History of Sexuality Volume 1: An Introduction*, trans. Graham Burchell (London: Allen Lane, 1979), 48.

3. Elisabeth Roudinesco, *Our Dark Side: A History of Perversion*, trans. David Macey (Cambridge: Polity, 2009), 53; Tim Dean, "The Frozen Countenance of the Perversions", *Parallax* 46 (2008): 93–114.

4. Roudinesco, *Our Dark Side*, 5.

5. It is no coincidence that the most vocal advocates of this position focus on the historical genesis of homosexuality, among them the author—and scion of the queer discursive family—whose title, *One Hundred Years of Homosexuality*, perfectly encapsulates this position: David Halperin. A complete and persuasive development of this argument falls outside the remits of this chapter, so I shall simply say that it is no coincidence because, in the "opposing camp" of psychoanalysis, homosexuality *was* seen as a perversion, before pragmatic and theoretical adjustments by liberals like Roudinesco (and many others) produced a thoroughly predictable renunciation, a renunciation which, with only a few exceptions, has nevertheless *not* seen fit to go back to texts where, for example, even the "Greek homosexuality" Halperin seeks to refute is unequivocally referred to as a perversion—namely, Lacan's eighth Seminar on Transference (but there are others). We *are not done with* perversion by simply reconfiguring the historically contingent—this much is obvious—criteria whereby homosexuality is or is not included within its conceptual reach and application. See David Halperin, *One Hundred Years of Homosexuality* (New York/London: Routledge, 1990), and Jacques Lacan, *Le Séminaire. Livre VIII. Le Transfert*, new ed. by Jacques-Alain Miller (Paris: Seuil, 2001).

6. The phrase in quotation marks is a parodic appropriation of Lacan's famous definition of sublimation as "raising the object to the dignity of the Thing". See Jacques Lacan, *The Ethics of Psychoanalysis, 1959–1960: The Seminar of Jacques Lacan, Book VII*, ed. Jacques-Alain Miller (New York: Routledge, 1992).

7. The three subheadings as well as the title of this chapter are all parodies, respectful—rather than mocking—perversions of song titles by the artist who in no small part inspired my efforts: Momus. The original song titles are as follows: "A Complete History of Sexual Jealousy (Parts 17–24)", "The Angels Are Voyeurs" and "I Was a Maoist Intellectual", all standing alongside "The Homosexual", with which my chapter closes, in Momus, *Tender Pervert* (Creation Records, 1988), which is, of course, the source of my title.

8. Michel Foucault, *Abnormal: Lectures at the Collège de France, 1974–1975*, trans. Graham Burchell (London: Verso, 2003), 32.
9. Foucault, *Abnormal*, 34.
10. If Foucault earlier spoke of the "strange notion" of *perversity*, whereas now he is concerned with the confluence between ideas of "the dangerous individual" and of *perversion*, could he be taken to task for eliding the difference between these two cognates? In my view, that would be fruitless, pedantic and entirely missing the point not only of Foucault's sexuality project but also of what is at stake in perversion itself. It is usually psychoanalysts, especially those for whom "perversion" must be raised to the dignity of a "psychic structure", who point a pedantic finger at such elisions between perverse substantives—for we also have the potentially significant distinction between monolithic, singular *perversion* and plural, often proliferating *perversions* to add to that between *perversion* and *perversity*, the latter customarily derived from the "polymorphous perversity" Freud attributes to infantile sexuality. Being pedantically stringent about such distinctions has certainly *not* helped the cause or the arguments of psychoanalysis, for the grey area opened up by them and covered over by their seemingly inevitable elision is intriguing precisely *because* it is a grey area where strip-lights and disciplinary surveillance come to falter.
11. Foucault, *Abnormal*, 40.
12. Foucault, *Abnormal*, 42.
13. Foucault, *Abnormal*, 49, 50, my emphasis.
14. Foucault, *A History of Sexuality*, 37.
15. Foucault, *Abnormal*, 131.
16. Foucault, *Abnormal*, 132.
17. Foucault, *Abnormal*, 279.
18. Foucault, *Abnormal*, 279 (my emphasis), 280.
19. For a more precise diagnosis, please consult the DSM.
20. Paul Verhaeghe, *On Being Normal and Other Disorders: A Manual for Clinical Psychodiagnostics* (London: Karnac, 2008).
21. Verhaeghe, *On Being Normal*, 10. The citation is from Thomas Szasz, *The Myth of Mental Illness: Foundations of a Theory of Personal Conduct* (New York: Harper & Row, 1972), 12.
22. Verhaeghe, *On Being Normal*, 10–11.
23. With the publication in 2013 of *DSM-5* came also the most prominent critiques of its clinical and practical bankruptcy as well as calls for its abolition. This fifth edition is far too "fresh" to be taken into account here, and all the sources I refer to, including Verhaeghe and Roudinesco, refer to the latest version prior to this—namely, the *DSM- IV-TR* (2000).
24. "Version" in perversion allows for more general takes on the meaning of the concept, something which has not gone unnoticed and has been richly and variously utilized by Lacan and his followers, to say the least. *Philia* in the scientific lexicon denotes a propensity for union, an "attraction to" or "love of" (for example, hydro-philia is the quality of some material to absorb water). Paraphilia in this sense is somewhat eccentric, as all the "attraction" is to the prefix "para" and not something concrete. Even paedophilia is at least more precise than that—except, of course, that if one is to get serious about etymology, one would have to consider how the Greek *philia* is certainly related to, but equally certainly something *other than*, sexual attraction. Admittedly, it is foolhardy to expect the authors of diagnostic manuals to have done extensive research on Plato, or "Greek Love" in general—except for someone like Lacan, who spends almost half of his eighth Seminar discussing, among others, the complexities arising from the distinction between *eros* and *philia*. I cannot help suspecting that it is precisely such an attention to nuance that partly helps raise Lacan's verdicts on the "paraphilias" safely over the ordinary taxonomer's morass.
25. Verhaeghe, *On Being Normal*, 399.
26. Verhaeghe, *On Being Normal*, 30.
27. Verhaeghe, *On Being Normal*, 31.

28. Roudinesco, *Our Dark Side*, 138.

29. For object-relations theory, the title of a seminal study must suffice: Robert Stoller's *Perversion: The Erotic Form of Hatred*—despite appearances, *far from* the disciplinary thesis one might expect. See Robert J. Stoller, *Perversion: The Erotic Form of Hatred* (New York: Pantheon, 1975). As for these "required" steps, they are taken in my forthcoming book-length study on perversion for which this chapter is a kind of introduction or stepping-stone. Another such stepping-stone, opting for the riskier strategy of diving headlong into the Lacanian deep, is: Hector Kollias, "'Another Man Is An Other': Impersonal Narcissism, Male Homosexuality, Perversion", *Textual Practice* 27, no. 6 (2013): 991–1011.

30. *Our Dark Side*, 4.

31. *Our Dark Side*, 136, translation modified. For a thorough critique, from a Lacanian position, of the various efforts to associate perversion with the ogre of fascism, see Antonios Vadolas, *Perversions of Fascism* (London: Karnac, 2009).

32. Slavoj Žižek, *The Ticklish Subject: The Absent Centre of Political Ontology* (London: Verso, 2000), 248.

33. *The Ticklish Subject*, 253.

34. Jean Laplanche and Jean-Bertrand Pontalis, *The Language of Psychoanalysis* (New York: Norton, 1973), 307.

35. I apologize for this sudden swell of Lacanian *aperçus*, which a more patient and serious exposition would have carefully teased out of an incredibly complex and varied thirty-year-long elaboration in Lacan's Seminars and writings themselves. If their hasty agglomeration helps make the style and tone of my argument here appear perhaps a tad Žižek-like, it will have served its purpose.

36. *The Ticklish Subject*, 247, my emphasis.

37. Jacques Lacan, *Le Séminaire, Livre VI: Le Désir et son interpretation*, ed. Jacques-Alain Miller (Paris: La Martinière, 2013), 571 (translation mine). Another chapter in the ongoing unfolding of what I called "the destiny or perversion" must surely be the momentous mobilization of the entire army of international Lacanianism responding to the diagnosis of the decline of Oedipus, with Miller as its leader and flag waver, and with flag perhaps the all-too-recent publication of the sixth Seminar, where we read, from Miller's pen, that Lacan sings the praises of perversion as resistance. The question of how all these different approaches to Lacan's legacy come to be united or differentiated, and of the critical role the question of perversion plays in all this, is yet another important thread I cannot develop here.

38. *The Ticklish Subject*, 251.

39. Momus, "The Homosexual", in *Tender Pervert*.

NINETEEN

Jouissance in Lacan's Seminar XX

Progelomena to a New Reading of Strange Enjoyment and Being an Angel

Lorenzo Chiesa

This chapter aims to provide elements for a new reading of Lacan's notion of *jouissance*, in particular woman's *jouissance*, as exposed in Seminar XX. In addition to treating feminine *jouissance* as an "Other enjoyment" which lies "beyond the phallus", Lacan also speaks of it in terms of a *jouissance* that is "strange" [*étrange*] and of woman's "being-an-angel" [*l'être ange*]. My claim is that, although these two terms are homophonous, and both irreducible to "Other enjoyment", they should not be confused: while the former refers to a *phallic jouissance* that is different from that of man, the latter points to an *asexual jouissance*, which Lacan denounces as the structural illusion of the being-One of a body that would no longer be subjected to the Other, and identifies with the basic fantasy of masculine totalization. The angel can in the end only be obliquely materialized in linguistic reality in the far less edifying body of the hysteric, in her striving for the "outsidesex" [*horsexe*] and her "pretending to be the man" [*faire l'homme*].

Right at the opening of Seminar XX, Lacan introduces *jouissance* in a particularly dense but enlightening lesson. His arguments presuppose here a line of enquiry he had been pursuing at least since Seminar XVII and its theory of discourses: psychoanalytic discourse—and its novel historical determination of the status of other discourses—starts off from fully assuming the empirical evidence, already intuitively experienced in our everyday lives,[1] that human sexuality is inextricably entwined with the absence of the sexual relationship [*rapport*]. More precisely, psychoanalysis acknowledges and formalizes the fact that it is impossible to enunciate this relationship as One, whereby one sex could relate to the other sex as *the* other one, thus achieving a unified couple, or two-as-One. Consequently, Lacan contends in Seminar XX, the sexual relationship as One, which *would be* the "*jouissance* of the body as such", can only be supposed in terms of the *asexual* being One of the body.[2] Psychoanalysis accepts the logical necessity of this supposition of the One but refuses to grant it any essence, which is regarded as purely mythical. In doing so, it claims to break with all classical philosophical discourses revolving around an ontology of substance,[3] in order to focus, in its theory and practice, exclusively on the *sexual jouissance* involved in the liaison between sexed human beings as beings of language.

The impossibility of establishing the sexual relationship as One follows from the fact that, although there are two sexes, *sexual jouissance* can only be mediated symbolically—for both men and women—by the contingent bodily inscription of the signifier in the image of the male sexual organ, or phallus. As Lacan has it, "woman's sex does not say anything" to man, or better, for him, "nothing distinguishes woman as a sexed being, but sex",[4] the phallic bodily enjoyment he obtains from her during intercourse. Of crucial importance, here, is the specification that sex—bluntly put, in the sense of "having sex"—ultimately differentiates woman as a *sexed* being there where the lack of an independent feminine symbol that would linguistically mediate her biological sexual characteristics fails to accomplish this.[5]

Lacan's argument is not that woman is less sexed than man—nor that she is even without sex, leaving aside the symbolic dominance of the phallus and man's having sex with her—as a precipitous reading of these pages could suggest ("nothing distinguishes woman as a sexed being"), but rather that, given the asymmetry of the phallic signifier, her sexuation—her sex as a sexed being of language—and following this her *jouissance* of man, is "strange"[6] [*étrange*]. In other words, *l'étrange* must by all means not be confused with *l'être ange*, "being-an-angel", although the two terms are homophonous. *L'être ange* in fact points to a projection onto woman of *asexual jouissance*, which Lacan resolutely denounces as the structural illusion connected to the being-One of a chimerical body, or, we may add, to the basic fantasy of man—that of totalizing *jouissance*.

Unsurprisingly, in the same context, Lacan also states that *jouissance* insofar as it is sexual can only be phallic, for both men and women: "Analytic experience attests precisely that everything revolves around phallic *jouissance*".[7] Having said this, at the same time, "woman is defined by a position that I have indicated as 'not-all' [*pas toute*] with respect to phallic *jouissance*". So, if, on the one hand, as we have just seen, the sex of woman says something to man only through his *jouissance* of her body, on the other, since she is not entirely contained within phallic *jouissance*, the latter will also be "the obstacle owing to which man does not come to enjoy woman's body": what he enjoys is rather the "*jouissance* of the organ". In other words, to the extent that *jouissance* is sexual, and hence phallic, it is marked for man by a hole; man never relates to Woman as a universal, which is why Lacan bars the determinate article, the "*La*", of "*La femme*".

Moving from this premise, it is now a question of seeing how this fault [*faille*] or gap [*béance*] within *jouissance* is confronted, or alternatively avoided, as such by a woman.[8] In the first lesson of Seminar XX, Lacan seems to provide two answers: it can either be realized as a *phallic jouissance* that is *étrange* in comparison with that of man, in the sense that it deals with the nontotalizability of *jouissance* (with its structural being deficient, or "in default") differently from masculine *jouissance*, or be evaded to attempt to achieve the totalization of *jouissance* by positing precisely what is lacking as the enjoyment of an asexual *être ange*—that is, by using the impasse of phallic *jouissance* as an alibi for a mythical (and ultimately man-oriented) desexualization. Note that up to this point Lacan has not yet invoked what, in later lessons, he will call "Other *jouissance*", a feminine *jouissance* that is neither phallic (it famously lies beyond the phallus, and consequently cannot be considered as sexual) nor angelical (and thus totalized), but rather mystical.

It can therefore be advanced that Seminar XX is concerned with four different kinds of *jouissance*: (a) masculine phallic *jouissance*, which in attempting to totalize enjoyment uncovers its very nontotalizability; (b) feminine phallic *jouissance*, or *jouissance étrange*, which is the nontotalization *inherent and immanent to* the thwarted process of totalizing enjoyment, as well as mutually dependent on it; (c) asexual and mythical *jouissance être ange*, which is the fantasy of masculine phallic *jouissance* as totalized (projected onto woman or adopted by her); (d) nonsexual but really existing feminine *jouissance stricto sensu*, which is a mystical supplement (or *en plus*) of phallic *jouissance*. In order to refrain from locating it on a transcendent level, we could also call it "*non*totalizability". Feminine *jouissance stricto sensu* is, for Lacan, beyond the phallus (and its inherent nontotalization) but this beyond does not exist without referring to the phallus.[9]

Let us now dwell on feminine phallic *jouissance*, the *jouissance étrange*. As Lacan has it, woman phallically "possesses" man just as much as man "possesses" woman; she is far from being indifferent to the phallus (i.e., in common parlance, to "her man"). Or, more accurately, "it is not because she is not-wholly in the phallic function that she is not there at all". Being not-all in the phallic function involves, rather than excludes, being "*not* not at all there", to the extent of being "in full" [*à plein*] in it where sexual *jouissance* is concerned; in other words, as I have already remarked, woman—her *pas-toute*—is not less sexed (that is, phallically engaged) than man.[10] What is then so "strange" about feminine phallic *jouissance*? In a few words, woman approaches, and sustains, the phallus in her own way; she complies with the "requirement of the One" [*l'exigence de l'Un*], which is inseparable from phallic *jouissance*, but replaces the "One of universal fusion" that underlies masculine phallic *jouissance* with a singular "one by one"[11] [*une par une*]. For woman, the other sex as masculine is phallically the one by which she makes herself be taken, but only one by one, not universally. To put it simply, woman is, at the same time, infinite and countable, or better, she becomes countable and able to count precisely insofar as she exposes the nontotalization of the count.

Lacan provides an example to depict this: the myth of Don Juan. He has women one by one; as long as he knows their names, he can list them and hence count them. Yet, if, at a given point in time, he has possessed, say, 1,003 women, this is only the 1,003rd instance—and obviously not the last—he counts again woman as one. This different form of counting, which inherently hinders even the semblance of successful totalization explains why—and this has not been emphasized enough by critics—Lacan unexpectedly calls Don Juan a "*feminine* myth".[12] The singular one-by-one as opposed to the attempted fusional One of masculine phallic totalization, albeit entwined with it, is a feminine—but nonetheless phallic—counting that shows "what the other sex, the masculine sex, is for women":[13] counting one more woman.[14]

We now need to take a step back and ask ourselves: Why does phallic *jouissance*—as the only possible sexual *jouissance*—underlie the "requirement of the One"—or its "one-by-one" feminine variation—in the first place? Lacan explains this by and large underrated yet fundamental issue right at the beginning of the first lesson of Seminar XX. Phallic *jouissance* depends on love as "the desire to be One"; *jouissance* is not a "sign of love", but it remains nevertheless "secondary"—that is, epiphenomenal, with respect to it.[15] The body of man and woman as beings of language is certainly sexed symbolically (in an asymmetrical way), but the sexual *jouissance* of the body of the Other "remains a question, because the answer it may constitute is *not necessary*. We can take this further still: it is not a sufficient answer either, because love demands *love*. It never stops demanding it. It demands it . . . *again* [*encore*]".[16]

Thus, against doxastic readings of the title of Seminar XX, "*encore*" does not primarily refer to the "I want it all and I want it now!" of masculine phallic *jouissance*, or to the "!!!" of feminine/mystical *jouissance* beyond the phallus,[17] but to the "I want more!" of love. The following argument should by now be clear: human sexuality issues from the primacy of the absence of the sexual relationship, from an impossibility, and *jouissance* should consequently be seen, first and foremost, as a "negative instance", which as such, "serves no purpose" or "is worth nothing"[18] [*ne sert à rien*]. That is to say, it is the demand for love as a desire to be One—which Lacan[19] specifies as thwarted, "impotent"—that in the end sustains human sexuality—and thus, indirectly, reproduction and the preservation of the species—as based on a relationship that is not One between the sexes. In other words, as Lacan[20] claims later in Seminar XX in a passage that should be read together with the one earlier, and equally refers to necessity, we eventually even manage "to give a shadow of life to the feeling known as love. This is *necessary, really necessary;* it is necessary that this goes on [*ça dure encore*]. It is necessary that, with the help of this feeling, this leads, in the end . . . to the reproduction of bodies".[21]

To put it very bluntly, as beings of language, we do not primarily make love because sex is instinctively "fun". Rather, we (strive to) have sex because we love, whatever our polymorphously perverse motivations for sleeping with the other may be. *Jouissance* is nothing else than a byproduct of the impossibility of the One necessarily desired by love. Here, we should pay particular attention to the fact that Lacan obliges us to thoroughly rethink the opposition between love and desire—on which he had insisted throughout his earlier Seminars—in the more general terms of an interaction, if not of a noneliminable presence of desire within love. Love can well be, as a passion, as the desire to *be One,* the "ignorance of desire"; yet this does not in the least involve a weakening of the *desire* to be One.[22]

Lacan's question "Is love about making one?", which at first sight seems to be redundant in a context that defines love as the desire to be One, cannot simply be answered affirmatively[23]: love as the *desire* to be One always goes together with the "again!" of the demand for love, hence it would be misleading to identify its aim with a final "making one" [*faire un*]. Love desires to be One, and fabricates in the process a fragile semblance of the One. In other words, the capitalized One at stake in love—as *desire* to be One—does not lead us back to what is allegedly a primordial unity, the "earliest of con-fusions"; it rather evidences the fissure of the One (the One of the enunciation "*Y a d' l'Un*", "There's such a thing as One") as the "essence of the signifier", a differential gap which alone allows us to propose a discourse on *jouissance* and being. If Lacan concludes the first lesson of Seminar XX by returning to the irreconcilability of "being absolute" with "being sexed", this is not just a reminder of the fact that angelical *jouissance,* the supposed being-One without fissure of a totalized body, would necessarily be asexual; it also means that we can finally think the existence of God—which as a hypothesis implicitly upholds all of our enunciations—as *insubstantial,* as a *non*-angelical One with a fissure, only on the basis of *sexual jouissance* and its nonphallic feminine supplement.

FROM *L'AUTRE SATISFACTION* TO *L'AUTRE JOUISSANCE*

It would not be an exaggeration to suggest that, in Seminar XX, Lacan goes as far as positioning love and desire on the side of the satisfaction of needs, as opposed to—albeit inseparable from—what he names the "other satisfaction" of—the uselessness of—phallic (masculine and feminine) *jouissance*. This topic is extensively covered in the fifth lesson. The sixth lesson then introduces feminine *jouissance* as "beyond the phallus". Contrary to what has been wrongly proposed by several critics, Lacan's efforts are aimed overall here at distinguishing *l'autre satisfaction*—that is, phallic (masculine and feminine) *jouissance*, from *l'Autre jouissance* (or, as it is also referred to, *la jouissance de l'Autre* and *jouissance radicalement Autre*[24]), an exclusively feminine non-phallic *jouissance*.[25]

The fifth lesson of Seminar XX revolves around the contrast between the satisfaction of needs and "another satisfaction". More specifically, against any naturalist-reductivist reading, the satisfaction of needs in the human being can only be grasped indirectly as what is lacking from, or does not fulfil [*fait défaut*] this other satisfaction, which "supports itself from language" and can be defined as "*jouissance*".[26] This specification thus entails an elaboration of the previous assumption according to which *jouissance* is epiphenomenal with regard to love-desire and their indirect fulfilment of the sexual biological function; to sum up: (a) love-desire makes reproduction possible; and (b) *jouissance* is a byproduct of their interaction with (sexual) need, if not replacement of it; yet, nevertheless, (c) (sexual) need as such, its "non-other" satisfaction (i.e., satisfaction *tout court*; compliance with the pleasure principle), can only be approached in a roundabout way by means of the other—unsatisfied—satisfaction of *jouissance*.[27]

Moving from this interweaving of need, love-desire, and *jouissance*, how should we decipher Lacan's crucial claim that other satisfaction "supports itself from language"? In a few words, this simply means that *jouissance* does not precede the institution of human reality as a linguistic reality.[28] Or, at least, that any discourse about an alleged prediscursive reality, and an associated *jouissance* which would not be deficient, is necessarily mythical for the reason that being a "man" or a "woman" does not per se denote anything like a pre-discursive reality; that "men [and] women . . . are . . . signifiers. . . . A woman seeks out a man qua signifier. A man seeks out a woman qua . . . that which can only be situated through discourse".[29] More generally, there is no such thing as a "human species" at the level of pre-discursive reality; the species—and its sexual *jouissance*—is only made possible "thanks to a certain number of conventions, prohibitions, and inhibitions that are the effect of language".[30] From this perspective, *jouissance*, love and desire all equally follow from *homo sapiens'* rupture from—what appears to be—the immanence of animal need.[31]

Having said this, the very fact that *jouissance* only sustains itself from language—that is, from the absence of the sexual relationship that determines human *sexuality* as deprived of essence—necessarily entails the supposition of a mythical substance that enjoys itself absolutely. If, on the one hand, there is "*jouissance* of a body" [*jouir d'un corps*]—in the subjective and objective sense—only insofar as this very body is symbolized—or, which is the same, insofar as the "signifier is the cause of *jouissance*"[32]—then, on the other hand, psychoanalysis presents itself as a discourse on how discourse as such is founded on an inevitable supposing of substance. In the first half of Seminar XX, Lacan also attempts to move beyond this fluctuation between the absence of the sexual relationship and the mythical substance by introducing a "new form of substance", a "*substance jouissante*",[33] through which *both* the myth of absolute *jouissance* and the *jouissance* of the symbolized body could dialectically be thought anew.[34] While with respect to substance as really existing body, and as such symbolized, *jouissance* can only be enjoyed *in part*, given that there is no sexual relationship as One,[35] with regard to substance as supposed by language, the logical necessity to posit it involves the positing of an imaginary absolute *jouissance*, the *jouissance* of the body as One. But, as Lacan has it in a later lesson of Seminar XX, necessity goes here together with impossibility, or better, that which is necessary in logic *qua* the founding exception to the rule is impossible in reality.[36]

We thus have to conceive of one mythical man, the Father (of the horde), for whom the phallic function that decrees the partiality of *jouissance*—that is, the fact that there is no sexual relationship, or, which is the same, that human sexuality equals the absence of the sexual relationship as One—is not valid. To put it differently, we have to *think* of an absolutely self-enjoying, and hence ultimately asexual, substance that embodies "the correlate of the *fact* that there's no such thing as a sexual relationship":[37] Lacan calls it "the substantial aspect of the phallic function" [*le substantiel de la function phallique*] and then proceeds to formalize it as $\exists x$–Φx (there exists an x, the Father, for which Φx, the phallic function, is negated) in his so-called formulas of sexuation.[38]

In his seminars of the early 1970s, Lacan puts forward strong arguments to deny any essence to the purely symbolic existence of the Father—that is, to detach the exceptional one which does not abide by the phallic function from its imaginary embodiment. And yet, in Seminar XX, he decides to ambiguously characterize such exception as the *substantial* aspect of the phallic function. We could argue that "Father", and his substantiality, stand here for the imaginary/phantasmatic reification into One of the logical existence of the one, where the latter is actually inextricable from its movement towards the not-One. This reading could be supported by Lacan's recurrent distinction between the living/self-enjoying God of religion and the God hypothesis of metaphysics.[39]

The broader issue at stake, however, cannot be settled so easily: for instance, is the embryonic theory of *substance jouissante* sufficient to reconcile the apparently antinomical claims for which, on the one hand, "psychoanalytic discourse . . . is lent support . . . by the fact that it *never resorts to any substance*, never refers to any being"[40] (and thus unmasks the substantial aspect of the phallic function as an imaginary mirage), while, on the other, "*the substance of the body*, on the condition that it is defined only as that which enjoys itself [*is*] *precisely what psychoanalytic experience assumes*"?[41] Should these two statements be read together as (1) "psychoanalytic discourse is lent support by the fact that it never resorts to any substance, never refers to any being" *as the being One of the body*, but, at the same time, (2) "the substance of the body, on condition that it is defined only as that which enjoys itself" *sexually*—that is, on condition that *the enjoying body is not One*—"is precisely what psychoanalytic experience assumes"?

The apparent contradiction could in this way be explained, yet, given that Lacan never expands on *substance jouissante*,[42] a more general tension remains: bluntly put, is there substance in the real or is there not? Is Lacan trying to sketch the para-ontology of a human corporeal *substance without essence* as not-One, which would be graspable *aside* from the logical existence of the God hypothesis—that is, of the oscillation between the one/One and the not-One? Should this be the case, does he successfully manage not to entail here any being as the being One of the body? What seems certain at this stage is that the very interpretation I am advancing is a palpable instantiation of the fact that, as Lacan[43] himself acknowledges, to speak about *jouissance* puts more than ever "this *the Other* [*ce* l'Autre]"[44]—the symbolic order as such—as *the* symbolic—into question. In brief, it is precisely an enquiry into substance from the standpoint of enjoyment that indicates how the really existing Other as not-One (always mediated through "the Other sex"—that is, woman) can ultimately be thought only against the background of an absolute mythical One and, vice versa, how any cogitation about the symbolic-imaginary phallic whole goes hand in hand with the nontotalizability of linguistic reality.

Let us now briefly focus on feminine *jouissance* as a different way of putting the consistency of the Other into question: instead of challenging it by means of language—that is, by uncovering the logical codependence of the one/One and the not-One in the symbolic order—as phallic (masculine and feminine) *jouissance* does, feminine *jouissance* rather indicates that this order is as such not entirely sayable. While the "other satisfaction" of phallic *jouissance* ultimately amounts always to a *jouissance* of speech,[45] in the sense that, as we have seen, it supports itself from language and can even be regarded as a (always deficient) satisfaction of the "blah-blah",[46] all that women can say about their *jouissance* beyond the phallus—as exemplified by the "sporadic" writings of mystics—is that "they feel it [*ils l'éprouvent*], but know nothing about it".[47] This contrasting reference to speech clearly corroborates the view that Lacan distinguishes *l'autre satisfaction* from *l'Autre jouissance*. To sum up, we could suggest that not only is there a satisfaction which is other with respect to the satisfaction of needs, but this very "other satisfaction", phallic *jouissance*, also has its Other. In spite of a general tendency among Lacanians to write this other otherness with a capital O, we should resist the temptation to turn it into a transcendent feeling. The "supplementary" nature of unspeakable feminine *jouissance* should rather be understood as that which, in "escaping" symbolization, nonetheless refers to it qua symbolization's inherent impasse, as the *not*-all of symbolization.[48] Conversely, woman's "beyond the phallus" or *"en plus"*—"Be careful with this 'more'—beware of taking it too far too quickly" Lacan warns us[49]—that is, her being not entirely contained within the phallic function, should be considered as a precondition for the symbolic as such, since the latter is also not-One, and can propose itself as a one/One only on this basis.

The best way to understand the nontranscendence of feminine *jouissance* with regard to the symbolic is by closely analysing its relation to sexual difference. As I have anticipated, Lacan does not hesitate to state that *jouissance*, insofar as it is sexual, is phallic. If feminine *jouissance* lies "beyond the phallus", it inevitably follows that it must be seen as somehow *non*sexual. Yet, at the same time, this does not entail that it is *a*sexual *sensu stricto*. Rather, *feminine* jouissance *is nonsexual within the sexual relationship (that is not One)*. Lacan[50] says that feminine *jouissance* derives from woman's "being the Other, in the most radical sense, *in the sexual relationship*" (as a *liaison* that is not a *rapport*); "Woman is that which has a relationship to *that* Other";[51] or woman is that which has a relationship to herself as Other in the most radical sense. In other words, feminine *jouissance* is the consequence of woman's unique opening onto the Other as barred, as not-One, or, better, onto the barred Other insofar as it is barred only as *marked by the signifier*:

> Woman has a relation to the signifier of that Other, insofar as, qua Other, it can but remain forever Other. I can only assume here that you will recall my statement that there is no Other of the Other. The Other, that is, the locus in which everything that can be articulated on the basis of the signifier comes to be inscribed, is, in its foundation, the Other in the most radical sense. That is why the signifier . . . marks the Other as barred: S(A̸).[52]

Woman's relation with the "most radical" Other when she experiences feminine *jouissance* is thus far from coinciding with the attainment of the alleged primordial unity of substance—that is, with the mythical end of sexual difference (and of the symbolic order along with it) which is instead evoked by the image of the *être-ange qua* the purely *a*sexual enjoyment of the body as One, fictitiously situated outside of the Other (or before/after it).

L'être-ange (being-an-angel) must therefore be opposed to both phallic *and* non-phallic feminine *jouissance*—both the *jouissance étrange* (strange enjoyment) through which woman relates to man by making herself be taken/counted "one by one" *and* the silent "impulses of fervour and passion" [*jaculations*][53] experienced by the mystics, which men cannot relate to. Furthermore, not only is the mystic's expressing *one* of the faces of God (or "one face of the Other"; an expression we cannot unravel here) not enough to turn her into an angel, but the angel can only be obliquely materialized in the far less edifying body of the hysteric. Lacan[54] advances this point concisely but effectively when he claims that hysteria aims at the "outsidesex" [*horsexe*], and for this reason stands for a "playing the part of", or "pretending to be the man" [*faire l'homme*]. To put it differently, as Lacan puns, the hysteric is a "*hommosexuelle*"—that is, a "man-sexual" rather than a homosexual woman[55]—in the sense that she attempts to embody the epitome of the masculine-phallic fantasy of overcoming sexual difference in an asexual being as corporeal being One.[56] In the end, the hysterical angel intends to occupy the mythical place of the desexualized partner of the noncastrated and hence fully enjoying Father, who would thus be himself asexual: in not facing an Other sex, in having all women as *The* angelical/hysterical woman, his sexuality—which can only be differential—would remain in fact undetermined.[57]

BIBLIOGRAPHY

André, Serge. *Que veut une femme?*. Paris: Seuil, 1995.

Balmes, François. *Dieu, le sexe, et la vérité*. Ramonville Saint-Agne: Érès, 2007.

Chiesa, Lorenzo. *Subjectivity and Otherness: A Philosophical Reading of Lacan*. Cambridge, MA: MIT Press, 2007.

———. "The World of Desire: Lacan between Psychoanalytic Theory and Evolutionary Biology". *Filozofski Vestnik*, 2009.

———. *The Not-Two: Logic, Love, and God in Lacan*. Cambridge, MA: MIT Press, 2015.

Cock, Édith De. "Encore, 1972–1973". In *Lacaniana. Les séminaires de Jacques Lacan, 1964–1979*, edited by Moustapha Safouan. Paris: Fayard, 2005.

David-Ménard, Monique. *Les constructions de l'universel. Psychanalyse, philosophie*. Paris: Presses universitaires de France, 1997.

Fink, Bruce. "Knowledge and Jouissance". In *Reading Seminar XX*, edited by Suzanne Barnard and Bruce Fink. Albany: State University of New York Press, 2002.

Lacan, Jacques. *Encore: The Seminar of Jacques Lacan. Book XX*. Translated by Bruce Fink. New York: W.W. Norton, 1998.

———. *Le séminaire. Livre XIX . . . ou pire*. Paris: Seuil, 2011.

Salecl, Renata. "Love Anxieties". In *Reading Seminar XX*, edited by Suzanne Barnard and Bernard Fink. Albany: State University of New York Press, 2002.

NOTES

1. And witnessed by colloquial language, in which, for example, "to fuck" refers to both copulation and something that does not work or go well (as in "fuck!" or "to fuck

up"). See Jacques Lacan, *Encore: The Seminar of Jacques Lacan. Book XX*, trans. B. Fink (New York: W.W. Norton, 1998), 32.

2. Lacan, *Encore*, 6–9.

3. Lacan, *Encore*, 11.

4. Lacan, *Encore*, 7. Translation modified.

5. Lacan's basic assumption is that woman's primary biological characteristic concerning sex, the vagina, is not as such symbolizable since, not "sticking out", it cannot imaginarily be associated with the differentiality of the signifier independently of the image of the male organ. This is a point he puts forward as early as Seminar III and is still valid in Seminar XX.

6. Lacan, *Encore*, 8.

7. Lacan, *Encore*, 7 and 9. Even more explicitly: "Jouissance, qua sexual, is phallic".

8. Lacan, *Encore*, 8.

9. For a clear association of feminine nonphallic *jouissance* with ex-sistence, see Lacan, *Encore*, 77.

10. Lacan, *Encore*, 73–74.

11. Lacan, *Encore*, 10.

12. This would need to be complicated further. See Monique David-Ménard, *Les constructions de l'universel. Psychanalyse, philosophie* (Paris: Presses universitaires de France, 1997), 139. David-Ménard is right in suggesting that there are two meanings of "not-all": "Not-all" [*pas tout*] refers to the logical fact that women do not form a whole. But when Lacan comments on what he writes, the "not-all" acquires a different meaning: it is not all of a woman that is bound to the phallic. In brief, I would add that it is as not-"all women" that *a* woman is singularly not-all phallic. Two entwined notions of infinity are here at stake: (1) that of the set that collects all the natural whole numbers, the smallest infinite which is as such countable although we cannot count up to infinity; (2) that of what, in Seminar XIX, Lacan calls a "non-numerable" [*non-dénombrable*] (see *Le séminaire. Livre XIX . . . ou pire* [Paris: Seuil, 2011], 205). Counting women one by one means counting the inexistence, or zero, of woman as one, yet this count also discloses as such an undecidable, Lacan says, situated "between 1 and 0" in each singular woman—which is what a woman relates to as feminine *non-phallic* enjoyment. I tackle this topic in the last part of my forthcoming book entitled *The Not-Two: Logic, Love, and God in Lacan* (Cambridge, MA: MIT Press, 2015).

13. Lacan, *Encore*, 10.

14. See Renata Salecl, "Love Anxieties", in *Reading Seminar XX*, ed. Suzanne Barnard and Brice Fink (Albany: State University of New York Press, 2002), 96 (my emphasis). Renata Salecl fails to make this point when she reduces the feminine myth of Don Juan as discussed by Lacan to a fantasy that "proves that there is at least one man who has it from the outset, who *always has it and cannot lose it*. . . . Since women often are concerned that a man may completely lose himself when he is with another woman, the fantasy of Don Juan reassures women that there is at least one man who never loses himself in a relationship". Salecl is right in identifying Don Juan with the primal Father of the horde as seen by woman (rather than by his sons)—that is, in associating him with the exception to the logic of castration, and to the hole of *jouissance* (as robbed by him) which sustains the phallic function. However, she mistakenly confines the singularizing count of the one-by-one (the feminine phallic function) *within* the count of the One of universal fusion (the masculine phallic function), instead of evidencing them as the two sides of the same coin: for Lacan, Don Juan is rather the primal Father as *unable* to have all women at once. Pertinently, as Serge André observes, "in many versions of the myth, the character of Don Juan mocks the father". See Serge André, *Que veut une femme?* (Paris: Seuil, 1995), 228–29.

15. Lacan, *Encore*, 5–6.

16. Lacan, *Encore*, 5–6. My emphases.

17. Speech conveys this ecstatic speechlessness in expressions such as "Oh my God!" The feminine mystics do after all *write* about what they feel but do not know.

18. Lacan, *Encore*, 3.

19. Lacan, *Encore*, 6.
20. Lacan, *Encore*, 46.
21. Lacan, *Encore*, 4. Translation modified; my emphasis.
22. Lacan, *Encore*, 4.
23. Lacan, *Encore*, 5.
24. See, for instance, Lacan, *Encore*, 4, 24, 83. Given the context, there is no doubt that when Lacan speaks of *la jouissance de l'Autre* in these passages of Seminar XX, he is referring to woman's nonphallic *jouissance* of the not-all. This phrase is, however, highly misleading, since Lacan had previously (in Seminar XVII) used it to designate knowledge and, clinically, perversion—that is, the *phallic jouissance* of enjoying for the consistency of the Other, the ideological *j'ouïs-sens* as "I enjoy hearing the sense" of the symbolic order, which thus corks the hole in structure. See my book on *Subjectivity and Otherness: A Philosophical Reading of Lacan* (Cambridge, MA: MIT Press, 2007), 185–86. The notion of the *jouissance* of the *barred* Other in Seminar XXIII seems to be trying to obviate this ambiguity.
25. Fink seems to be taking for granted the equation between "another satisfaction" and "Other jouissance" as a peculiarly feminine form of enjoyment. As he promptly acknowledges, this leads him to an impasse, since the former is defined by Lacan as a "satisfaction of speech" while the latter is seen as unspeakable: he even says at one point in the seminar that it (Other jouissance) is "the satisfaction of speech". How that is compatible with the notion that it is an *ineffable* experience where the bar between signifier and signified does not function, I do not profess to know. Shortly after, he opts to attribute this apparent contradiction to Lacan's own inconsistent arguments: "We need not assume that there is some sort of complete unity or consistency to his work, for he adds to and changes things as he goes along". Bruce Fink, "Knowledge and Jouissance", in *Reading Seminar XX*, ed. Suzanne Barnard and Bruce Fink (Albany: State University of New York Press, 2002), 40. The reading I propose, whereby the "other satisfaction" amounts to phallic *jouissance* as different from Other/feminine/nonphallic *jouissance*, shows that Lacan is far from contradicting himself.
26. Lacan, *Encore*, 51 (translation modified).
27. Lacan, *Encore*, 50. In this light, it is not a coincidence that the last sentence of the fourth lesson announces that the main theme of the fifth concerns the point "where love and sexual jouissance meet up".
28. Lacan, *Encore*, 55.
29. Lacan, *Encore*, 33.
30. Lacan, *Encore*, 33.
31. See my article entitled "The World of Desire: Lacan between Psychoanalytic Theory and Evolutionary Biology", in *Filozofski Vestnik*, 2009.
32. Lacan, *Encore*, 24.
33. Lacan, *Encore*, 23.
34. Lacan is, however, cautious: what he has so far put forward in terms of *jouissance*, he says, "perhaps" involves a *substance jouissante*.
35. Lacan, *Encore*, 23: "One can only enjoy a part of the Other's body, for the simple reason that one has never seen a body completely wrap itself around the Other's body, to the point of surrounding and phagocytizing it. . . . We must confine ourselves to simply giving it a little squeeze, like that, taking a forearm or anything else—ouch!"
36. Lacan, *Encore*, 59.
37. Lacan, *Encore*, 59. My emphasis.
38. Lacan, *Encore*, 59, 79–80.
39. See Balmes, François *Dieu, le sexe, et la vérité* (Ramonville Saint-Agne: Érès, 2007).
40. Lacan, *Encore*, 11.
41. Lacan, *Encore*, 23. Translation modified; my emphases.
42. He very briefly returns to it in *Seminar XXI. "Les non-dupes errent". 1973–1974* (unpublished), lesson of 3 December 1974.
43. Lacan, *Encore*, 38.
44. Translation modified.

45. Lacan, *Encore*, 64.

46. Lacan, *Encore*, 56. At this point, Lacan goes as far as explicitly associating this satisfaction with Freud's pleasure principle. That is, here, the supposed satisfaction of needs is directly equated with "other satisfaction"; in the case of the human animal, the former can only be mediated by language and its logorrheic *jouissance*.

47. Lacan, *Encore*, 76. Translation modified. Note that Lacan uses the masculine, or mixed, personal pronoun *ils*, not the feminine *elles*, which shows that "woman" (and her specific mystical *jouissance*) is not limited to biological females.

48. Lacan, *Encore*, 24, 33.

49. Lacan, *Encore*, 74.

50. Lacan, *Encore*, 81.

51. My emphases.

52. Lacan, *Encore*, 81.

53. Lacan, *Encore*, 76. Translation modified.

54. Lacan, *Encore*, 85.

55. To highlight this "man-sexuality" of the hysteric, Lacan adds an extra "m" to the French *homosexuelle*, which designates a homosexual woman.

56. Édith De Cock speaks of a "*hors sexe*" in relation to feminine *jouissance*. This is confusing in light of Lacan's restricted application of the term in question to hysteria. It also completely misses our explanation of the difference between feminine *jouissance*'s nonsexuality within the sexual relationship as opposed to the angelic mirage of *a*sexuality that enchants the hysteric. Furthermore, De Cock states that Other *jouissance* "touches in equal measure man and woman, and can be located on both sides" of the graph of sexuation: this is a clear misunderstanding of Lacan's argument, which rather advances that both biological females and males can symbolize themselves as "woman" (i.e., on the right-hand side of the graph) and, consequently, experience feminine *jouissance*. See É. De Cock, "Encore, 1972–1973", in *Lacaniana. Les séminaires de Jacques Lacan, 1964–1979*, ed. Mohamed Safouan (Paris: Fayard, 2005), 307.

57. The sexual indetermination of the Father of the horde has repeatedly been noted by Žižek.

Index

abortion, 176
Actor Network Theory (ANT), 136, 138–140
advertising, and desire, 251
affect theory, 138
Agamben, Giorgio, 20n59, 35, 39
ageing: epigenetic process of, 205–207; free radical theory of, 200–201; and longevity, 205–207; as normative biological process, 199–200; as normative self-destructive process, 201–204; and physiological constraints, 203; and repulsive constraints, 203; and vicious circles of destruction, 201–205
agency, 116
Alain (Emile Chartier), 154–156, 160, 164n37
Alainists, 159–160
allures de vie, 164n37
Althusser, Louis, 11, 12, 104
American Medical Association, 25
anatomo-politics of the body, 272
animal rights movements, rise of, 19n54
Ansell-Pearson, Keith, 75n6
anthropophagy, 231
anti-production, 269
anti-"specism" movements, rise of, 19n54
Aristotle, 2, 66–67, 70, 89–90, 253
Aron, Raymond, 159
asexual *jouissance*, 295, 296, 297
assemblages, 139, 140
attachment, 52, 54
Audard, Catherine, 228
autonomy, 168, 185–186, 187, 203

Bachelard, Gaston, 2, 66
Badiou, Alain, 89
Balibar, Etienne, 119n30
Balvert, Paul, 131n11
Barbaras, Renaud, 5
bare life, 55
Bataille, Georges, 255–257
Bauplan, 4
becoming, reciprocal, 194
becoming-able concept, 188–189
behavior, regulating through discipline, 34, 36, 37
Bénézé, Georges, 159–160, 164n37
Bentham, Jeremy, 231, 249
Bergson, Henri, 2; on ageing, 207; influence on Canguilhem, 75n6, 76n11; vitalism of, 5, 56–57, 77n34
Bergsonian intuitionism, 69, 70
Bernard, Claude, 80, 82, 153, 154, 157
Bertrand, François, 229–230, 232–233
Bichat, Marie François Xavier, 161
bioethical maximalism/minimalism, 55
bioethics, 8–11; archaeology of, 28–29; ecological bioethics, 25–30; medical bioethics, 25–30; and medicine as clinic approach, 27–28; as not having earlier history, 25–27; origins of, 27, 29–30; and technological advances, 8. *See also* biopolitics
biological order, 6
biological value of physical measurements, 80–81
biophilosophy, 6–7, 63–65
biopolitics: anecdotes on, 33–35, 42; anecdotes on, relationship among, 36–37; and bioethics, 9–11; and biopower, 39–40; and definition of life, 41–42; Foucauldian interpretation of, 38–39; interpretive branches of, 39–40. *See also* bioethics
bios: principle of, 33; relation to *zoe*, 38

bios politikos vs. oikos, 243–244
Biran, Maine de, 160–161
Birre, Andre, 38
Birth of the Clinic (Foucault), 8, 26, 52
Black Skins, White Masks (Fanon), 123
body: as network, 263–265; utopian body, 262
body-without-organs concept, 269–270, 275n4
Bouglé, Célestin, 154, 156
Brandon, Robert, 19n37
Broussais, François Joseph Victor, 153–154
Brunschvicg, Léon, 4
Butler, Judith, 99, 110, 118n1, 118n12, 119n14, 196n14, 289

Caldwell, Lynton K., 44n19
Calvin, John, 228
Canguilhem, Georges: and Bergson, 64–65, 67–68, 75n6, 76n11; on biophilosophy, 6–7, 64–65; on Biran, 161; on cognitive neuroscience, 12–13; on conception of value, 156; definition of medicine/therapeutics, 96; on health and pathology, 80–82; influence of Alain on, 154–156; influence of Goldstein on, 127; influence on Fanon, 131n11; on infraction/irregularity, 111; on Le Senne, 157; on life as generating norms, 88–89, 92n1, 96; on mechanism, 67–68; on medicine as clinic approach, 27–28; on negative values, 97; on norms and techniques, 155–157; on pain and pathology, 151–153; on physiological constraints, 203; and race/racism, 124–125, 128; on relation of organism and society, 125–126; on relation of vitalist biology to totalitarianism, 126; on repulsive constraints, 203; theory of organism of, 151–153; vitalism of, 7, 63–67, 71–73, 97–99, 100, 143; on vital norms/biological values, 83–85
cannibalism, 233
capitalism: and desire, 11, 252, 268–269; global neoliberal capitalism, 140; limits of, 269; and nihilism, 268–270
capital surplus absorption problem, 256
care of self, 273
Carlier, Félix, 238n31
Cartesian mechanism, 4, 65
Casper, Johann Ludwig, 238n29
Cassirer, Ernst, 4, 126, 216
Castel, Robert, 119n30
Cavell, Stanley, 176
cell theory, 125
Certeau, Michel de, 111
Cézanne, Paul, 217, 220
Chamberlain, Houston Stewart, 127
Changeux, Jean-Pierre, 12
Chartier, Emile. *See* Alain
Chicago School of economics, 249
Churchland, Patricia, 12
Churchland, Paul, 12
citizenship, 186; right to, 33–34, 42; social, 117, 119n30
Civilization and Its Discontents (Freud), 10–11
classical vitalism, 68, 73
Cohen, Hermann, 4
Comte, Auguste, 4, 128, 153, 154
conceptual knowledge, 69–70
concrete operations stage, 218–219
consent for medical treatment, 27
constructivism, 271
continental tradition, 14
conventional morality, 218
Cornier, Henriette, 282–283
The Courage of Truth (Foucault), 170–172
creative destruction, 220
Creative Evolution (Bergson), 56, 67–68
Crick, Francis, 1, 17n5
cultural capital, 250
Cynicism, 170

Darwin, Charles, 8
Dasein, 4
David-Ménard, Monique, 305n12
Dean, Tim, 280
death: legal limits to, 34–35, 36, 42; penalty, 232; subjective perspective on dying, 169–170
de Boismont, Brierre, 231

De Cock, Édith, 307n56
de Grey, Aubrey, 220
Deists, 216
de Lauretis, Teresa, 279
Deleuze, Gilles: on bodies as networks, 263–265; on body as own agent of subjection, 266–267; on capitalism/desire, 252, 268–269; and concept of environment, 126; definition of power, 264; on desire as mode of vitality, 262; on ethics of care and face transplants, 188–191, 194; on power, 261–262, 265–268; and reactive type, 267; on representation, 267, 276n14; on subjectification, 276n23; and transcendence of power, 265
de Martigny, Collard, 231
democracy, and ethics, 187–188
Dennet, Daniel, 12
Descartes, René, 2
desexualization, 297
desire: active mode of, 266; advertising's role in, 251; ascetic *vs.* new regime of, 242; biopolitical image of, 271–274; and birth of liberalism, 244–249; and body-without-organs concept, 269–270; and capitalism, 11, 252, 268–269; companies as systems of, 252; as cultural construction, 189; economic regime of, 242–244; and immunological paradigm, 244; and libidinal energy, 251; love *vs.*, 299–300; as mode of vitality, 262; as natural/vital, 242–243; and neoliberalism, 249–253; and new sovereignty, 253–257; and *oikos vs. bios politikos*, 243–244; power as, 267–268; reactive mode of, 266; repression of, 11; suppression of, 11; and techniques of subjectivism, 252; traditional desire, 262. *See also* pleasure
Diagnostic and Statistical Manual of Mental Disorders (DSM), 235, 285–287, 288, 293n23
dialectical materialism, 4
disposable soma theory, 206
divergence/infraction, and norms, 101
divine law conception of ethics, 165, 172
Dollimore, Jonathan, 279
Don Juan, feminine myth of, 298, 305n14
dressage, 252
Driesch, Hans, 5, 66, 128
DSM. *See Diagnostic and Statistical Manual of Mental Disorders*
DuBois-Reymond, Emil, 214
Durkheim, Emile, 154, 156, 158

Eccles, John, 12
Echo Objects (Stafford), 136–137
Edmunds, June, 212
effort, Biranian concept of, 160–161
Einstein, Albert, 213, 215
élan vital, 66, 68, 71, 77n34
elastin/laminin receptor, 204–205
Ellis, Havelock, 230
enjoyment. *See jouissance*
entelechy, 5
epigenetic process of ageing, 205–207
epithumology, 244
eros vs. philia, 293n24
erotic capital, 250
error-theorists, 83, 84
Esposito, Roberto, 20n59, 122–123, 244
Esquirol, Jean, 231
essentialization, and vitalism, 55–56
ethics of care and face transplants, Deleuze on, 188–191, 194; and becoming-able concept, 188–189; and care as not natural/moral obligation, 189–190; and care of one's self, 193–195; face as politics, 191–193; face transplants, 190. *See also* subjectivity
ethics of care and face transplants, Levinas on, 182–186; and call of the face, 184–185; criticism of, 188; ethics and liberal democracy, 187–188; face as ethical resistance, 183; face as transcendent, 182–183; and justice/power relations, 186–187; origins of ethics in desire, 183–184; Other as face, 182; and suffering, 185. *See also* subjectivity

ethology, 51–52, 56, 65
Euclidean geometry, 216–217
eugenics, 127
euthanasia, 36
evolution theory, 3
exigence, 71
existence, hostility of, 100
existentialism, 65, 70
explanatory/heuristic vitalism, 66
extracellular matrix (ECM), 204–205

facelessness, Butler on, 196n14
face transplants. *See* ethics of care and face transplants
faciality (*visagéité*), 191
Fanon, Frantz: influence of Canguilhem on, 123–124, 125, 131n11; influence of Lacan on, 131n11; on normal and abnormal, 123–124, 129
fascism, 3, 11, 154–155, 265
Faury, Gilles, 205
feminine phallic *jouissance* (*étrange*), 297–299
Fink, Bruce, 306n25
Fletcher, Joseph, 8
Foot, Philippa, 89–90, 91
forensic psychiatry, 233
Foucault, Michel, 2, 3, 6; on biopolitics, 28, 38–39, 41; on biopower and racism, 121–122; on bodies as networks, 263–264; on Canguilhem, 67; on crime and punishment, 228; on Cynicism, 170, 174; on desire, 11, 262, 271–274; on division of political and scientific domains, 36–37; on economy, 250; on history-writing, 142–143; on liberty, 227; on mortalism, 52; on nominalism, 104; on perversion, 281–284, 286, 289, 293n10; on pleasure, own difficulty in experiencing, 262, 275n2; on power, 9, 95, 261, 264; on primary moral work, 166, 169–171; on utopian body, 262
Franck, Philipp, 5
free radical theory of ageing, 200–201
Freud, Sigmund, 4, 10–11, 286, 293n10
Freudo-Marxism, 10, 11

Gayon, Jean, 75n6
Gehlen, Arnold, 4, 126
generational change: as central problem in sociology of knowledge, 211–212; as cognitive rejuvenation, 216–217; and logical positivism, 215; and Newtonian mechanics, 213–215, 216; in sacred and secular democracies, 217–219; and social contract, 217–218; and transhumanism, 219–221; and virtual reality machines, 221
genetic capital, 250
genetics, 2, 71, 125
genocide, 20n59
genome, discovery of, 1–2
geocentrism, 66, 71
Georget, Etienne-Jean, 230–231
Gestalt theory, 65
global neoliberal capitalism, 140
Gobineau, Arthur de, 125
Goethe, Johann, 214
Goldstein, Kurt, 7, 126, 127, 153
Gould, Stephen Jay, 214
Green, Richard, 235
Gretter, Lystra, 25
Gross, Walter, 126–127
Guarente, Leonard, 206–207
Guattari, Félix: on bodies as networks, 263–264; on care of one's self, 193–195; on desire and capitalism, 252, 268–269; on desire and power, 274; on desire as mode of vitality, 262; on face as politics, 191–193; on power, 261–262, 267–268; on transcendence of power, 265

Hacking, Ian, 230
Haldane, J. B. S., 4
Halévy, Elie, 159
Halperin, David, 292n5
Hamelin, Octave, 157, 160
Hardt, Michael, 40
Harman, Denham, 200
Harrington, Anne, 128
Harvey, David, 256
"having a sake", 91
Hayflick, Leonard, 201
Hegel, Georg, 70

Heidegger, Martin, 2, 4, 18n23, 123, 126, 184, 215
Helvétius, Claude Adrien, 247
Hermeneutics of the Subject (Foucault), 170, 172
History of Sexuality (Foucault), 11, 281–282, 287, 289
history of the body, 135–136
history-writing: and Actor Network Theory, 136, 138–140; and Foucault's history of the present, 142–143; and need for new theories, 140–142; and neuroscience, 136–137
Hobbes, Thomas, 228, 232
Holle, H. G., 128
homo economicus, 250
homosexuality, 238n29; coining of, 234; epistemological instability of, 234–236; historical genesis of, 292n5; *jouissance* located in name *homosexual*, 291
hors sexe, 307n56
Hull, David, 5
human capital, 250–251
human will, 157
Hume, D., 83, 246–247
Husserl, Edmund, 5
hybridity, 125
Hyppolite, Jean, 159
hysteria, 290, 291, 304

immunological paradigm, 244
infant baptism, 217
Infinite, 196n6
institutional racism, 20n59
institutions, and power of bodies, 264–265
invisible hand, 248
irrationalism, 4
irreducibility of living being to matter, 6

Jacob, François, 1, 3, 6, 201
Jefferson, Thomas, 217–218
Jim Crow, 129
Jonas, Hans, 5
jouissance (enjoyment): feminine (Other enjoyment), 295, 302–304; feminine phallic (*jouissance étrange*), 297–299; *jouissance être ange*, 295, 297; *jouissance stricto sensu*, 297; *l'être ange* (being-an-angel), 295, 296, 304; and love *vs.* desire, 299–300; male phallic *jouissance*, 295, 296–297, 298–299; and myth of Don Juan, 298; and other satisfaction, 300–301; psychoanalytic view on, 295–296, 302
Joyce, Patrick, 140–142

Kaan, Heinrich, 284
Kahane, Ernest, 4
Kant, Immanuel, 70, 213–214, 215–216
Kauffman, Stuart, 203
Kertbeny, Karl-Maria, 234
Kirkwood, Thomas, 206
Kohlberg, Lawrence, 218
Korsgaard, Christine, 90–91, 92
Krafft-Ebing, Richard von, 229, 230, 233, 234, 284, 286
Krieck, Ernst, 127
Kuhn, Thomas, 211–212, 213

Labat-Robert, Jacqueline, 204
Lacan, Jacques, 4; discourse theory of, 11; on Don Juan, 305n14; influence on Fanon, 131n11; on *jouissance*, 295–304; on love *vs.* desire, 299–300; on "not-all" concept, 305n12, 306n24; on paraphilia, 293n24; on perversion, 283, 284, 288; on sublimation, 292n6
Lachelier, Jules, 155, 160, 161
lactose operon, 201
Lagneau, Jules, 160
Lalande, André, 160
Lamarck, Jean-Baptiste, 126
land ethic, 30
Latour, Bruno, 139–140, 142
Lavelle, Louis, 159
Lebensphilosophie, 7
Lebenswelt, 5
Le Blanc, Guillaume, 8–9, 99, 111, 113
Lecourt, Dominique, 77n54
Léger, Antoine, 231
Legrand, Stéphane, 102–104
Leopold, Aldo, 30
Leriche, René, 26–27, 151, 153

lesbian, gay, bisexual, transgender (LGBT) communities, 176
Le Senne, René, 157–159, 160
l'être ange (being-an-angel), 295, 296, 304
Levinas, Emmanuel, 182–186, 196n14
Lévi-Strauss, Claude, 2
Lévy-Bruhl, Lucien, 154, 156
Leys, Ruth, 138
liberal democracy, and ethics, 187–188
liberalism: emphasis on liberty, 227; and modern culture, 167–168; optimistic view of humanity, 228; and problem of radical evil, 228–229; and sexuality, 229–233
life, concept of: birth and death of, 1–3; reasons for disappearance of, 3–5; rebirth of, 13–14
life and death, legal limits to, 34–35, 36, 42
linguistic-cum-somatic turn, 135–136
Locke, John, 228, 244–245, 246
logical positivism, 5, 215
Lordon, F., 252
love, 197n23; *vs.* desire, 299–300
lumpenproletariat, 34
Lwoff, André, 1, 3, 6, 201
Lyotard, Jean-François, 11, 259n40
Lysenko, Trofim, 125
Lysenko affair, 4

Mach, Ernst, 214–215
Macherey, Pierre, 101–102
machination, biological philosophy of, 67–68
Maffini, M., 204
Malabou, Catherine, 13
male phallic *jouissance*, 295, 296–297, 298–299
Malthus, Thomas, 8
mania without delirium, 230, 231–232, 233
Mannheim, Karl, 212
Marcuse, Herbert, 10–11
Marx, Karl, 248
materialist eliminative approach, 12
Matthey, André, 231–232
Maupertuis, Pierre Louis, 125
Maurras, Charles, 128
Mayr, Ernst, 5
McCloskey, Michael, 214
McCormmach, Russell, 213
meaning (*sens*) of life, 64–65
medical care, institutional context of, 189–190. *See also* ethics of care and face transplants
medicine: biological value of physical measurements, 80–81; Canguilhem's definition of medicine/therapeutics, 96; as clinic, 27–28; institutional context of care, 189–190; medical vitalism, 65; of normality, 26–27; as power, 28–29
Mendel, Gregor, 125
Merck, Mandy, 279
Merleau-Ponty, Maurice, 5, 126
metaphors, use in biology, 6
metapsychology, 4, 286
Meyerson, Emile, 4
Mill, John Stuart, 167, 168–169, 174, 228, 250
Miller, Jacques-Alain, 294n37
Miller, Peter, 251
Mirowski, Philip, 249
mobility of the self, 173
modernity, clash with tradition, 165–167
Monod, Jacques, 1, 3, 6, 201
monomania, 230, 231, 232
Montagu, Ashley, 122
moral perfectionism, 176
moral philosophy, 156
moral realists, 83, 84
Morals and Medicine (Fletcher), 8
moral statements, 85
Morange, Michel, 2, 13
Morin, E., 38
mortal, meaning of, 58n3
mortalism, and vitalism, 52–55
Moser, Charles, 235

Nagel, Ernst, 5
Natrop, Paul, 4
Nazis: concept of life of, 122–123; ideology of, 18n23, 20n59, 126–127; and Mendelain theory, 125; and Nuremberg Code, 27
negative values, 97

Negri, Antonio, 40
neoliberalism, 144, 250
network, body as, 263–265
neurohistory, 136, 137, 144
neuroscience, 12–13, 136–137
New Reflections on the Normal and the Pathological (Canguilhem), 100–101, 103
Newtonian mechanics, 213–215, 216
Nicolle, Charles, 153
Nietzsche, Friedrich, 2, 3, 123, 141, 142, 268
Nietzsche and Philosophy (Deleuze), 265–266
nihilism, and capitalism, 268–270
nominalist theory of norms, 102–103
noncognitivists, 83, 84
nonintentionality, 137
The Normal and the Pathological (Canguilhem), 6, 80, 121, 126, 128–129, 151–152, 154, 155
normality, defining, 7
normativity: and emotivists, 83; and error-theorists, 83, 84; and "having a sake", 91; and life-based norms and values, 85–86; and living *vs.* nonliving entities, 90–91; and moral realists/objectivists, 83, 84; and noncognitivists, 83, 84; norms and values as subjectively/objectively grounded, 85; and objectivity of life-based values, 88; and quasi-realists, 83, 84; and reductionism, 88; and relationality of life-based values, 86–88; and subjectivists/skeptics, 83, 84–85
norms: and divergence/infraction, 101; immanent causality of, 101–102; nominalist theory of, 102–103; polemical nature of, 100–101; and systemization, 103–104; vital, 81–82, 83–85
norms, critical philosophy of: and clinical process, 112–113, 115; critical/political stakes of, 115–117; multiple and conflictual normative systems, 100; nonnaturalistic vitalism, 99; norms as essential to life, 99; norms as mediated process, 99–100; and precariousness concept, 113–114, 115, 116; and problem of recognition, 114; and social exclusion/devaluation, 111–114; summary of main features of, 105; vital normativity, 96–99
norms, philosophy of, 6; and social sciences, 6, 113–114; and subjectification, 109–111
Nuremberg Code, 8, 27

objectivists, 83, 84
oikos vs. bios politikos, 243–244
ontological vitalism, 5, 77n54
ordered liberty, 227
The Order of Things (Foucault), 3
organicism, 4
organic representation, 266
organic unconscious, 161
organic *vs.* animal sensation, 161
organisms, structural properties of, 84
organizational closure, 201, 202
organology, 68
overdetermination, 104

pacifism, 154, 155
paedophilia, 293n24
pain: and Biranian concept of effort, 160–161; consciousness of, 158–159; and pathology, 151–153; as source of human action, 246–247, 248
Papavoine, 232
paraphilia, 235, 239n34, 285–287, 293n24
parrhesia (truth-saying), 170
Paugam, Serge, 113, 119n19
Percival, Thomas, 25
performativity, 118n5
personality disorder, 286
perversion: classification of, 284–287; Foucault on, 281–284, 286, 289, 293n10; history of, 281–284; modalities of, 288; as not sexual, 289–290; paraphilias, 235, 239n34, 285–287, 293n24; polymorphous perversity, 293n10; as third category, 289; use of term, 279–280, 287; "version" in, 293n24; Žižek on, 288–291

phallic *jouissance*. *See* feminine phallic *jouissance*; male phallic *jouissance*
philia vs. eros, 293n24
phlogiston, 66, 71
Piaget, Jean, 214, 218–219
Pinel, Philippe, 230
Planck, Max, 211
plane of immanence, 270
plane of transcendence, 191–192
Planet, Camille, 160
pleasure: Foucault on own sense of not knowing what pleasure is, 262, 275n2; as source of human action, 246–247, 248. *See also* desire
Plessner, Helmuth, 4
political economy, 247–249
Politzer, Georges, 4
polymorphous perversity, 293n10
Pope, Alexander, 220
Popper, Karl, 5
The Postmodern Condition (Lyotard), 259n40
Potter, Van Rensselaer, 8, 29, 30
power, 261, 264; as domination, 264; fascination for, 265–268; medicine as, 28–29; as production, 95; right to live as, 9; sovereign, 9, 33, 36, 39; transcendence of, 265
power (*pouvoir*), 193
powers (*puissances*), 190
precariousness concept, 113–114, 115, 116
preconsciousness, 142, 143
Prenant, Marcel, 4
primary moral work, 169–170
proetarian science, 4
professional integration, 119n19

quality of life, 168
quasi-realists, 83, 84

racism: Foucault on biopower and, 121–122; and genetics, 125; institutional racism, 20n59; and Nazi concept of life, 122–123; systemic, 122, 123
radical evil, 228–229, 232
Railton, Peter, 87–88
Ramsay, Paul, 29
rationalism, 70; and mechanism, 68; and vital, 65
Rauh, Frédéric, 157
realism, 4
reciprocal becoming, 194
recognition, problems of, 114
reductionism, 65; and normativity, 88; reductionist character of sociology, 157; and vital, 56
Regnault, Elias, 232, 236
Reich, Wilhelm, 10
relationality *vs.* relativity, 87
relational worldview, 215
representation, 267, 276n14; organic, 266; relationship of effort to, 160
Revel, Judith, 40
Ribot, Theodule, 124, 154
Rickert, Heinrich, 123
Ricoeur, Paul, 13
right to citizenship, 33–34, 42
rigid designator, 2
Robert, Ladislas, 204–205
Roger, Jacques, 64
Rose, Nikolas, 20n59, 122, 143, 251
Rosen, Robert, 201, 202
Roudinesco, Élisabeth, 280, 287
Rousseau, Jean-Jacques, 54
Ruffié, Jacques, 122
Russell, Bertrand, 5
Ruyer, Raymond, 7, 63, 76n11

sacred life, 55
sadism, 229, 233, 235
Salecl, Renata, 305n14
The Savage Mind (Lévi-Strauss), 2
Schaeffer, Jean-Marie, 5
Scheler, Max, 4
Schiller, Joseph, 64
Schlick, Moritz, 5
Schrödinger, Erwin, 2
Schuhl, Pierre-Maxime, 157
Schultz, Theodore W., 250
science, relationship to technology, 155–156
scientia sexualis, 242, 271, 272
scientific knowledge, as essentialist/transhistorical, 137–138
seatbelt law, 34, 36, 37

sexuality: and emergence of sexual instinct concept, 230–233; first case of sexual pervert, 229–230, 232–233; and liberalism, 229–233; meaning of concept of, 238n29; and monomania, 230, 231, 232; psychoanalytic view on, 295–296, 302. *See also* homosexuality; *jouissance*
Silverstein, Charles, 235
Simmel, Georg, 113
Simondon, Gilbert, 51, 63, 190
Singer, Peter, 166, 167, 168
Sirtuins, 206
slaves, 276n14
Smail, Daniel, 137
Smith, Adam, 248
social citizenship, 117, 119n30
social Darwinism, 4
social exclusion, 109
social integration through work, 119n19, 266
social positivism, 4
sociology of morals, 156
sodomy, 233, 234, 235
Somit, Albert, 38
Sonnenschein, C., 204
Soto, A., 204
sovereign power, 9, 33, 36, 39
Spencer, Herbert, 4
Spengler, Oswald, 127
Spinoza, Baruch, 268, 270
Staatsbiologie (Uexküll), 126–127, 128
Stafford, Barbara Maria, 136–137
"Statement on Race" (UNESCO), 122
state racism, 10, 20n59
subject formation, as repetitive process, 119n16
subjectification, 109–111, 276n23
subjectivity: and Cynicism, 174; and desire, 252; and divine law conception of ethics, 165, 172; and experience of dying, 169, 174–175; and mobility of self, 173; and modes of self-transformation, 172–173; and moral rules and codes, 172–173; and primary moral work, 169–173; and procreation, 174–175; and sexuality, 174–176. *See also* ethics of care and face transplants
sublimation, defining, 292n6
substantival vitalism, 66, 71, 73
systemic racism, 122, 123
Szasz, Thomas, 286, 288

technology: and bioethics, 8; relationship to science, 155–156. *See also* medicine
teleological anthropology, 165
teleonomy, 6
thermodynamics, 200, 201, 203, 249
Thompson, Judith Jarvis, 166
Thorson, Thomas, 38
Thrift, Nigel, 138
tigridomanie, 232
Tosquelles, François, 131n11
transhumanism, 219–221
Turner, Bryan S., 212
The Two Sources of Morality and Religion (Bergson), 56–57

Uexküll, Jacob von, 4, 65, 126–127
Ulrichs, Karl Heinrich, 234
unconscious, organic, 161
Unitarians, 216
urgence, meanings of, 58n9
utilitarianism, 166, 248
utopian body, Foucault on, 262

values: biological, 83, 85; judgements of, 156–157, 158; negative, 97; objectivity of life-based, 88; relationality of life-based, 86–88; as subjectively/objectively grounded, 85
van Gogh, Vincent, 217, 220
van Horn, Rob, 249
Verhaeghe, Paul, 284–285, 286, 287
vicious circles of destruction, 201–205
vicious circles of work, 203
Vienna Circle, 5, 66, 215
virtual reality machines, 221
vital/vitalism, 2, 4, 17n5; of Bergson, 5, 56–57, 77n34; and Bergsonian intuitionism, 69, 70; Bergson *vs.* Cangulilhem, 68–71; biology, relation to totalitarianism, 126; and biophilosophy, 63–65; of

Canguilhem, 7, 63–67, 71–73, 97–99, 100, 143; classical, 68, 73; conceptualist, 70; and condition/ principle of exteriority and interiority, 48–49; continuation aspect of, 50–51; defining, 50–52; and desire, 242–243, 262; dual nature of, 64–65; *élan vital*, 66, 68, 71, 77n34; emergency, 53, 55; and essentialization, 55–56; and existentialism, 65; explanatory power of, 71; as historical/ metaphysical, 65–67; individuation aspect of, 51–52, 54, 56; and life-technique connection, 97–99, 100; medical, 65; and the minimum and the extensive, 47–48, 50–55; and mortalism, 52–55; nonnaturalistic, 99; normativity, 96–99; norms/ biological values, 81–82, 83–85; ontological, 5, 77n54; paradoxes of, 47–50; and the positive and the negative, 49; and preservation/ development of life, 97, 99; and rationalism, 65; and reductionism, 56, 65; relational aspect of, 52, 53–54, 56, 57; substantival, 66, 71, 73

Watson, James, 1
Weber, Max, 212
Weil, Simone, 159
Weismann, August, 206
Westphal, Carl, 234
Whitehead, Alfred North, 5
Wilkins, Maurice, 1
The Will to Knowledge (Foucault), 271
Wilson, Edward O., 38
Wöhler, Friedrich, 66
Wolfe, Charles T., 12
work: social integration through, 119n19, 266; vicious circles of, 203

Žižek, Slavoj, 288–291
zoe, relation to *bios*, 38
zoepolitics, 248

About the Contributors

Miguel de Beistegui is Professor of Philosophy at the University of Warwick. He was the Principal Applicant of the Leverhulme Research Project in Bioethics and Biopolitics. He is the author of books and articles in the areas of German and French Phenomenology (especially Heidegger), twentieth-century French philosophy, Proust and Aesthetics. He is currently researching a book entitled *The Subject of Desire: Between Genealogy and Ontology*.

Robert Bernasconi is Edwin Erle Sparks Professor of Philosophy and African American Studies at Penn State University. He is the author of two books on Heidegger, a book on Sartre and numerous essays on such figures as Kant, Hegel, Levinas, Derrida and Fanon. He is coeditor of the new journal *Critical Philosophy of Race* and he has published extensively on the history of race thinking and its contemporary dimensions. He recently published *Race* (2001) and *Race and Racism in Continental Philosophy* (2003), and he edited, with Tommy Lee Lott, *The Idea of Race* (2000).

Giuseppe Bianco was a Research Assistant for the Leverhulme Research Project in Bioethics and Biopolitics at the University of Warwick. He is now a fellow at Paris' Institute pour les Etudes Avancées, where he is working on a project on Canguilhem and social sciences. His work focuses on the history of the relation between philosophy and social sciences. He is the author of *Après Bergson. Portrait de groupe avec philosophe* (2014), *La signification du concret. Psychologie et philosophie chez Georges Politzer* (2014), *Jean Hyppolite entre structure et existence* (2013) and, with Tzuchien Tho, *Badiou and the Philosophers* (2013).

Florence Caeymaex is a F.R.S.-FNRS researcher at the University of Liège, codirector of the ARC/fructis project "Contemporary Politics of Nature" and member of the Belgian Advisory Committee on Bioethics. She specialized on Bergson and phenomenology (*Sartre, Merleau-Ponty, Bergson: Les phénoménologies existentialistes et leur héritage bergsonien*, 2005), and is now working on the relations between life, norms and politics to develop a "critical philosophy of norms".

Lorenzo Chiesa is Professor of Modern European Thought, Head of Italian and the codirector of Centre for Critical Thought at the University of

Kent, UK. He has published numerous journal articles and book chapters on Lacanian theory and contemporary French philosophy, and serves as production editor for the *Journal for Lacanian Studies*. He is the author of *Subjectivity and Otherness: A Philosophical Reading of Lacan*, and the editor, with Alberto Toscano, of *The Italian Difference: Between Nihilism and Biopolitics*.

Roger Cooter, now retired, was formerly professor of the history of medicine at University College London. He is the author with Claudia Stein of *Writing History in the Age of Biomedicine* (2013) and is currently coauthoring with her *The Spectacle of Hygiene: Capitalism, History and Visual Culture in Britain and Germany 1880s–1920s*. His latest article, "Neural Veils and the Will to Historical Critique: Why Historians of Science Need to Take the Neuro-turn Seriously", appeared in *Isis* in March 2014.

Piergiorgio Donatelli is Professor of Philosophy at Sapienza University of Rome. He is the author of several books and many articles and chapters in Italian, English and French. He has worked extensively on the philosophy of Wittgenstein, J. S. Mill, Michel Foucault, and on issues tied to the notion of human life. His recent books include *La vita umana in prima persona* (2012) and *Manières d'être humain* (2014). He is the editor of the journal *Iride*.

Steve Fuller is Auguste Comte Professor of Social Epistemology in the Department of Sociology at the University of Warwick, UK. Originally trained in the history and philosophy of science, he is best known for his foundational work in the field of "social epistemology", which is concerned with the normative grounds of organized inquiry. "Social Epistemology" is also the name of a quarterly journal that he founded in 1987 as well as the first of his twenty books. His recent books include *Preparing for Life in Humanity 2.0* (2012) and (with Veronika Lipinska) *The Proactionary Imperative: A Foundation for Transhumanism* (2014).

Marjorie Gracieuse was Research Assistant for the Leverhulme Research Project in Bioethics and Biopolitics at the University of Warwick. She has worked on the problem of hierarchy and desire in Deleuze's philosophy for her PhD thesis and her research interests revolve around the possibility of rethinking the distinction and classification of differences in the context of a political ontology of immanence. She's now an affiliate member of the ERRAPHIS research laboratory (Toulouse, France) and teaches philosophy and French literature at the *Institut Florimont* in Geneva, Switzerland.

Hector Kollias is lecturer in French at King's College London. He has written extensively on the articulation between philosophy psychoanalysis and queer theory. He is currently director of the Queer@King's Research Centre and was recently awarded an AHRC-funded project on Queer Theory in France.

Guillaume Le Blanc is professor of philosophy at the University of Bordeaux Michel de Montaigne. He is editor in chief of the collections *Pratiques théoriques* (Presses Universitaires de France) and *Diagnostics* (Bord de l'eau) and member of the editorial boards of the journals *Esprit* and *Raison Publique*. His work deals with the philosophy of norms, social philosophy and contemporary French philosophy. He has recently published *Vies ordinaires vies précaires* (2007), *Les maladies de l'homme normal* (2007), *L'invisibilité sociale* (2009), *Dedans, dehors. La condition d'étranger* (2010), *Que faut-il faire de notre vulnérabilité?* (2011) and *La philosophie comme contre-culture* (2014).

Paul-Antoine Miquel is professor in contemporary philosophy at the University of Toulouse. He has worked extensively on Bergson and the philosophy of sciences, with a particular focus on the epistemology of biology and the problem of ageing. He is a member of Toulouse ERRAPHIS Research Laboratory. His recent books include *Qu'est-ce que la vie?* (2007) and *Le Vital, aspects physiques, aspects métaphysiques* (2011). He's currently writing a book on the concept of Nature.

Nicolae Morar is a visiting assistant professor in the departments of philosophy and biology at the University of Oregon. His research interests are at the intersection of biology, ecology, ethics and biopolitics. He is a Co-PI on the grant *Between Deleuze and Foucault* and a Co-PI on the grant *Biodiversity at Twenty-Five: The Problem of Ecological Proxy Values*, which will provide a critical assessment of the normative role of biodiversity. His coedited volume, *Biopower: Michel Foucault and Beyond*, will be published in spring 2015.

Luca Paltrinieri is a researcher at the Chambre de Commerce Paris Ile-de-France and at the Collège international de philosophie. He specializes in the work of Michel Foucault and is now working on a genealogy of the notion of "company" and on actual entrepreneurial practices. He is the author of *L'expérience du concept: Michel Foucault, entre épistémologie et histoire* (2012) and of *La sanction éducative: Une recherche-action* (2012).

Julien Pieron is a lecturer at the University of Liège (Belgium). He is the author of a book on Heidegger, *Pour une lecture systématique de Heidegger* (2010). He has also worked on French historical epistemology, focusing on the interconnection between *life*, *norms* and *politics*. His present research focuses on the articulation between metaphysics, epistemology and politics and he is currently completing a research project on Bruno Latour.

Patrick Singy is adjunct professor at Union College in Schenectady, New York. His research interests include the history of medicine and sexuality, the historiography of science and the history and philosophy of psychology and psychiatry. He has published essays in several American and international journals, including *Representations, Journal of Modern History, Journal of the History of Sexuality* and *Journal of the History of Medicine and Allied Sciences*. He is the author of *L'Usage du sexe. Lettres au Dr Tissot, auteur de* L'Onanisme *(1760)* (2014).

Claudia Stein is Associate Professor in the History Department at the University of Warwick and a co-applicant of the Leverhulme Research Project in Bioethics and Biopolitics. She specializes in the history of science and medicine from 1500 to today. Together with Roger Cooter, she recently published, *Writing History in the Age of Biomedicine* (2013). She is currently finishing a book project, titled *Spectacles of Hygiene: Capitalism, Visual Culture and Medicine in Germany and Great Britain, 1880–1930* (with Roger Cooter), which investigates biopolitical technologies and strategies in the German and British Empires. Her second book project, *The Birth of Biopower in Enlightenment Germany*, aims to test Foucault's ideas on sovereign and biopower by applying them to concrete historical examples in late eighteenth-century Germany.

Sander Werkhoven is a doctoral student in philosophy at the University of Warwick, UK. Sander has a professional background in physiotherapy for which he obtained a BA of health in physiotherapy at the University of Applied Science Utrecht. His PhD thesis concerns the nature of health and the possibility of developing a naturalistic ethics on a conception of health.

Charles T. Wolfe is a research fellow in the Department of Philosophy and Moral Sciences and Sarton Centre for History of Science, Ghent University and an Associate of the Unit for History and Philosophy of Science, University of Sydney. He works primarily within history and philosophy of the early modern life sciences, with a particular interest in materialism and vitalism. He has edited volumes including *Monsters and Philosophy* (2005) and *Brain Theory: Essays in Critical Neurophilosophy* (2014).

Andy Wong is a doctoral student in philosophy at the Université de Liège, Belgium. He is currently writing a dissertation on the concept of life according to Bergson and Canguilhem.

Frédéric Worms is professor in French and Moral Philosophy at Paris' Ecole normale supérieure. He is the director of the Centre international pour l'étude de la philosophie française. Editor in chief of the *Annales bergsoniennes*, he has published extensively on Bergson, including *Bergson ou les deux sens de la vie* (2004) and *Bergson Biographie* (with Philippe Soulez, 2002). He is also the author of a history of French philosophy, *La philosophie en France au XXème siècle. Moments* (2009) and of several publications concerning life and care: *Revivre: Eprouver nos blessures et nos ressources* (2012; *Soin et politique* (2012); *La vie qui unit et qui sépare* (2013).

Lightning Source UK Ltd.
Milton Keynes UK
UKOW02n1059300316

271168UK00012B/260/P